Studies in Big Data

Volume 181

Series Editor

Janusz Kacprzyk, Polish Academy of Sciences, Warsaw, Poland

The series "Studies in Big Data" (SBD) publishes new developments and advances in the various areas of Big Data- quickly and with a high quality. The intent is to cover the theory, research, development, and applications of Big Data, as embedded in the fields of engineering, computer science, physics, economics and life sciences. The books of the series refer to the analysis and understanding of large, complex, and/or distributed data sets generated from recent digital sources coming from sensors or other physical instruments as well as simulations, crowd sourcing, social networks or other internet transactions, such as emails or video click streams and other. The series contains monographs, lecture notes and edited volumes in Big Data spanning the areas of computational intelligence including neural networks, evolutionary computation, soft computing, fuzzy systems, as well as artificial intelligence, data mining, modern statistics and Operations research, as well as self-organizing systems. Of particular value to both the contributors and the readership are the short publication timeframe and the world-wide distribution, which enable both wide and rapid dissemination of research output.

The books of this series are reviewed in a single blind peer review process.

Indexed by SCOPUS, EI Compendex, SCIMAGO and zbMATH.

All books published in the series are submitted for consideration in Web of Science.

Hien Thu Thi Nguyen · Hai Hong Phan ·
Van-Nam Huynh
Editors

Data Science in Finance and Accounting

Springer

Editors
Hien Thu Thi Nguyen
School of Finance and Accounting
Industrial University of Ho Chi Minh City
Ho Chi Minh City, Vietnam

Hai Hong Phan
Industrial University of Ho Chi Minh City
Ho Chi Minh City, Vietnam

Van-Nam Huynh
School of Knowledge Science
Japan Advanced Institute of Science
and Technology
Nomi, Japan

ISSN 2197-6503 ISSN 2197-6511 (electronic)
Studies in Big Data
ISBN 978-3-032-06178-2 ISBN 978-3-032-06179-9 (eBook)
https://doi.org/10.1007/978-3-032-06179-9

This Springer imprint is published by the registered company Springer Nature Switzerland AG
The registered company address is: Gewerbestrasse 11, 6330 Cham, Switzerland

Preface

The rapid evolution of data science technologies has transformed the landscape of finance and accounting, offering new capabilities for prediction, analysis, and decision support. This volume *Data Science in Finance and Accounting* brings together a selected collection of chapters that explore the intersection of advanced analytical techniques and real-world applications within the realms of accounting and finance.

The chapters in this book reflect a rich diversity of perspectives, methodologies, and use cases, contributed by researchers and practitioners across different disciplines. The opening chapters provide theoretical underpinnings for explainable data science and elucidate key constructs such as the Shapley value and skew-normal distributions–offering foundational insights relevant to both economic modeling and machine learning.

Subsequent chapters focus on applied studies that leverage machine learning for financial forecasting tasks, including financial distress prediction, post-IPO performance analysis, and corporate bankruptcy detection. These contributions not only demonstrate the efficacy of techniques such as LSTM networks, sentiment analysis, and ensemble learning, but also emphasize issues of interpretability, model validation, and practical implementation.

Several contributions examine explainable artificial intelligence (XAI) in the financial domain, highlighting its potential to enhance transparency, trust, and accountability in automated decision-making. Complementing these are case studies and empirical investigations into the roles of social media, textual data, ICT adoption, and corporate governance in shaping financial behavior and outcomes.

We extend our sincere appreciation to all the contributing authors, reviewers, and editorial collaborators who have made this volume possible. We would also like to thank Prof. Janusz Kacprzyk (Series Editor) and Dr. Thomas Ditzinger (Senior Editor, Engineering/Applied Sciences) for their support and cooperation in this publication.

It is our hope that this book will serve as a valuable resource for academics, professionals, and graduate students working at the frontier of data science and financial decision-making.

Ho Chi Minh City, Vietnam
July 2025

Hien Thu Thi Nguyen
Hai Hong Phan
Van-Nam Huynh

Contents

On Explainable Data Science

Hung T. Nguyen

Abstract Viewing statistics as the oldest and the most influencial science of current data science, including AI and machine learning, we elaborate upon some significant developments related to the pursuit of credibly empirical applications of data science with emphasis on demystifying hypothesis testing, relating coalitional games to the explanation of black box models, and evoking useful contributions of optimal transport theory to data analysis.

1 Introduction

This paper aims simply at providing some "significant" closer looks at current efforts in the realm of research on credible methods in data science, especially in explainable models for prediction.

In Sect. 2, inspired by the (rare) thought provoking chapter "A closer look at tests of significance" in Freedman et al. [1], and the Lecture 26 in Cosma Shalizi's course "Modern Regression" [2] on covariate selection in standard (traditional) linear regression models (where it was stated, and analyzed, out loud that "Why variable selection using p-values is a bad idea?"), we elaborate shortly again the "p-value crisis in statistical inference" [3–7] and demystify hypothesis testing, as well as dig a bit into explainable statistics. The main message could be "While it is still so hard to get rid of p-values [8, 9], it should be clear that testing using p-values should no longer be used in feature selection of regression models" in traditional statistical practices of empirical research.

In Sect. 3, we elaborate on the unexpected beautiful marriage between data analysis and game theory, especially for explainable prediction models, exemplified in [10–12].

H. T. Nguyen (✉)
New Mexico State University, Las Cruces, USA

Chiang Mai University, Chiang Mai, Thailand
e-mail: hunguyen@nmsu.edu

H. T. T. Nguyen et al. (eds.), *Data Science in Finance and Accounting*, Studies in Big Data 181, https://doi.org/10.1007/978-3-032-06179-9_1

In Sect. 4, we embark on surprising contributions to data analysis, especially for data on manifolds, from Optimal Transport Theory in Mathematics [13, 14] with emphasis on the old pioneering program of M. Frechet where Wasserstein metrics and Waserstein data now surface as a useful setting for data science.

2 Explainable Statistics

Any science is relied upon at least two "ingredients", namely its rationality and explanation. As data science is subject to uncertainty, it must provide its rationality and its explanation (for the users and regulators). Prediction under uncertainty is everywhere in data science, starting of course with statistical science (although a joke could be "Statisticians can predict anything, except the future"). Talking about statistical science, let's look at it closely. Why should we do that? Well, we all agree that statistical estimation is rational since we can explain it by mathematics (consistency of estimators, limiting distributions, based on two-valued logic). How about "statistical inference"? Its main "tool" is (frequentist) testing of hypotheses. It is the most mysterious procedure.in statistical inference. Its rationale is somewhat hidden in Fisher's "mathematics of the Lady tasting tea". It is an "argument" (not a mathematical proof) analogous to "a proof by contradiction" in mathematics under two-valued logic. The problem is this. The proof by contradiction (called modus tollens in logic) is only valid in two-valued logic. However, inference in statistics is in the realm of "*probability logic*" (a multi-valued logic) where modus tollens is not valid, see e.g. [5, 6, 15]. Why no statisticians see it? until the "p-value crisis" [3, 4, 7–9].

Remark 1 (i) There is a joke in the literature: What p stands for in p-value? p stands for "problems" that they give us!
(ii) In terms of *Fuzzy Logics*, see [16].

In a rare text book on Statistics [1], after a careful exposition of almost all topics in Statistics (for applications), the authors ended their text by the last Chap. 29 entitled "A Closer Look at Tests of Signficance". Why another chapter on testing while they aleady wrote three chapters on Tests of Significance (based on p-value)?

What are things that they want to look at closely? Are there something which bothers them? or not "explainable" to them? noting that text books in Statistics did not explain clearly the testing procedure to reach decisions, they just told the readers what to follow. Also, it seems that teachers of statistics courses also do not bother to explain to students, perhaps they do not understand why themselves? This is precisely the missing problem of "explainable statistics". Without explaining statistical inference based on the notion of p-value is the reason of [4]!

The Chap. 29 in [1] is "significant"! It pointed out that the rationale of using p-values to reject or accept a hypothesis (i.e., making decisions under uncertainty) is an argument of proof by contradiction, although it did not go a little further to prove that proof by contradiction is not valid in probability logic. Instead, it says this: "Nowadays, tests of signficance are extremely popular. This sounds so impressive, and there is so much mathematical machinery clanking around in the background, that tests seem truly scientific- even when they are *complete nonsense*. St. Exupery understood this kind of problem very well: When a mystery is too overpowering, one dare not disobey (The Little Prince)".

Remark 2 (i) We may wonder why statisticians did not look at the mysterious testing procedure suggested by Fisher? Perhaps it is not easy to understand it? Note that Fisher's Maximum Likelihood procedure for estimation is rational and clear since it has the same rationale of Bayesian decision-making. Perhaps probabilists turned statisticians are more curious than trained statisticians? This is somewhat similar to G. Choquet when he considered capacities to solve a problem in Potential Theory that experts in Potential Theory considered as difficult, because Choquet did not know Potential Theory! A probabilist turned statistician could ask, as in Chap. 26 (tests of significance) [1] "If there were any justice in the world, the p-value would be the probability of the null hypothesis given the data. Even worst, according to the frequency theory, there is no way to define the probability of the null hypothesis being right (because H_o is not considered as a random event), the p-value is not the chance of the null hypothesis being right"!

(ii) As far as the "p-value crisis in science" is concerned (as a major tool to make decisions under uncertainty), we can describe its present status as follows. A lot of statisticians, either unaware of it, or just cannot get out from 'tradition', continue to use p-values in inference in their empirical research, some arguing that "misuse does not mean wrong", and relying on the fact that most Journals still accept p-values. Well, time will tell. However, remember a typical situation such as this from Kenneeth Rothman (Editor of *Amer. Journal of Public Health,* and of *Epidemilogy*):
"All references to statistical hypothesis testing and statistical significance should be removed from the paper. I ask that you delete p-values as well as comments about statistical significance. If you do not agree with my standards (concerning the inapropriateness of significance tests), you should feel free to argue the point. or simply ignore what you may consider to be my misguided view, *by publishing elsewhere*".

Important Note: A new *proof,* using *conditional event algebra* [17], of the fact that the statistical inference using (Fisher) p-value is *wrong* is given in an *Appendix* of this paper.

Now, for practical purposes, while we cannot "solve" the p-value problem in general, at least we should ask a modified question "Where and when hypothesis testing with p-values is used in traditional statistics?".

What you are going to find out will surprise you! Recently, by following the literature on advances of regression analysis in statistics, I happened to run into a course with the title "Modern Regression" of Cosma Shalizi at CMU [2]. Reading his Lecture 26 shocked me with an unexpected pleasure! The pleasure comes from the fact that, as he put it upfront, "Why variable selection using p-values is a *bad* idea?"! His unusual (rare) deep analysis of the traditional covariate selection in linear regression destroys the "traditional" use of testing using p-values.

Just like the special closer look at tests of significance in Freedman et al. [1], Shalizi's lecture 26 is a real closer look at how covariate selection was performed until now (i.e., explaining the procedure), a so important and frequent problem in credible applied statistics, neglected by statisticians. In fact, after providing all details concerning "Why variable selection using p-values is a *bad* idea?", Shalizi had this to say with respect to the routine of using hypothesis testing for covariate selection in linear regression:

"This pattern of reasoning shows up over and over again among users of regression, including, I *am ashamed* to say, not a few statisticians."

For your convenience, here is the traditional covariate selection in linear regression, followed by Shalizi's take.

Consider the problem of predicting a real-valued random variable Y from the observation of a random vector $X = (X_1, X_2, \ldots, X_k)^t$ with values in $\mathbb{R}^k$.

In traditional statistical analysis, having in mind the mean square error as the "favorite" prediction error, the best prediction of Y based on X (i.e., the relation between Y and X) is $E(Y|X)$. For that to make sense, "standard assumptions" are imposed such as Y, X_j's have finite second moments.

Suppose there is enough "evidence" to suggest a linear model for $Y = \varphi(X_1, X_2, \ldots, X_k)$, say

$$Y = \sum_{j=1}^{k} \beta_j X_j + \varepsilon$$

with ε_i independent, distributed as $\varepsilon \sim N(0, \sigma^2)$, and $E(X|\varepsilon) = 0$.

To specify the model for applications, we need data to estimate the "population parameters" $\beta_j's$. Suppose data are of the form of i.i.d. observations $(X_{1i}, X_{2i}, \ldots, X_{ki}, Y_i)$, $i = 1, 2, \ldots, n$, from (X, Y).

The traditional analysis proceeds as follows (see any statistical text books on regression). First, estimate the coefficients $\beta_j's$ by OLS, providing consistent estimators and their distributions, from the given assumptions. The next step is "inference about the parameters", i.e., conducting tests for hypothesis such as $H_o : \beta_j = 0$ versus $H_a : \beta_j \neq 0$ (or nested models). The cook books say this. The reason for interest in such testing is to find out whether there is or not a linear association between Y and X_j. You could ask: in the previous step, we have obtained consistent estimates of the β_j's, so why we still need to test?

Remark 3 Of course, unlike "tests" in physical science, tests in social sciences are subject to (probabilistic) uncertainty. The explanation of frequentist testing is about "how to carry out tests?", noting that Bayesian testing is logical!

But let's follow cook books. Statistical inference using tests (of significance) with p-values is a process of covariate selection (finding out which "features" X_j are really affecting Y). ANOVA is investigated to provide test statistics. The covariate selection in regression is carried out by testing $H_o : \beta_j = 0$ versus $H_a : \beta_j \neq 0$. If H_o is rejected then the feature (covariate) X_j is declared "significant" (in Fisher's language), i.e. should be part of the checked model (X_j does affect Y, without specifying its quantitative contribution). Rather than just following the theory of (frequents) hypothesis testing, let's look really close to that theory, i.e., trying to explain it.

Under all necessary assumptions imposed, the test is carried out as follows. Let T be the test statistic when H_o is true (in fact, T is defined also under all assumptions imposed on the model and data!) Specifically,

$$T = \frac{\hat{\beta}_j}{SE(\hat{\beta}_j)}$$

which is distributed as a $t-$distribution, and

$$SE(\hat{\beta}_j) = \frac{\hat{\sigma}}{\sqrt{nS^2_{X_j}}}, \quad S^2_{X_j} = \frac{1}{n}\sum_{i=1}^{n}(X_{ji} - \overline{X}_j)^2$$

Let $p = P(|T| \geq |T_{ob}||H_o)$ (called the *p-value* of the test statistics T) be the probability of observing a value greater than the actual observed value T_{ob} of T. Testing procedure: Reject H_o if p is very small, say, $p \leq 0.05$ (just a threshold for decision-making). Why? Well, it is an "argument by contradiction": p is small for observing T_{ob}, but we do observe T_{ob}, then H_o is not true!

Without evoking logic, let examine the credibility of this testing procedure for covariate selection (it is a "procedure" and not a mathematical theorem like in statistical estimation). Here is a summary of Shalizi's take by looking at the formula for the p-value.

First clearly, the p-value depends on the sample size n of the data. $\hat{\beta}_j$ larger makes β_j more significant! As $\hat{\sigma_n}$ decreases will make all β_j more significant. Increasing the sample size n will increase all test statistics resulting in β_j significant. As $\hat{Var}(X_j)$ increases, β_j becomes more significant. More correlation among features will make β_j less significant.

With the above facts, Shalizi concluded "None of testing using p-values is even much help in answering the question which covariates help us predict the response? let alone which covariates help us explain the response", resulting in "why testing using for covariate selection in regression is a bad idea", in fact a wrong procedure to use.

The above explanation of testing procedure of traditional statistics should prevent statisticians from using it from now on and should turn to more rational methods for covariate selection as well as for explainable data science.

Shalizi's analysis ends the traditional way of doing covariate selection using testing with p-values, although without attacking the "hidden" fact that we should also abandon p-values (and hence tests of significance) in statistical inference as a whole, including nonlinear regression models such as "regression kink with an unknown threshold" [18]. But since regression analysis is the workhorse (Bread and Butter) of applied statistics, eliminating p-values from it is like avoiding the use of p-values in a big chunck of a cake! i.e., avoiding providing uncredible empirical results in applied statistics. Thus, whether you are not sure about getting rid of p-values, at least you won't make mistakes in your research. Teachers of fundamental courses in statistics should elaborate Shalizi's findings to students.

After clearly and completely explaining why statisticians should no longer use p-values in covariate selection of linear models, Shalizi suggested to replace it by cross validation. We will, instead, elaborate on a more popular method for covariate selection in the next Section.

3 The Hidden Thing in Plain Sight

How to do credible covariate selection in regression models (in fact, in black box models of machine learning) as well as explaining them, when "using testing with p-values is a bad idea"? noting that covariate selection is an important step in applied econometrics since rough models need to be checked before using them.

Continuing with the covariate selection problem in regression let's us examine a related problem of variance-based sensitivity analysis in a general setting.

It is interesting to note that in the context of functional ANOVA of traditional statistics (for testing purpose) there is something unexpected hidden that will change our way of doing covariate selection !

Consider a black box model $Y = \varphi(X_1, X_2, \ldots, X_k)$, then under suitable conditions (e.g., Y has finite variance, and the features $X_j's$ are independent), the functional ANOVA decomposition yields

$$Y = \varphi(X_1, X_2, \ldots, X_k) = \sum_{A \subseteq I} \varphi_A(X_A)$$

where $I = \{1, 2, \ldots, k\}$, $X_A = (X_j : j \in A)^t$, $\varphi_A(.)$ is a function just depending on X_A, so that $Var(Y) = \sum_{A \subseteq I} Var[\varphi_A(X_A)]$ for variance-based sensitivity analysis.

Since, denoting $|A|$ to be the cardinality of the set A,

$$\varphi_A(X_A) = \sum_{B \subseteq A} (-1)^{|A \setminus B|} E(Y|X_B)$$

we have, by Mobius inversion,

$$E(Y|X_A) = \sum_{B \subseteq A} Var[\varphi_B(X_B)]$$

Thus if we let $v(.) : 2^I \to \mathbb{R}^+$, $v(A) = Var[E(Y|X_A)]$, then $v(.)$ is a bona fide characteristic function of a coalitional game since $v(\varnothing) = 0$, and $v(.)$ satisfies a stronger property than supper additivity (See [19, 20]).

Remark 4 A "belief function" or equivalently a probability distribution function $F(.)$ of a non empty random set S on a finite set I is a special characteristic function of a coalitional game. Specifically, let $S(.) : (\Omega, \mathscr{A}, P) \to 2^I$ with distribution $F(.) : 2^I \to [0, 1]$, $F(A) = P(\omega : S(\omega) \subseteq A)$ which is characterized as

(i) $F(\varnothing) = 0$,
(ii) For any $A_i, i = 1, \ldots, n$ in 2^I,

$$F(\cup_{i=1}^n A_i) \geq \sum_{\varnothing \neq B \subseteq \{1,2,\ldots,n\}} (-1)^{|B|+1} F(\cap_{i \in B} A_i)$$

It follows that $F(.)$ is supper-additive.

The question we may like to ask is: What good to have a characteristic function of a coalitional game in the above variance-based sensitivity analysis? Well, first, the random variable $E(Y|X_A)$ is the prediction of Y based on X_A, its variance $v(A) = Var[E(Y|X_A)]$ can be interpreted as the "payoff" of the coalition A. Next, from this characteristic function $v(.)$, we obtain its solution as the Shapley values of "players", i.e., an individual contribution of each player. Does that ring the bell?

Well, while $v(A) = Var[E(Y|X_A)]$ indicates the payoff of the coalition A, so that $V(I) = Var[E(Y|X)]$ is the payoff of the grand coalition I (all "players" join it to produce the total payoff $Var[E(Y|X)]$ (the variance of the prediction $E(Y|X)$, when using mean square error), a priori, we still do not know how to divide fairly (i.e., according to players'contributions) that total amount $Var[E(Y|X)]$ to each player. We need to know such divisions since they correspond precisely to what we are looking for, namely individual contributions of features to our response Y when we use all of features (for predicting the response Y). In traditional statistical analysis, such concerns were not considered, i.e., the correct explanation of the model was not seriously considered.

Remark 5 Understanding the topology of a network is a main concern since it helps to get a better insight of the system it represents. A (traditional) major approach to achieve that goal is hypothesis testing (!). To set up hypothesis testing (when p-value is used, without knowing the p-value crisis, of course!) we need test statistics. That is why functional ANOVA is needed. In other words, functional ANOVA (e.g., Hoeffding decomposition) is the basis for getting tests, see also [10]. With background from statistics, engineers just proceed to tests to explain to themselves how to design their systems, but not explaining their systems to the users or regulators (which should be the answer to a different question, namely "What are the individual feature contributions to the response output of interest?").

Now, inspired from the important issue of explainability in machine learning (or AI in general), the same issue is also a must in statistics. The connection between a characteristic function $v(.)$ of a coalitional game and a fair division of payoffs to players in a grand coalition is provided by the well known notion of *Shapley values* in game theory. Thus, variance-based sensitivity analysis in, say, regression models (or more generally for black box models) using functional ANOVA reveals the "thing" we are looking for, *hidden in plain sight* (!): Shapley values of coalitional games are individual feature (covariate) contributions to output response, i.e., explaining the model under investigation. In fact, Shapley values "kill two birds with one stone": They explain the model and do covariate selection (logically/ replacing the illogical p-value, see especially [3]). The marriage of Statistics and Game Theory (althought existed previously) at this new junction is really the best glory of Game Theory, unexpected from J. von Neumann, namely "Shapley's values explain statistics" (in fact, they explain data science).

Basic references on the use of Shapley values for explainable regression models (linear or not, e.g., [21]) or black box models are [10–12, 18, 22–24].

Remark 6 How machine learning different from statistics, as both fields seem to have the same goal: prediction? or what makes machine learning a new sub-field of data science, besides algorithmic approach and learning (generalization) process ? Is there an analogy with traditional mathematics and *constructive mathematics*? (e.g., explaining "there exists" as "we can construct").

4 The Contributions of Optimal Transport

In the context of explainable data science, it is impossible for not mentioning the significant contributions of the mathematical theory of Optimal Transport (OT) [14].

In one hand, OT provides a host of new tools to data analysis and inference, as exemplified in [13], such as the long waited notion of *multivariate quantile functions* (for, e.g., multivariate quantile regression). On the other hand, as we will spell out in this section, OT provides also additional *explanations* for data analysis of statistics and machine learning.

Here is a short tutorial on what is OT and how it contributes to explainable data science.

In general data analysis, as laid out by the well known pioneering work of Maurice Frechet, we need a metric to assess the variation of data points in a sample space. In "modern" applications (e.g., in ML) data need not be euclidean (e.g., histograms, point clouds). For example, our sampling space could be a space of probability measures.

Consider, for example, the situation where histograms (as data "points") are sitting in the euclidean space $\mathbb{R}^2$. They "move" not like points of $\mathbb{R}^2$, so that the euclidean metric of $\mathbb{R}^2$ does not "measure" their "true" variations. Histograms move according to different geodesics! To formalize this, we should put the histograms (or general random probability measures) on their own (sampling) space (of probability measures on the metric space $(\mathbb{R}^2, ||.||)$). Within this non-euclidean space, data points (e.g., histograms) move (i.e., change) differently which cannot be captured by euclidean metrics.

How they move? Well, each data point is a probability measure μ at a location, and it "moves" to another probability measure ν at another location (on the underlying metric state space). As such, a natural metric in OT becomes apparent, recalling that in OT we are interested in finding the optimal transport plan to move a mass μ from a location $\mathscr{X} \subseteq (\mathbb{R}^2, ||.||_2)$ to a mass ν at location $\mathscr{Y}$ when a cost function $c(x, y)$ is given, such as $c(x, y) = ||x - y||_2^2$, which is

$$T^* = \arg\min\{\int_{\mathscr{X}\times\mathscr{Y}} ||x - y||_2^2 d\lambda(x, y) : \lambda \in \Pi(\mu, \nu)\}$$

where $\Pi(\mu, \nu)$ denotes the set of joint probability measures on $\mathscr{X} \times \mathscr{Y}$ having μ, ν as marginals.

It turns out that

$$W_2(\mu, \nu) = [\inf\{\int_{\mathscr{X}\times\mathscr{Y}} ||x - y||_2^2 d\lambda(x, y) : \lambda \in \Pi(\mu, \nu)\}]^{\frac{1}{2}}$$

is a bona fide metric on a probability measures space, called a *Wasserstein metric*.

The upshot of the use of a Wasserstein metric is that it does take into account of the geometry of the underlying space $\mathbb{R}^2$ (or, more generally, of Riemannian differential manifolds), capturing the variations of data points. As such, Wasserstein metrics arising from OT *explain* their rational uses in, e.g., unsupervised machine learning such as Wassertein Generative Adversary Networks/ WGAN). In view of this explanation aspect, a new field of statistics, namely *Statistics on Wasserstein Space*, has recently emerged.

Remark 7 Copula Theory, as Abe Sklar developed in his Ph.D. Thesis in 1959, was motivated by Maurice Frechet's question concerning joint distribution functions with given marginals. How copulas are useful for Statistics? In fact, Frechet, ahead of his times, was generalizing the foundations of statistics on euclidean spaces to arbitrary metric spaces in

Les elements aleatoires de nature quelconque dans an espace distancie'
Frechet, M.
Ann. Inst. H. Poincare' (10), 215-310, (1948)
(Random elements of arbitrary nature in a metric space)

i.e., considering an arbitrary metric space $(\mathscr{S}, d)$ as a sample space. A Frechet mean of a random element Y with values in $(\mathscr{S}, d)$ is a solution of minimizing $E[d(Y, s)^2]$ over $s \in \mathscr{S}$. However, such optimization problems in general spaces (e.g., polish spaces) do not have unique minimizer. Instead, we get a set of minimizers, called the Frechet mean set. As such, we face the problem of "random set estimation".

The use of Wasserstein metrics in financial risk analysis is justified by the robustness of quantile-based risk measures.

An interesting "explanation" of Wasserstein metric in Machine Learning is WGAN.

It is so since essentially any statistical problems (treating in ML) involve the measure of a distance between probability measures.

Without going into details of unsupervised learning using Generative Adversary Networks/algorithms (GAN), it suffices to mention the following.

In GAN, KL-Divergence (or more general divergence measures) is used to compare probability measures. But these divergence measures (inspired from Statistics) are applicable only for probability density functions (i.e., for absolutely continuous probability measures) in one hand, and on the other hand, exhibit some undesirable properties with regard to computations and interpretations.

It turns out that another metric can replace divergences to avoid these undesirable properties, and it is precisely a Wasserstein metric, leading to the WGAN!

This is a significant new application of Wasserstein metrics in unsupervised machine learning, since the training algorithm involes a distance between probability measures, for comparison. It illustrates again the need to consider, not only a metric between probability measures, but a good one!.

Appendix: Explainable Logic By CEA

In discussing explainable data science it would be incomplete if we do not include XLogic as logic is the backbone of reasoning under uncertainty , especially the logic of "If...Then..." rules in intelligent systems.

The logic of *Conditional Event Algebra* (CEA) [17] was investigated precisely to explain reasoning with conditional data, and as a by-product, to shed light on the invalidity of "proof by contradiction" in probability logic, noting that the concept of "conditional events" and their associated logic are lacking from probability theory.

In two-valued logic, the "If...Then..." rule is interpreted as a material implication (a connective on propositions), i.e.,

$$(B \to A) = A^c \cup B$$

so that $(B \to A)$ is an event (a proposition). In this context of two-valued logic, the proof by contradiction is valid since

$$(A^c \to B^c) = (B^c)^c \cup A^c = B \cup A^c = A^c \cup B = (B \to A)$$

In the setting of reasoning under uncertainty assigned probability to a conditional data "If B then A" is the conditional probability (when $P(B) > 0$)

$$P(A|B) = P_B(A) = \frac{P(A \cap B)}{P(B)}$$

As such "If B then A" cannot be modeled as a material implication $\rightarrow$ since (for $P(B) > 0$):

$$\begin{aligned} P(B \rightarrow A) = P(A^c \cup B) &= P[B^c \cup (A \cap B)] = P(B^c) + P(A \cap B) \\ &= P(B^c) + P(A|B)P(B) \\ &= P(B^c) + P(A|B)[1 - P(B^c)] \\ &= P(A|B) + P(B^c)[1 - P(A|B)] \\ &= P(A|B) + P(B^c)P(A^c|B) \geq P(A|B) \end{aligned}$$

unless either $P(A^c|B) = 0$, or $P(B) = 1$, a rather trivial case (Lewis' Triviality Result implying that there is no binary operation $\rhd$ on a $\sigma-$ field $\mathscr{A}$ of subsets of Ω such that, for all probability measures P on $\mathscr{A}$, and all A, B in $\mathscr{A}$ with $P(B) > 0$, $P(A \rhd B) = P(A|B)$).

Thus, material implication $\longrightarrow$ is not compatible with conditional probability evaluation, and "If B then A" should be some other implication $\Longrightarrow$ outside on $\mathscr{A}$.

The fundamental question is "What should be $\Longrightarrow$" so that we can assign rigorously conditional probability $P(A|B)$ to $(B \Longrightarrow A)$? noting that a mathematical modeling for $(B \Longrightarrow A)$ is needed for data analysis with "conditional events", i.e., to provide an algebra (logic) to manipulate rules, e.g., to combine rules such as $(B \Longrightarrow A)$ and $(D \Longrightarrow C)$ into a rule of the form $[(B \Longrightarrow A)$ and $(D \Longrightarrow C)]$ which we will call a CEA [7].

It should be noted that Fisher's testing procedure using p-values falls somewhat into the above setting where "If...Then..." rules are of the form

"If H_o is true then T_{ob} (likely) does not occur"
"If T_{ob} occurs then (likely) H_o is not true"

It turns out that $(B \Longrightarrow A)$ is a bona fide "event", not in $\mathscr{A}$, but in an associated higher $\sigma-$ field $\mathscr{A}^*$, and for any probability measure P on $\mathscr{A}$, there is a probability measure P^* on $\mathscr{A}^*$ such that for any $A, B \in \mathscr{A}$ with $P(B) > 0$, we have $P^*(B \Longrightarrow A) = P(A|B)$ [17].

For ease of reference, here is the proof.

Theorem. For any measurable space $(\Omega, \mathscr{A})$, there is an associated measurable space $(\Omega^*, \mathscr{A}^*)$, and a mapping $\alpha(.,.) : \mathscr{A} \times \mathscr{A} \rightarrow \mathscr{A}^*$ such that for any probability measure P on $\mathscr{A}$, there is a probability measure P^* on $\mathscr{A}^*$such that for any $A, B \in \mathscr{A}$, with $P(B) \neq 0$, we have $P^*(\alpha(A, B)) = P(A|B)$, i.e., the "conditional event" $(A|B) = (B \Longrightarrow A)$ is in fact an ordinary event $\alpha(A, B)$ but in a "higher boolean space", whose P^* probability is compatible with initial conditional probability.

Proof. Take Ω^* be the infinite (but countable) cartesian product space $\prod_{n\geq 1} \Omega_n$ where $\Omega_n = \Omega$ for all n. Next, equip this product space with the usual product σ−field $\mathscr{A}$, i.e., the σ−field of subsets of Ω^* generated by cylinders that are subsets of $\mathscr{A}^*$ of the form

$$A_1 \times A_2 \times \cdots \times A_n \times \Omega \times \Omega \times \Omega \times \cdots$$

for any $n \geq 1$, and $A_1, A_2, \ldots, A_n$ in $\mathscr{A}$.

Finally, define a map $\alpha(.,.) : \mathscr{A} \times \mathscr{A} \rightarrow \mathscr{A}^*$ as

$$\alpha(A, B) = \cup_{n\geq 0}(B^c \times B^c \times \cdots B^c(n \text{ times}) \times (A \cap B))$$

where $\cup$ denotes union among subsets of Ω^*, and the term $A \cap B$ is the standard short hand for $A \cap B \times \Omega \times \Omega \cdots$ which is a subset of Ω^*, and for $n = 0$, the first term is simply $A \cap B$.

For P on $\mathscr{A}$, we let P^* to be the infinite product probability measure on $\mathscr{A}^*$with identical components P, i.e., P^* is constructed from

$$P^*(A_1 \times \ldots \times A_n \times \Omega \times \Omega \times \Omega \times \cdots) = \prod_{i=1}^{n} P(A_i)$$

Since the subsets (of Ω^*) $B^c \times \ldots \times B^c(n$ times$) \times (A \cap B)$, for $n \geq 0$, are pairwise disjoint in Ω^*, we have

$$\begin{aligned}
P^*(\alpha(A, B)) &= P^*[\cup_{n\geq 0}(B^c \times B^c \times \cdots B^c(n \text{ times}) \times (A \cap B))] \\
&= \sum_{n=0}^{\infty} P^*[(B^c \times B^c \times \cdots B^c(n \text{ times}) \times (A \cap B))] \\
&= \sum_{n=0}^{\infty} P(A \cap B)[P(B^c)]^n = P(A \cap B)\sum_{n=0}^{\infty}[P(B^c)]^n \\
&= \frac{P(A \cap B)}{1 - P(B^c)} = \frac{P(A \cap B)}{P(B)} = P(A|B)
\end{aligned}$$

With the basics of CEA, it should be clear that reasoning with proof by contradiction (Modus Tollens) in probability logic is not valid, *explaining* that testing using p-values is not credible. It is so simply because

$$P^*(B \Longrightarrow A) = P(A|B) \neq P(A^c|B^c) = P^*(A^c \Longrightarrow B^c)$$

References

1. Freedman, D.A., Pisani, R., Purves, R.: Statistics. W.W. Norton (2007)
2. Shalizi, C.: Modern regression (2015). https://www.stat.cmu.edu/~cshalizi/mreg
3. Gelman, A., Robert, C.P.: The statistical crisis in science. Am. Sci. **102**(6), 560–465 (2014)
4. Ioannidis, J.P.A.: Why most published research findings are false. PLoS Med. **2**(8), e124 (2005)
5. Nguyen, H.T.: On evidence measure of support for reasoning with integrated uncertainty: a lesson from the ban of p-values in statistical inference. In: Lecture Notes in Artificial Intelligence (LNAI), vol. 9978, pp. 3–15. Springer, Berlin (2016)
6. van der Pas, S.L.: Much ado about p-value. Bachelor Thesis, Mathematisch Institute, Universiteit Leiden (2010)
7. Wasserstein, R.L., Lazar, N.A.: The asa's statement on p-values: context, process, and purpose. Am. Stat. **70**, 129–133 (2016)
8. Goodman, S.N.: Why is getting rid of p-values so hard? musings on science and statistics. Am. Stat. **73**(S1), 26–30 (2019)
9. Hubard, R., Lindsay, R.M.: Why p-values are not a useful measure of evidence in statistical significance testing. Theory Psychol. **18**(1), 69–99 (2008)
10. Cohen, S., Ruppin, E., Dror, G.: Feature selection based on the shapley value. In: Proceedings of the 19th International Joint Conference on Artificial Intelligence (IJCAI-05), pp. 665–670 (2005)
11. Lundberg, S.M., Lee, S.-I.: A unified approach to interpreting model predictions. In: Advances in Neural Information Processing Systems, vol. 30, pp. 4765–4774 (2017)
12. Strumbelj, E., Kononenko, I.: An efficient explanation of individual classifications using game theory. J. Mach. Learn. Res. **11**, 1–18 (2010)
13. Galichon, A.: Optimal Transport Methods in Economics. Princeton University Press (2016)
14. Villani, C.: Topics in Optimal Transportation. American Mathematical Society (2003)
15. Yalcin, S.: A counterexample to modus tollens. J. Philos. Log. **41**, 1001–1024 (2012)
16. Nguyen, H.T., Walker, C.L., Walker, E.A.: A First Course in Fuzzy Logic, 4th ed. Chapman & Hall/CRC (2019)
17. Goodman, I.R., Nguyen, H.T.: A theory of conditional information for probabilistic inference in intelligent systems II: product space approach. Inf. Sci. **76**, 13–42 (1994)
18. Gramegna, A., Giudici, P.: Shapley feature selection. FinTech **1**, 72–80 (2022)
19. Nguyen, H.T.: On random sets and belief functions. J. Math. Anal. Appl. **65**, 531–542 (1978)
20. Nguyen, H.T.: An Introduction to Random Sets. Chapman & Hall/CRC (2006)
21. Hansen, B.E.: Regression kink with an unknown threshold. J. Bus. Econ. Stat. **32**(2), 228–240 (2017)
22. Chen, H., Lundberg, S.M., Lee, S.-I.: Algorithms to estimate Shapley value feature attributions. Nat. Mach. Intell. **5**, 590–601 (2023)
23. Owen, A.B., Prieur, C.: On Shapley value for measuring importance of dependent inputs. J. Uncertain. Quantif. **5**(1) (2017)
24. Verdinelli, I., Wasserman, L.: Feature importance: A closer look at Shapley values and loco (2023). https://doi.org/10.48550/arXiv.2303.05981

Why Shapley Value and Its Generalizations Are Effective in Economics and Finance, Machine Learning, and Systems Engineering

Miroslav Svitek, Niklas Winnewisser, Michael Beer, Olga Kosheleva, and Vladik Kreinovich

Abstract In many practical situations, it is necessary to fairly divide the joint gain between the contributors. In the 1950s, the Nobelist Lloyd Shapley showed that under some reasonable conditions, there is only one way to make this division. The resulting Shapley value is now actively used in situations that go beyond economics and finance—and in which Shapley's conditions are not always satisfied: in machine learning, in systems engineering, etc. In this paper, we explain why Shapley value can be applied to such situations, and how can we generalize Shapley value to make it even more adequate for these new applications.

1 Formulation of the Problem

Practical problem with which all this started. This story starts with an important practical problem: we have a group of n people working together on a project. This project is a success. What is the fair way to divide the resulting gain between the participants?

M. Svitek
Faculty of Transportation Sciences, Czech Technical University in Prague, Praha, Czech Republic
e-mail: miroslav.svitek@cvut.cz

N. Winnewisser · M. Beer
Institute for Risk and Reliability, Leibniz University Hannover, Hannover, Germany
e-mail: winnewisser@irz.uni-hannover.de

M. Beer
e-mail: beer@irz.uni-hannover.de

O. Kosheleva
Department of Teacher Education, University of Texas at El Paso, El Paso, TX, USA
e-mail: olgak@utep.edu

V. Kreinovich (✉)
Department of Computer Science, University of Texas at El Paso, El Paso, TX, USA
e-mail: vladik@utep.edu

H. T. Nguyen et al. (eds.), *Data Science in Finance and Accounting*, Studies in Big Data 181, https://doi.org/10.1007/978-3-032-06179-9_2

What information can we use to make a fair division? To decide who contributed more to the success, a reasonable idea is to take into account, for all possible subsets S of the set $N \stackrel{\text{def}}{=} \{1, 2, \ldots, n\}$ of all the participants, the amount $v(S)$ that they could gain if they worked on their own, without others. Based on this function v—that transforms a subset S into a real number $v(S)$—we need to decide how to distribute the amount $v(N)$ that they earned together between the participants, i.e., which amount $x_i(v)$ to allocate to each participant so that $x_1(v) + \cdots + x_n(v) = v(N)$.

For example, if Participant i did not contribute to the gain, i.e., if we have

$$v(S \cup \{i\}) = v(S)$$

for all S, then clearly this participant should not get anything: we should have

$$x_i(v) = 0.$$

What do we mean by fair? The solution to this practical problem was proposed, in the early 1950s, by the future Nobelist Lloyd Shapley; see, e.g., [5–10]. He showed that under some reasonable conditions—including the above condition that a non-contributor should not get anything—there is only one way to solve this problem. This way is now called the *Shapley value*. Here is how Shapley formalized fairness.

First, fairness means that the result should not depend on how we order the participants. In precise terms, if we perform any permutation $\pi : N \mapsto N$, then for the resulting function $v'(S) \stackrel{\text{def}}{=} v(\pi(S))$, the gains $x_i(v')$ should be the same for the same participants, i.e., we should have $x'_{\pi(i)}(v') = x_i(v)$.

Second, it means that the distribution should only depend on this particular situation and not on anything else. In other words, if the same group participates in two projects, with gains $u(S)$ and $v(S)$, then, whether we view them as two project, or a single project with gains $w(u) = u(S) + v(S)$, each participant's overall gain should be the same: $x_i(u + v) = x_i(u) + x_i(v)$. This condition is called *additivity*. In particular, when we combine identical situations, we get $x_i(m \cdot v) = m \cdot x_i(v)$ for all m.

Final formula for Shapley value. Shapley has shown that these two conditions—plus the condition that a non-contributor should not gain anything—uniquely determine the distribution of gain. The resulting formula is

$$x_i(v) = \sum_{S:\, i \notin S} \frac{|S|! \cdot (n - |S|)!}{n!} \cdot (v(S \cup \{i\}) - v(S)), \tag{1}$$

where $|S|$ denotes the number of elements in a set S, and $n!$ is the factorial:

$$n! \stackrel{\text{def}}{=} 1 \cdot 2 \cdot \ldots \cdot n.$$

Shapley value is now effectively used in other applications, but why? In the last decade, Shapley value has been effectively used in many other applications, e.g., in machine learning, where it is used to describe which features are most important in a classification problem. Why Shapley value successfully works in such applications is a big mystery, because, e.g., in machine learning, one of the possible applications is to the case when $v(S)$ is the frequency of cases in which, if we only use features from the set S, we still get a correct classification.

In this application, adding probabilities does not make any sense, so we do not have additivity condition—one of the important conditions that lead to Shapley value. However, Shapley value seems to work well in this application.

What we do in this paper. In this paper, we explain:

- why the original Shapley value works well in these applications, and
- how to modify Shapley value so that it will work even better in these new applications.

2 Analysis of the Problem and the Resulting Explanation of Why Shapley Value Can Be Applied to Machine Learning etc.

What are the new problems to which Shapley value is now applied: analysis. Let us recall different applications to which Shapley value is now applied.

Case of machine learning. Let us start with the application to machine learning. In this application, we have n features based on which we have a reasonably good classification—e.g., classification of images into images or cats and dogs, or, to give a more serious example, a classification of X-ray breast images into cancerous and benign. In many such cases, the full use of all these features requires a lot of processing, more than is possible to implement in each medical facility. So, a natural idea is to try to use fewer features.

To decide which features to use, ideally, we need to know, for each set S of features, what will be the probability $p(S)$ that using only these features will still lead to a correct classification. For n features, there are 2^n possible sets $S \subseteq \{1, \ldots, n\}$. When n is large, the number 2^n is astronomical—e.g., for $n = 300$, the number of possible subsets is larger than the number of particles in the Universe. We therefore cannot test every set S. Instead, a reasonable idea is to have some few-parametric approximation of the actual function $p(S)$. Specifically, what people do in machine learning applications is use the following linear approximation:

$$p(S) \approx v_{\approx}(S) \stackrel{\text{def}}{=} c_0 + \sum_{i:\, i \in S} c_i, \tag{2}$$

for some coefficients c_i—and the coefficients c_i are calculated by using the Shapley value formula.

Other applications—possible and actual. There are many other problems in which it is desirable to find a few-parameter approximation to a function of sets:

- When we have n companies that are potentially interested in merging, it is desirable to predict the gain $v(S)$ (or loss) if companies from a set $S \subseteq \{1, \ldots, n\}$ merge together; see, e.g., [4, 12–14].
- When we have n researchers, it is desirable to predict, for each subset S, what will be the productivity gain when researchers from a set S start actively collaborating. Sometimes, collaboration works, sometimes, it does not, it would be nice to be able to predict when it will work; see, e.g., [4, 12–14].
- When we have a system consisting of n components, and these components are not 100% reliable, it is desirable to predict the system's productivity $v(S)$ in situations when only components from the set S remain working; see, e.g., [16].

How is the success of Shapley value is explained now. A possible explanation for the success of Shapley value in situations like machine learning (when additivity does not seem to be applicable), was provided in [2] (see also [1]). Specifically, if we use the usual least squares approach (see, e.g., [11]) to find the coefficients that best fit the approximate formula (2), then we get linear expressions for the coefficients Δv_i in terms of the values $v(S)$. In particular, one of the options is the Shapley value, but more general expressions are also possible—depending on what standard deviations we use for different values of $|S|$.

Need to go beyond linear approximations. In many practical cases, a linear approximation (2) makes perfect sense. However, when we are interested in studying potential mergers or researcher collaboration, the whole purpose is to make sure that the joint effect is larger than the sum of the original contributions. To capture this additional joint effect, we need to go beyond the linear approximation.

How can we go beyond linear approximations? To answer this question, let us reformulate the right-hand side of the expression (2) in the following equivalent form:

$$v \approx (S) = c_0 + \sum_{i=1}^{n} c_i \cdot m_i(S), \tag{3}$$

where $m_i(S) = 1$ if $i \in S$ and $m_i(S) = 0$ otherwise. In this form, this is clear that the resulting approximation is a linear function of the variables $m_i(S)$.

What can we do if we have an approximate linear dependence, and we want to come up with a more accurate approximation? Such a situation is typical in physics; see, e.g., [3, 15]. A usual physics approach is to take into account that most real-world dependencies $f(x_1, \ldots, x_n)$ are smooth, they can be expanded in Taylor series

$$f(x_1, \ldots, x_n) = c_0 + \sum_{i_1=1}^{n} c_{i_1} \cdot x_{i_1} + \sum_{i_1=1}^{n} \sum_{i_2=1}^{n} c_{i_1 i_2} \cdot x_{i_1} \cdot x_{i_2} + \cdots \tag{4}$$

In general, the linear terms are the largest, quadratic terms are second largest, etc. So, if we are not satisfied with an accuracy of the linear approximation, a natural idea is to also take into account quadratic terms. If this will not lead to a sufficiently accurate approximation, we can take cubic terms into account, etc. In general, we thus get the following approximate formula

$$f_{\approx}(x_1, \ldots, x_n) = c_0 + \sum_{i_1=1}^{n} c_{i_1} \cdot x_{i_1} + \sum_{i_1=1}^{n} \sum_{i_2=1}^{n} c_{i_1 i_2} \cdot x_{i_1} \cdot x_{i_2} + \cdots +$$

$$\sum_{i_1=1}^{n} \ldots \sum_{i_k=1}^{n} c_{i_1 \ldots i_k} \cdot x_{i_1} \cdot \ldots \cdot x_{i_k}. \quad (5)$$

It is reasonable to apply the same approach to our case as well—and this was indeed proposed, e.g., in [4, 12–14, 16]. The only difference from the general formulas (4)–(5) is that in our case, each value $m_i(S)$ is equal either to 0 or to 1, $(m_i(S))^k = 1$ for all $k \geq 1$ and thus, it does not make sense to separately consider terms like $(m_i(S))^2$, etc. So, we arrive at the following formula:

$$v_{\approx}(S) = c_0 + \sum_{i_1} c_{i_1} \cdot m_{i_1}(S) + \sum_{i_1 < i_2} c_{i_1 i_2} \cdot m_{i_1}(S) \cdot m_{i_2}(S) + \cdots +$$

$$\sum_{i_1 < i_2 < \cdots < i_k} c_{i_1 \cdots i_k} \cdot m_{i_1}(S) \cdot \ldots \cdot m_{i_k}(S). \quad (6)$$

Actually, in many cases, there is additivity. In cases of gain—like companies merger or researcher collaboration—we do have the same additivity idea as in the original applications of Shapley value.

In the machine learning case, it does not make sense to add probabilities, but we *can* add gain coming from the increased probability of success. This gain is proportional to probability of success, so additivity of gain naturally translates into additivity of probabilities. *This, by the way, explains why the Shapley value is so effective in machine learning.*

Similarly, in the systems engineering case, we can measure the results of only-some-components-functioning in terms of productivity; in this case, we also have a natural idea of adding productivity of both cases—so we can also formulate a natural additivity condition.

The desired approximation should be exact in some cases. For the usual Shapley value, the linear approximation is exact for the case when S is the empty set: if no one does anything, there is clearly no gain. For the proposed Taylor-series-type generalization, we have more parameters, so we can add more sets for which this approximation is exact. We can start with this exactness requirement for sets S consisting of a single element, then formulate a similar requirement for sets consisting of two elements, etc.

In the next section, we will show that with this additional requirement, there is a unique Taylor-type expression that is additive and symmetric.

3 Our Main Result

Definition. *Let $n \geq 2$ and $k \geq 1$ be integers.*

- *By a* set function *we mean a function that assigns, to each set $S \subseteq N \stackrel{\text{def}}{=} \{1, \ldots, n\}$, a number $v(S)$ so that $v(\emptyset) = 0$.*
- *By a k-*approximation*, we mean an expression of the type*

$$v_{\approx}(S, v) = c_0(v) + \sum_{i_1} c_{i_1}(v) \cdot m_{i_1}(S) + \sum_{i_1 < i_2} c_{i_1 i_2}(v) \cdot m_{i_1}(S) \cdot m_{i_2}(S) + \cdots +$$

$$\sum_{i_1 < i_2 < \cdots < i_k} c_{i_1 \ldots i_k}(v) \cdot m_{i_1}(S) \cdot \ldots \cdot m_{i_k}(S), \tag{7}$$

for some continuous functions $c_{i_1 \ldots}(S)$.
- *We say that the k-approximation is* dummy-fair *if whenever for some set A, we have $v(S \cup A) = v(S)$ for all S, we should get $v_{\approx}(S \cup A, v) = v_{\approx}(S, v)$ for all S. In particular, for $S = \emptyset$, we should get $v_{\approx}(A, v) = v_{\approx}(\emptyset, v)$.*
- *We say that the k-approximation is* symmetric *if for every permutation $\pi : N \mapsto N$, we have $v_{\approx}(\pi(S), \pi(v)) = \pi(v_{\approx}(S, v))$, where $(\pi(v))(S) \stackrel{\text{def}}{=} v(\pi(S))$.*
- *We say that the k-approximation is* additive *for every two set functions u and v, we have $v_{\approx}(S, u + v) = v_{\approx}(S, u) + v_{\approx}(x, v)$.*
- *We say that the k-approximation is* exact for small sets *if $v_{\approx}(S, v) = v(S)$ for all sets S for which $|S| < k$.*

Proposition. *For each n and k, there exists exactly one k-approximation which is dummy-fair, symmetric, additive, and exact for small sets:*

$$v_{\approx}(S, v) = \sum_{T:\ |T| \leq k} \left(\frac{w_v(T)}{|T|} \cdot \prod_{i:\ i \in T} m_i(S) \right) +$$

$$\sum_{T:\ |T| > k} \left(\frac{k! \cdot (|T| - k)!}{|T|!} \cdot \sum_{U:\ U \subset T\ \&\ |U| = k} \left(\frac{w_v(U)}{|U|} \cdot \prod_{i:\ i \in U} m_i(S) \right) \right), \tag{8}$$

where

$$w_v(T) \stackrel{\text{def}}{=} \sum_{S:\ S \subseteq T} (-1)^{|T| - |S|} \cdot v(T). \tag{9}$$

Comment. For $k = 1$, we get a linear expression for which the coefficients c_i are exactly the Shapley values.

Example beyond Shapley value. If $n = 3$, and we get some result only when at least 2 people work together, i.e., if $v(S) = 0$ for all 1-element sets and $v(S) = 1$ for all 2-element sets and for the whole set N, then:

- For $k = 1$, we get a linear approximation based on the Shapley values:

$$v_{\approx}(S) = \frac{1}{3} \cdot m_1(S) + \frac{1}{3} \cdot m_2(S) + \frac{1}{3} \cdot m_3(S). \tag{10}$$

  This approximation is exact only for the empty set $S = \emptyset$.
- For $k = 2$, we get the following quadratic approximation:

$$v_{\approx}(S) = \frac{1}{3} \cdot m_1(S) \cdot m_2(S) + \frac{1}{3} \cdot m_1(S) \cdot m_3(S) + \frac{1}{3} \cdot m_2(S) \cdot m_3(S). \tag{11}$$

  This approximation is exact for all 1-element sets S.
- For $k = 3$, we get the following cubic approximation:

$$v_{\approx}(S) = m_1(S) \cdot m_2(S) \cdot m_3(S). \tag{12}$$

  This approximation is exact for all 2-elements sets S—and, it so happens that it is also exact for the whole 3-element set $N = \{1, 2, 3\}$.

Proof 1°. First, one can see that additivity implies that each coefficient function $c_{i_1 \ldots i_\ell}(v)$ is additive, i.e., that for each such coefficient function, we have

$$c_{i_1 \ldots i_\ell}(u + v) = c_{i_1 \ldots i_\ell}(u) + c_{i_1 \ldots i_\ell}(v). \tag{13}$$

In particular, if we add m identical set functions, we get

$$c_{i_1 \ldots i_\ell}(m \cdot v_{\approx}) = m \cdot c_{i_1 \ldots i_\ell}(v_{\approx}). \tag{14}$$

In particular, for any integer q, for the set function v/q, we get

$$c_{i_1 \ldots i_\ell}(v) = q \cdot c_{i_1 \ldots i_\ell}\left(\frac{v}{q}\right) \tag{15}$$

hence

$$c_{i_1 \ldots i_\ell}\left(\frac{1}{q} \cdot v\right) = \frac{1}{q} \cdot c_{i_1 \ldots i_\ell}(v). \tag{16}$$

By applying the formula (14), we conclude that for every positive rational number m/q, we have

$$c_{i_1 \ldots i_\ell}\left(\frac{m}{q} \cdot v\right) = m \cdot c_{i_1 \ldots i_\ell}\left(\frac{1}{q} \cdot v\right).$$

So, due to (16), we get:

$$c_{i_1 \ldots i_\ell}\left(\frac{m}{q} \cdot v\right) = \frac{m}{q} \cdot c_{i_1 \ldots i_\ell}(v). \tag{17}$$

Since the function $c_{i_1 \ldots i_\ell}(v)$ is continuous, and every non-negative real number a can be represented as a limit of positive rational numbers, we conclude that, in the limit when $m/q \to a$, the same equality (17) holds for every non-negative real number a:

$$c_{i_1 \ldots i_\ell}(a \cdot v) = a \cdot c_{i_1 \ldots i_\ell}(v). \tag{18}$$

Together with additivity, this means that for all possible linear combination of set functions with non-negative coefficients, we have

$$c_{i_1 \ldots i_\ell}(a_1 \cdot v_1 + \cdots + a_t \cdot v_t) = a_1 \cdot c_{i_1 \ldots i_\ell}(v_1) + \cdots + a_t \cdot c_{i_1 \ldots i_\ell}(v_t). \tag{19}$$

By combining these equalities for all the coefficient functions $c_{i_1 \ldots}(v)$, we conclude that

$$v_{\approx}(S, a_1 \cdot v_1 + \cdots + a_t \cdot v_t) = a_1 \cdot v_{\approx}(S, v_1) + \cdots + a_t \cdot v_{\approx}(S, v_t). \tag{20}$$

2°. Let us show that the equality (20) holds also when some coefficients a_i are negative—but the expression

$$v \stackrel{\text{def}}{=} a_1 \cdot v_1 + \cdots + a_t \cdot v_t \tag{21}$$

is still positive and thus, makes sense. Indeed, in this case, if, in the equality (21), we move all the terms for which a_j is negative to the left-hand side, we get the following equality:

$$v + \sum_{j:\, a_j < 0} |a_j| \cdot v_j = \sum_{j:\, a_j \geq 0} a_j \cdot v_j. \tag{22}$$

On both sides, the coefficients are non-negative, so we can apply the formula (20) to both sides and get the following equality:

$$v_{\approx}(S, v) + \sum_{j:\, a_j < 0} |a_j| \cdot v_{\approx}(S, v_j) = \sum_{j:\, a_j \geq 0} a_j \cdot v_{\approx}(S, v_j) \tag{23}$$

Now, we can move the terms corresponding to $a_j < 0$ back into the right-hand side and get

$$v_{\approx}(S, v) = \sum_j a_j \cdot v_{\approx}(S, v_j), \tag{24}$$

i.e., indeed, the formula (20) holds for all possible coefficients a_j.

3°. To complete the proof, we will use th fact—that is used in the original Shapley's proof—that for every set function $v(S)$, we have:

$$v(S) = \sum_T w_v(T) \cdot v_T(S), \tag{25}$$

where the sum is over all sets T, and $v_T(S)$ denotes the following set function:

- we have $v_T(S) = 1$ when $T \subset S$ and
- we have $v_T(S) = 0$ otherwise.

In other words, the set function $v_T(S)$ means the following:

- if all the participants from the set T get together, then they gain 1 unit—irrespective of what others do;
- otherwise, if not all participants from the set T get together, no one gains anything.

Because we proved that formula (20) holds for all linear combinations, with coefficients of any sign, we can conclude that

$$v_{\approx}(S, v) = \sum_T w_v(T) \cdot v_{\approx}(S, v_T). \tag{26}$$

So, to find the desired value $v_{\approx}(S, v)$, it is sufficient to describe the values $v_{\approx}(S, v_T)$ corresponding to all possible sets T.

4°. For each function v_T, participants outside the set T do not affect the value of the set function, and thus—due to dummy-fairness, adding these participants should not change the approximate value $v_{\approx}$:

$$v_{\approx}(S, v) = v_{\approx}(S \cap T, v). \tag{27}$$

Thus, it is sufficient to consider the values $v_{\approx}(S, v)$ for subsets S of the set T.

By definition of the set function v_T, we have $v_T(S) = 0$ for all S for which $|S| < |T|$. When $|T| \le k$, then, due to the exactness requirement, this means that we should also have $v_{\approx}(S, v) = 0$ for all S. So, we cannot have any products of fewer than $|T|$ values $m_i(S)$. So, the only possible term is the product of all $|T|$ values $m_i(S)$ corresponding to $i \in T$.

When $|T| > k$, we should also have $v_{\approx}(S, v) = 0$ for all sets S with $|S| < k$. Thus, we also cannot have any products with fewer than k values. So the only remaining option is to have products of k values. The set function $v_T(S)$ does not change if we perform any permutation of the elements of the set T. Thus, due to symmetry, once we have one such product, we should have, with the same coefficient, *all* such products corresponding to all possible subsets U of size k. From the fact that the sum of all these terms should be equal to 1, we deduce the coefficient at each such product—it is 1 divided by the number of such subsets U.

Substituting the resulting expressions for $v_{\approx}(S, v_T)$ into the formula (26), we get the desired expression (8). The proposition is proven.

4 Conclusions: How Can These Results Be Applied to Finance/accounting Domain?

General topic of this paper. The main purpose of this paper is to analyze financial aspects of collaboration—be it collaboration between companies, between subdivisions within the same company, between individuals within the same subdivision, etc.

What is known—the starting point of our research. Once collaboration is in place, and this collaboration has been successful, a natural question is: what is the fair way to distribute the resulting gain between collaborating entities?

A solution to this problem was proposed by the Nobelist Lloyd Shapley. Specifically, Shapley showed that if we impose several reasonable conditions—that describe the notion of fairness in precise terms—then there is exactly one way to distribute the gain that satisfies all these conditions. The resulting distribution is known as the *Shapley value*.

First natural question, how we answered this question, and how this answer can be used in financial domain. Shapley value assumes that the companies (departments, people, etc.) have already decided who collaborates with whom. But how can they make such a decision? To make this decision, participants need to be able to predict, based on the experience of previous collaborations (between companies, departments, people, etc.) what will be the results of different groupings.

It is known that the Shapley value technique can be used to provide an approximate solution to this question—but this solution often provides only a crude estimate, an estimate that does not properly take into account that collaboration often leads to an increase in productivity. In this paper, we show that a natural modification of Shapley's fairness conditions leads to a more general model that provides a more accurate prediction of collaboration results.

This new model can be used to make the desired predictions more accurately.

Second natural question, how we answered this question, and how this answer can be used in financial domain. A second natural question is related to the fact that in the last decade, Shapley value has been used in the area different from economics and finance: namely, in machine learning. At present, after appropriate training, deep learning systems get spectacularly good results—in particular, results in economics and finance, where machine learning algorithms are actively used, e.g., in trading of stocks, bonds, futures, derivatives, and other financial instruments. The problem is that the corresponding training takes a very long time—and making real-time decisions requires so many computational steps that only special high-performance computers can do it.

One way to speed up computations is to take into account that, while the system starts with all the features that could possibly affect the prices, later on it turns out that effect of many of these features is negligible. Thus, we can get almost perfect predictions—and much faster—if we do not use such irrelevant features at all.

To decide which features to use, we need to be able to predict how effective will our prediction be if we only use some of these features. From the commonsense viewpoint, this problem is similar to collaboration problem, so researchers tried to use Shapley value here—and it worked, not perfectly, but it worked. A natural question is: yes, in several examples this idea worked, but will it always work?

This question was not raised for the economic and financial applications of Shapley value—because in these applications, Shapley value was uniquely determined by natural fairness conditions. These conditions, however, do not apply to machine learning: while fairness towards participating people makes sense, there is no sense to require fairness with respect to features—all we care about is the final result.

In this paper, we provide an alternative explanation for the Shapley value, an explanation in which the Shapley value is still uniquely determined by several conditions, but this time, the conditions do not use fairness—they are appropriate in the context of machine learning.

Since this way, the use of Shapley value in machine learning is justified, we can therefore be more confident in the results of this use. Similar justifications show that we can use, in the machine learning case, generalizations of the notion of the Shapley value—generalizations that were developed as an answer to the above-described first question.

Acknowledgements This work was supported in part by the National Science Foundation grants 1623190 (A Model of Change for Preparing a New Generation for Professional Practice in Computer Science), HRD-1834620 and HRD-2034030 (CAHSI Includes), EAR-2225395 (Center for Collective Impact in Earthquake Science C-CIES), and by the AT&T Fellowship in Information Technology.

It was also supported by a grant from the Hungarian National Research, Development and Innovation Office (NRDI), by the Institute for Risk and Reliability, Leibniz Universitaet Hannover, Germany, by the Deutsche Forschungsgemeinschaft (DFG, German Research Foundation) Focus Program SPP 100+ 2388, Grant Nr. 501624329, and the European Union under the project ROBOPROX (reg. no. CZ.02.01.01/00/22 008/0004590).

The authors are thankful to Van Nam Huynh for important suggestions.

References

1. Autchariyapanikul, K., Kosheleva, O., Kreinovich, V.: Shapley value under interval uncertainty and partial information. In: Kreinovich, V., Yamaka, W., Leurcharusmee, S. (eds.) Data Science for Econometrics and Related Topics. Springer, Cham, Switzerland, to appear
2. Bokati, L., Kosheleva, O., Kreinovich, V., Thach, N.N.: Why Shapley value and its variants are useful in machine learning (and in other applications). In: Kreinovich, V., Sriboonchitta, S., Yamaka, W. (eds.) Machine Learning for Econometrics and Related Topics, pp. 169–174. Springer, Cham, Switzerland (2024)
3. Feynman, R., Leighton, R., Sands, M.: The Feynman Lectures on Physics. Addison Wesley, Boston, Massachusetts (2005)
4. Galindo, O., Svitek, M., Kreinovich, V.: Quantum (and more general) models of research collaboration. Asian J. Econ. Bank. **4**(1), 77–86 (2020)
5. Luce, R.D., Raiffa, R.: Games and Decisions: Introduction and Critical Survey. Dover, New York (1989)
6. Myerson, R.B.: Game Theory: Analysis of Conflict. Harvard University Press, Cambridge, Massachisetts (1997)
7. Owen, G.: Game Theory. Emerald Publishers, Bingley, UK (2013)
8. Roth, A.E., (ed.): The Shapley Value: Essays in Honor of Lloyd S. Cambridge University Press, Cambridge, Shapley (1988) 0-521-36177-X. https://doi.org/10.1017/CBO9780511528446
9. Shapley, L.S.: Notes on the n-Person Game - II: The Value of an n-Person Game, Research Memorandum RM-670. RAND Corporation, Santa Monica, California, USA (1951)
10. Shapley, L.S.: Value for n-person Games. In: Kuhn, H.W., Tucker, A.W. (eds.) Contributions to the Theory of Games, Annals of Mathematical Studies, vol. 28, Princeton University Press, Princeton, New Jersey, USA, 1953, pp. 307–317. https://doi.org/10.1515/9781400881970-018. ISBN 9781400881970
11. Sheskin, D.J.: Handbook of Parametric and Nonparametric Statistical Procedures. Chapman and Hall/CRC, Boca Raton, Florida (2011)
12. Svitek, M., Kosheleva, O., Kreinovich, V.: When is it beneficial to merge two companies? When is it beneficial to start a research collaboration? In: Ceberio, M., Kreinovich, V., (eds.) Uncertainty, Constraints, AI, and Decision Making. Springer, Cham, Switzerland, to appear
13. Svitek, M., Kosheleva, O., Kreinovich, V., Nguyen, T.N.: Why quantum (wave probability) models are a good description of many non-quantum complex systems, and how to go beyond quantum models. In: Kreinovich, V., Thach, N.N., Trung, N.D., Thanh, D.V. (eds.) Beyond Traditional Probabilistic Methods in Economics, pp. 168–175. Springer, Cham, Switzerland (2019)
14. Svitek, M., Kosheleva, O., Kreinovich, V., Phuong, N.H.: To which interdisciplinary research collaborations should we pay more attention? In: Phuong, N.H., Chau, N.T.H., Kreinovich, V., (eds.) Explainable Artificial Intelligence and Other Soft Computing Techniques: Biomedical and Related Applications. Springer, Cham, Switzerland, to appear
15. Thorne, K.S., Blandford, R.D.: Modern Classical Physics: Optics, Fluids, Plasmas, Elasticity, Relativity, and Statistical Physics. Princeton University Press, Princeton, New Jersey (2021)
16. Winnewisser, N., Beer, M., Kosheleva, O., Kreinovich, V.: How Shapley value and its generalizations can help in the analysis of complex engineering systems and what next. In: Huynh, V.-N., Honda, K., Le, B.H., Inuiguchi, M., Huynh, H.T., (eds.) Integrated Uncertainty in Knowledge Modelling and Decision Making. Proceedings of the 11th International Symposium, IUKM 2025, Ho Chi Minh City, Vietnam, March 17–19, 2025, Part II, Springer Lecture Notes in Computer Science, 2025, vol. 15586, pp. 85–97. https://doi.org/10.1007/978-981-96-4603-6_8

Why Skew-Normal Distributions and How They Are Related to ReLU Activation Function in Deep Learning

Damian Lorenzo Gallegas Espinosa, Olga Kosheleva, and Vladik Kreinovich

Abstract Normal distributions are ubiquitous, but many actual distributions are different from normal—for example, they are skewed. To describe such distributions, it is desirable to have a few-parametric family that extends the family of normal distributions. Several such families have been proposed. Empirically, the most effective among them is the family of so-called skew-normal distributions first proposed by A. Azzalini. In particular, this family is effective in econometrics. In this paper, we provide a theoretical explanation for this empirical success. This explanation is similar to an explanation of what ReLU activations functions are most effective in deep learning.

1 Why Skew-Normal Distributions: a Challenge

Normal distributions and why they are ubiquitous: a brief reminder. In order to understand our problem, let us first recall what are normal distributions and why they are ubiquitous.

D. L. Gallegas Espinosa · V. Kreinovich (✉)
Department of Computer Science, University of Texas at El Paso, El Paso, TX, USA
e-mail: vladik@utep.edu

D. L. Gallegas Espinosa
e-mail: dlgallegose@miners.utep.edu

O. Kosheleva
Department of Teacher Education, University of Texas at El Paso, El Paso, TX, USA
e-mail: olgak@utep.edu

H. T. T. Nguyen et al. (eds.), *Data Science in Finance and Accounting*, Studies in Big Data 181, https://doi.org/10.1007/978-3-032-06179-9_3

In many practical situation, we have many small independent factors affecting the desired quantity. In such cases, according to the Central Limit Theorem (see, e.g., [4]), the resulting distribution is close to *Gaussian* (*normal*), i.e., a distribution with the following probability density function:

$$f(x) = \frac{1}{\sqrt{2\pi} \cdot \sigma} \cdot \exp\left(-\frac{(x-m)^2}{2\sigma^2}\right),$$

where m is the distribution's mean and σ is its standard deviation.

In line with the Central Limit Theorem, many real-life distributions are well-described by normal distributions.

Comments.

- The formula for the normal distribution can be described in the following equivalent form:

$$f(x) = f_0\left(\frac{x-m}{\sigma}\right),$$

  where

$$f_0(x) = \frac{1}{\sqrt{2\pi} \cdot} \cdot \exp\left(-\frac{x^2}{2}\right)$$

  is the probability density function of the *basic* normal distribution, with mean 0 and standard deviation 1.
- A normal distribution is uniquely determined by its first two moments: the mean value m and the standard deviation σ. So, when the empirical distribution is normal, we can easily find the exact shape of this distribution:
  - first, based on the sample, we estimate the mean m and the standard deviation σ, and
  - then, we use the normal distribution with these values of the mean and the standard deviation.

  The resulting distribution is symmetric relative to the mean m—and, as a result, its third central moment is equal to 0.

Need to go beyond normal distributions. Although normal distributions are ubiquitous, not all empirical distributions are normal. Many empirical distributions are *skewed* (asymmetric), i.e., have non-zero third central moment.

It is therefore desirable to have a few-parametric generalization of the class of normal distributions that would allow to consider skewness.

Skew-normal distributions: a brief reminder. Many skew-related generalizations of the class of normal distributions are possible, and many such families have been proposed.

Empirically, one of such families—called *skew-normal* (see, e.g., [1])—has been most successful. The basic skew-normal distribution has the form

$$s_0(x) = f_0(x) \cdot F_0(\alpha \cdot x),$$

where $F_0(x)$ is the cumulative distribution function corresponding to the basic normal distribution, and α is some real number. The probability density function $s(x)$ of a general skew-normal distribution can be obtained from the pdf of the basic one:

$$s(x) = s_0\left(\frac{x - m}{\sigma}\right)$$

for some m and σ.

In particular, skew-normal distributions are very effective in econometrics; see, e.g., [3, 5, 7] and references therein.

Resulting challenge. How can we explain why this particular generalization of the family of normal distributions turned out to be the most successful?

What we do in this paper. In this paper, we provide a possible explanation for the success of skew-normal distribution.

It turns out that this explanation is similar to one of the possible explanations of another empirical success—the success of rectified linear activation functions in deep learning.

2 Why Skew-Normal: an Explanation

We need a distribution that can be used in simulations. One of the main objectives of determining the actual distribution of the analyzed phenomena is that we will be able to simulate this phenomenon—by appropriate simulating this distribution.

From this viewpoint, it is necessary to recall how different probability distributions are simulated.

Two usual ways to simulate a distribution. Most computers have built-in random number generators—hardware or software implemented—that generate either the uniform distribution on the interval [0, 1] or the basic normal distribution, with mean 0 and standard deviation 1. To generate different distributions, we usually use these random number generators.

There are two main ways to do it:

- we can either apply some easy-to-compute function to the results of a random number generator,
- or we compute an easy-to-compute (e.g., linear) combination of several such results.

Let us briefly recall both approaches.

First way to simulate a distribution. Let us start with the technique of applying an easy-to-compute function to the results of the standard number generator. This is, for example, how computers simulate a generic normal distribution, with mean m and standard deviation σ:

- they use the standard random number generator to generate a random number X distributed according to the basic normal distribution, and
- then they compute $Y = m + \sigma \cdot X$.

One can easily check that the resulting random variable Y indeed has the desired distribution.

The use of this technique for simulating distributions other than normal comes from the easy-to-prove fact that all continuous distributions on a finite or infinite interval—i.e., all distributions with continuous probability density functions—can be obtained from each other by a continuous re-scaling of the interval. In precise terms:

- if X is a random variable that is distributed according to the first distribution,
- then, for an appropriate increasing function $g(x)$, the value $Y = g(X)$ is distributed according to the second distribution.

The desired function $g(x)$ is easy to find. Indeed:

- Let us denote the cumulative distribution function of the first distribution by
$$F_X(x) \stackrel{\text{def}}{=} \text{Prob}_X(X \le x).$$
- Let us denote the cumulative distribution of the second distribution by
$$F_Y(y) \stackrel{\text{def}}{=} \text{Prob}_Y(Y \le y).$$

Then, for any strictly increasing function $g(x)$, we have $X \le x \leftrightarrow g(X) \le g(x)$. Thus, the cumulative distribution function for $g(X)$ has the property

$$F_{g(X)}(g(x)) = \text{Prob}(g(X) \le g(x)) = \text{Prob}_X(X \le x) = F_X(x).$$

So, for any y, if we, as usual, denote the x for which $g(x) = y$ by $g^{-1}(y)$, then we get $F_{g(X)}(y) = F_X(g^{-1}(y))$. Thus, to get the desired distribution for $g(X)$, we need to make sure that we always have $F_Y(y) = F_X(g^{-1}(y))$. In other words, for $x = g^{-1}(y)$, for which $y = g(x)$, we need to have $F_Y(g(x)) = F_X(x)$. Thus, if we apply the inverse function F_Y^{-1} to both sides of this equality, we get the explicit formula for the desired re-scaling function:

$$g(x) = F_Y^{-1}(F_X(x)).$$

Second way to simulate a distribution. When all we have is a random number generator for the uniform distribution, then one of the possible ways to simulate a normal distribution is to use the above-mentioned Central Limit Theorem, according to which the distribution of the sum of a large number of independent random variables is close to normal. Thus, we can form the sum of several uniformly distributed random variables—and we will get a good simulation of a normal distribution.

For independent normal random variables, their sum—and, more generally, their linear combination—is also normally distributed.

Main idea behind our explanation. We need a non-normal distribution. Since normal distributions are ubiquitous, all computer systems already have a normal distribution. So a natural idea is to apply some function to the normal random variable.

As we have mentioned, this does not restrict the class of distributions, since all continuous distributions can be obtained from each other by applying some function—this is one of the usual ways to simulate different distributions.

Applying a linear function will still keep the resulting distribution normal. So, to simulate a probability distribution that is different from normal, we need to us apply a nonlinear function.

From the practical viewpoint, the faster-to-compute this nonlinear function, the better. Thus, we need ot use the fastest-to-compute nonlinear functions.

So what are the fastest-to-compute nonlinear functions?

What are the fastest-to-compute nonlinear functions? To answer this question, let us recall which computer operations are the fastesr. In a computer, the fastest possible operations are:

- unary minus—it changes just one bit, the sign bit, and
- min and max—that require, on average, 2 bit operations.

Indeed:

- In 1/2 of the cases, the two numbers have different first bits, i.e., 0 and 1. So, by applying only one bit operation—namely, by comparing the first bits of the two numbers—we already know which number is larger.
- In 1/4 of the cases, the first bits are equal, but the second bits are different. In this case, we need 2 bit comparisons to decide which number is larger.
- In general, in $1/2^k$ cases, the first $k-1$ bits are equal, but the k-th bits are different. In this case, we need k bit comparisons to decide which number is larger.

So, the average number b of bit operations needed to compute min or max is equal to the following:

$$b = 1 \cdot \frac{1}{2} + 2 \cdot \frac{1}{2^2} + \ldots + (k-1) \cdot \frac{1}{2^{k-1}} + k \cdot \frac{1}{2^k} + (k+1) \cdot \frac{1}{2^{k+1}} + \ldots \quad (1)$$

To compute the value b, let us divide both sides of the formula (1) it by 2. Then we get:

$$\frac{b}{2} = \frac{1}{2^2} + 2 \cdot \frac{1}{2^3} + \ldots + (k-1) \cdot \frac{1}{2^k} + k \cdot \frac{1}{2^{k+1}} + (k+1) \cdot \frac{1}{2^{k+2}} + \ldots \quad (2)$$

If we subtract (2) from (1) term by term, subtracting the terms proportional to the same factor $1/2^k$, we will get

$$\frac{b}{2} = \frac{1}{2} + \frac{1}{2^2} + \ldots + \frac{1}{2^k} + \frac{1}{2^{k+1}} + \ldots \quad (3)$$

The sum in the right-hand side is a geometric progression, so, in principle, we can use the known formula for its sum. But for simplicity, instead of using this general formula, let us use the same trick that we have just used. First, we divide both sides of the equality (3) by 2, resulting in the following:

$$\frac{b}{4} = \frac{1}{2^2} + \frac{1}{2^3} + \ldots + \frac{1}{2^k} + \ldots \quad (4)$$

If we subtract (4) from (3), we conclude that

$$\frac{b}{2} - \frac{b}{4} = \frac{b}{4} = \frac{1}{2}. \quad (5)$$

Thus, indeed

$$b = 4 \cdot \frac{1}{2} = 2. \quad (6)$$

In addition to the above two operations, we can also have constants like 0 that are also easy to generate.

Next in complexity are addition and subtraction. They require as many bit operations as there are bits—i.e., 64 on most computers. Multiplication and division require even more bit operations.

So, what are the fastest-to-compute nonlinear functions? The fastest functions are the ones that only use the fastest computer operations:

- unary minus that computes $-x$ and
- minimum and/or maximum.

The function $x \mapsto -x$ is linear. So, the only way to get a nonlinear functions is to apply minimum or maximum to x and $-x$. This leads either to $|x| = \max(x, -x)$ or to $-|x| = \min(x, -x)$.

Comment. It would be even faster to take $\max(0, x)$ or $\min(0, x)$—this would require only one fastest computer operations—but this would not lead to a continuous random variable :-(

This explains the efficiency of skew normal distributions. Interestingly, if we apply each of these two functions: $|x|$ and $-|x|$, to the basic normal distribution, we get two particular cases of skew normal distributions.

And if we use the second way of simulating random distributions and consider all possible linear combinations of $|x|$ (for a normal random variable x) and independent normal random variables, then we get exactly all skew-normal distributions.

This explains the empirical success of this family of distributions.

3 How Is This Related to ReLU?

What we do in this section. In this section, following [6], we will show that similar ideas explain the empirical success of Rectified Linear (ReLU) activation functions in machine learning (see, e.g., [2]). Later in this section, we will explain what is an activation function in general and what is ReLU activation function.

What are neural networks: a brief reminder. In a neural network, we have several units (called neurons) that perform some transformations: they take several values $v_1, \ldots, v_n$ and return the value

$$v = s(w_1 \cdot v_1 + \ldots + w_n \cdot v_n - w_0),$$

where:

- the values w_i are numerical constants, and
- the function $s(x)$ is a function known as *activation function*.

Some neurons process inputs to the algorithm, others process results of the previous neurons, etc. The output of one of the neurons is then returned as the computation result.

Need for non-linear neurons. Some neurons have a linear activation function. For such neurons, the dependence of v on v_i is linear.

However, If all neurons were linear, then we would only get linear functions, and many real-life dependencies are nonlinear. Thus, to be able to describe real-life dependencies, we need to add neurons with nonlinear activation functions.

Which nonlinear activation functions should we choose?

Which nonlinear activation functions should we choose? In many practical situations, we cannot compute the desired quantity since computations take too long. From this viewpoint, it is desirable to perform computations as fast as possible.

In relation to neurons, this means that we need to select the fastest-to-compute activations functions.

This explains ReLU activation function. Similar to the above analysis of re-scaling functions $g(x)$, we can conclude that the fastest-to-compute are the activation functions that consist of the fastest computer operations—i.e., changing the sign, minimum, and maximum. Simuilar to the above case, we conclude that we should get one of the following functions:

$$|x| = \max(x, -x), \quad -|x| = \min(x, -x), \quad \max(0, x), \quad \min(0, x),$$

$$-\max(0, x), \quad -\min(0, x), \quad \max(0, -x), \quad \min(0, -x),$$

$$-\max(0, -x), \text{ and } -\min(0, -x).$$

In this case—in contrast to the case analyzed in the previous sections—all these options are possible. Of these options, the fastest are the ones that use only one fastest computer operations, i.e.,

$$\max(0, x) \text{ and } \min(0, x).$$

From the viewpoint of the future linear transformations in a neural network, the activation function $\min(0, x)$ is equivalent to $\max(0, x)$; indeed:

- we have since $\min(0, x) = -\max(0, -x)$ and,
- similarly, $\max(0, x) = -\min(0, -x)$.

Since, from the neural network viewpoint, these two functions are equivalent, we can therefore conclude that the fastest-to-compute activation function is

$$s(x) = \max(0, x).$$

This is exactly the rectified linear (ReLU) activation functions that is so effective in machine learning.

4 Conclusions and Future Work

Let us summarize what we learned, how this can be used, and what are the remaining problems and challenges.

Uncertainty is important in financial applications. To make adequate financial decisions, we need to be able to predict financial consequences of different possible investments and of different possible financial policies.

Historically, numerical predictions of future events started with physics, where it is often possible to take all the related effects into account—and thus, to make very accurate predictions. For example, already in antiquity, astronomers were able to make very good predictions of the positions of planets on the night sky—and even of solar eclipses.

In contrast, in such a complex field as finance, it is not realistically possible to take all the factors into account. As a result, financial predictions come with uncertainty. No matter how much knowledge we have, we cannot predict the exact future value of a financial instrument. At best, we can predict the probabilities of different future values—i.e., a frequency with which these values will appear in different similar situations. How can we describe these probabilities?

In financial applications, we often need to go beyond normal distributions. Again historically, the study of probabilities started with physics. When we toss a coin, we cannot predict on which side it will fall, we can only predict the probability of heads and tails. In general, when we use a measuring instrument to measure a quantity, we may get slightly different results if we use a different instrument—or even if we use the same instrument again. The difference between the measurement result and the actual value—known as *measurement error*—is usually normally distributed. In general, many real-life quantities are normally distributed. Because of this, normally distributed random variables form the basis of most widely used statistical methods. Many of these methods use the fact that to fully describe a normal distribution, we need to know only two parameters—mean and standard deviation, parameters that describe the first and the second central moments.

In contrast to physics, in finance, many distributions are not normal, they are skewed—i.e., have a non-zero third central moment. The classical example is the distribution of people by income. To process such data, we need to find a good way to describe such distributions. In precise terms, we need to have a few-parametric family that will allow us to accurately describe such distributions.

Resulting challenge, how we resolve it, and how our resolution can be used. For describing skewed probability distributions, many different families have been proposed and tested. It turned out that out of all these families, the family of so-called skew-normal distributions leads to the most accurate description of the actual distributions in economics and finance.

The resulting challenge is that researchers have tried only a few families, and there are, potentially, infinitely many possible families. So a natural question is: are skew-normal distributions really the best—or are there other not-yet-tried families that will provide an even more accurate description of financial processes?

To answer this question, we reformulate it in precise terms, and show that in this formulation, the skew-normal family is indeed the best. We are reasonably confident in our conclusion—since a similar reformulation of the question about selecting an activation function in deep neural network shows, in perfect accordance with empirical studies, that the best choice is the rectified linear (ReLU) function $\max(0, x)$.

So our recommendation is to use skew-normal distributions when processing financial data—and not to waste time trying other families.

Future work. The family of skew-normal distributions has three parameters—that correspond to the three central moments. But actual distributions also differ in their fourth central moment—kurtosis. For example, some actual distributions have *heavy tails*—i.e., larger values of this moment. It is therefore desirable to find an extension of the skew-symmetric family that would allow us to describe different values of the kurtosis as well.

Several such families have been proposed, but this time, it is not clear which of them should be selected as the best—this selection remains an important open problem.

Acknowledgements This work was supported in part by the National Science Foundation grants 1623190 (A Model of Change for Preparing a New Generation for Professional Practice in Computer Science), HRD-1834620 and HRD-2034030 (CAHSI Includes), EAR-2225395 (Center for Collective Impact in Earthquake Science C-CIES), and by the AT&T Fellowship in Information Technology. It was also supported by a grant from the Hungarian National Research, Development and Innovation Office (NRDI), and by the Institute for Risk and Reliability, Leibniz Universitaet Hannover, Germany.

The authors are greatly thankful to all the participants of the Workshop on Mathematics, Computer Science, and Computational Science organized by New Mexico State University and University of Texas at El Paso (Las Cruces, New Mexico, April 12, 2025), especially to Tonghui "Tony" Wang, for valuable discussions, and to Van Nam Huynh for important suggestions.

References

1. Azzalini, A.: The Skew-Normal and Related Families. Cambridge University Press, Cambridge, UK (2013)
2. Goodfellow, I., Bengio, Y., Courville, A.: Deep Learning. MIT Press, Cambridge, Massachusetts (2016)
3. Ma, Z., Wang, T., Choy, S.T.B., Wei, Z., Zhu, X.: Extending the A Priori Procedure for estimating location parameter under multivariate skew normal settings. In: Thach, N.N., Kreinovich, V., Ha, D.T., Trung, N.D. (eds.) Optimal Transport Statistics for Economics and Related Topics, pp. 107–116. Springer, Cham, Switzerland (2024)

4. Sheskin, D.J.: Handbook of Parametric and Nonparametric Statistical Procedures. Chapman and Hall/CRC, Boca Raton, Florida (2011)
5. Tong, T., Chen, X., Wang, T., Trafimow, D., Choy, S.T.B., Wang, C.: The A Priori Procedure (APP) for estimating standardized regression coefficients in multiple linear model with skew errors. In: Kreinovich, V., Sriboonchitta, S., Yamaka, W. (eds.) Machine Learning for Econometrics and Related Topics, pp. 143–160. Springer, Cham, Switzerland (2024)
6. Urenda, J.C., Kreinovich, V.: Why rectified linear activation functions? Why max-pooling? a possible explanation. In: Castillo, O., Melin, P. (eds.) New Perspectives on Hybrid Intelligent System Design based on Fuzzy Logic, Neural Networks and Metaheuristics, pp. 459–463. Springer, Berlin (2023)
7. Wang, Z., Wang, T., Trafimow, D., Choy, S.T.B., Chen, X.: Extending a priori procedure for simultaneously estimating location and scale parameters in the context of skew normal distributions. In: Thach, N.N., Trung, N.D., Ha, D.T., Kreinovich, V., (eds.) Artificial Intelligence and Machine Learning for Econometrics: Applications and Regulation (and Related Topics). Springer, Cham, Switzerland, to appear

A Study of Machine Learning Models for Financial Distress Prediction

Jiaying Ni, Hung Le, Hien Thu Thi Nguyen, and Van-Nam Huynh

Abstract Predicting corporate financial status is one of the key research directions in financial risk management. Machine learning methods, on the other hand, have shown tremendous ability and potential in the domain of finance, particularly in financial distress prediction. The lack of access to financial data, the ambiguous definition of financial distress, and the differences in the methods used in the literature, however, have contributed to the confusion of researchers who are interested in the field. In this study, we revisit the concepts of financial distress and review recent papers talking about predictors, datasets, and models to construct early warning systems for financial distress prediction. We identify the current trends, present our comparative results of the main approaches, and discuss the advantages and disadvantages of each approach. While much effort was spent on predicting the likelihood of distress, we find that there are still some research opportunities. We conclude our study by discussing the need to develop explainable methods and the potential of using multiple sources of information. Through this work, we provide an overview of the field and categorize our understanding of common methodologies to tackle the challenges in distress prediction research.

J. Ni (✉) · H. Le
Japan Advanced Institute of Science and Technology, Ishikawa, Japan
e-mail: jiaying@jaist.ac.jp

H. Le
e-mail: lehung@jaist.ac.jp

H. T. T. Nguyen
School of Finance and Accounting, Industrial University of Ho Chi Minh City, Ho Chi Minh City, Vietnam
e-mail: nguyenthuhien@iuh.edu.vn

V.-N. Huynh
Japan Advanced Institute of Science and Technology, Ishikawa, Japan
e-mail: huynh@jaist.ac.jp

School of Finance and Accounting, Industrial University of Ho Chi Minh City, Ho Chi Minh City, Vietnam

H. T. T. Nguyen et al. (eds.), *Data Science in Finance and Accounting*, Studies in Big Data 181, https://doi.org/10.1007/978-3-032-06179-9_4

1 Introduction

Economic depression is an unpleasant experience for all involved parties. Corporate stakeholders lose money, people lose jobs, and society as a whole becomes unstable. The depression started when a few companies went bankrupt, and then a few more that depended on them. Bankruptcy is contagious, as declared by [1]. If we can detect the companies that are going to be bankrupt, we can isolate them; or even better, if we can detect the companies that will go through financial difficulties, we can make the changes to save them; then, significant social and economic costs could be avoided. In fact, the attempt to build a corporate failure early warning system has been of great interest since the 1960s classical work of [2, 3].

A failed company's life cycle starts at its establishment and ends at its declaration of bankruptcy. During this life cycle, there are periods of prosperity and difficulties. Financial distress is a term used to describe difficult periods, which could eventually lead to bankruptcy. Thus, predicting financial distress is currently considered a very important direction in economic and financial research. Generally, financial distress refers to the likelihood that a corporation will be unable to repay its debt [4]. In other cases, scholars define financial distress as a state where cash flow is low, and where the corporation has suffered losses but is not yet insolvent [5]. Financial distress prediction (FDP) is the task of discriminating between healthy companies and financially distressed companies. Accurate prediction of financial distress is important to help investors make informed decisions and reduce risk. Banks and other financial institutions are also interested in establishing early warning systems to avoid serious financial crises [6]. Traditionally, the main line of research used to address FDP problems focused on accounting-based models using statistical methods, for example, [2, 7, 8]. In recent years, however, the emergence of large and diverse datasets and advances in machine learning (ML) techniques have given rise to another approach: applying ML models for the task of FDP.

There are several existing studies that review and systematically categorize bankruptcy and financial distress identification methods. [9] is one of the earliest reviews of bankruptcy prediction models that include artificial intelligence tools. Since then, an increasing number of studies have been dedicated to developing ML models for FDP, which in turn has called for more updated surveys. In [10], the authors reviewed the bankruptcy prediction literature related to privately held firms in underdeveloped countries. Their review revealed that in the majority of cases, the classical Z-score models are the preferred methods. Another systematic review of studies between 2017 and 2022, [11], provided a wide overview of artificial intelligence methods for addressing financial distress. While their work is extensive, there is a need to go deeper into the working mechanism of the models and to conduct comparative studies from a real-life dataset.

ML methods can efficiently manage large datasets preprocessed with the proper pipeline, and the prediction accuracy can be improved [12]. However, several challenges remain in the field of FDP research. The first major challenge is the inconsistency of the financial distress definition. This challenge arises from the fact that financial distress covers a wide range of stages and degrees, from temporary liquidity problems to severe financial problems leading to bankruptcy. Therefore, the question "*what is distressed?*" needs to be clearly understood and differentiated. Another significant challenge is the difficulty in accessing high-quality financial data. In modern times, financial distress forecasts often rely on multiple data sources, such as financial statements, market data, news articles, and social media posts. These data sources vary in format, quantity, quality, and unique identificators, so integrating them is a complex process. Effective preprocessing techniques (e.g., data cleaning, feature selection) are essential to handle these diverse data types and ensure the reliability of the predictive models. The various methods that exist in the literature may cause confusion among researchers. Therefore, a systematic survey is needed to categorize and compare these approaches to provide a clearer understanding of the field.

This study aims to address these challenges by revisiting the concept of financial distress, reviewing the relevant literature, and presenting a comprehensive overview of the predictor variables, datasets, and models used to construct early warning systems for financial distress prediction. We identify current trends and present comparative results of the main approaches, highlighting their strengths and weaknesses. Despite considerable efforts in predicting financial distress, our study reveals several unexplored research opportunities. We hope to contribute to a deeper understanding of FDP and highlight the use of advanced machine-learning techniques on various data sources to increase forecast accuracy and reliability. Through this work, we aim to provide an overview of the field and categorize our understanding of common approaches to address the challenges in FDP research. This survey aims to be a valuable resource to guide future research and the development of practical applications in financial risk management. The dataset, preprocessing technique methods, and machine learning model source code used in this study are publicly available at https://github.com/fdp. In summary, the main contributions of this work are as follows:

- Papers related to the prediction of financial distress and present the concepts of distress (Sect. 2), datasets (Sect. 3.1), models (Sect. 3.2), and evaluation methods (Sect. 3.4) are systematically reviewed for this task.
- A comparative study to compare the performance of five machine learning models has been conducted on the task of distress prediction with different sampling methods (Sect. 4). The results show that XGBoost outperforms the other models and that SMOTE sampling does not significantly improve the models' performance compared with random sampling.
- Potential future research directions are presented and their feasibility has been discussed (Sect. 5).

2 Concepts of Financial Distress

Financial distress is a term used to describe a period in the lifetime of a company, usually associated with its negative performance. Since this term is broad and vague, there is no clear agreement on when a company is said to enter its distress stage. A general understanding of this period is when obligations cannot be paid when due [13]. Some examples of obligations are debts and dividends. Bankruptcy and bond default are two main events that are publicly known and hence traditionally used as proxies to determine the distress stage of a company. However, much of the value of a company is lost in the year preceding these events and not in the years after; therefore, some research has also proposed operational definitions in the hope of detecting early stages of financial distress. In this section, these definitions of financial distress and the studies in which they were used are presented.

In the United States, a company is said to be bankrupted when it files for corporate bankruptcy reorganization (Chap. 11 of the Bankruptcy Code) or liquidation (Chap. 7 of the Bankruptcy Code). Since the date from which a company files for bankruptcy is publicly known, much work has focused on predicting whether a firm will go bankrupt [14–17]. The year before the bankruptcy year is labeled the distressed year and the preceding years before that year are labeled normal (or healthy) years. A less extreme scenario than bankruptcy is the time at which a company is delisted from a major stock exchange. Similar to the bankruptcy situation, the year before the year a company is delisted from the exchange is labeled as a distressed year, and the preceding years are labeled normal years. That said, a company may be delisted from a stock exchange for many reasons that are not related to the performance of the company itself. For example, a company may opt for voluntary delisting to be taken private or to merge with another company to strengthen the business. In the literature, the list of delisted companies classified as distressed is usually taken from Wharton Research Data Services (WRDS) with the delisted codes (reasons for being delisted) ranging between 300 and 700.

While bankruptcy filings and delisting from major stock exchanges (with conditions described in the previous paragraph) are definite signs of poor corporate performance, the assumption that the preceding year before the year these events occur is the only distressed year is not fully accurate. A company may go through the distress period long before it files for bankruptcy, and in other cases, some companies may get themselves together and get out of the distress period without having to file for bankruptcy. Much of the value of a company is also lost before the year of bankruptcy, so predicting when a company will go bankrupt may not provide much value. For these reasons, researchers usually aim to detect signs of distress as early as possible. For example, [18] defined that three years before the delisting year is the distress period. This is known as the early stages of distress. Predicting early stages of distress has multiple important implications. The most obvious reason is that this information helps investors and creditors adjust their risks and cut losses early if necessary. Second, perhaps more importantly, the distressed corporation would have

Table 1 Various definitions of financial distress in the literature

Authors	Definition of financial distress
Beaver et al. [13]	Financial distress refers to a company's inability to repay its debts, with operational definitions are bankruptcy and bond default
Gandhi et al. [18]	Financial distress is defined as three years before the year the company is delisted from a stock exchange
Farooq et al. [21]	Define financial distress as three stages: profit reduction, mild liquidity, and severe liquidity
Mai et al. [16]	A company is a bankruptcy case if it files for either Chap. 7 or Chap. 11 bankruptcy protection code
Qian et al. [20]	Define a distressed company as a company with two conservative years having negative net profit or the net assets of the most recent fiscal year are lower than the registered capital
Huang et al. [19]	ST or *ST labels (assigned to tickers by stock exchanges) are used as indicators of companies that are in distress

more time to adjust itself (for example, a change in management style) to improve its performance, in turn, reducing the probability of bankruptcy.

Owing to the shortcomings of the bankruptcy definition approach, some researchers choose to use the other operational definitions of distress in their work. Studies about distress in the Chinese market focus on predicting the "special treatment" (ST) labels. A stock is marked as an ST preceding its ticker if its net profit is negative for two consecutive years, and *ST if its net profit is negative for three consecutive years [19]. In [20], the criteria for a distressed company are that its net profit for two conservative years is negative or that net assets of the most recent fiscal year are lower than the registered capital. In [21], the authors proposed a three-stage dynamic model of financial distress. In this three-stage dynamic model, the distress level of a company gradually becomes more serious, starting from profit reduction (PR), then mild liquidity (ML), and then severe liquidity (SL). SL is defined as negative net worth where total liabilities exceed total assets. ML is defined as a firm that has an interest coverage ratio of less than 1 for two consecutive years or less than 0.8 for a particular year with a positive net worth. Finally, a firm is denoted as PR if its documented net loss for a particular year or its profits are reduced for two consecutive years and if it has not yet entered the ML and SL stages. [22] later called for the development of multistage FDP models in addition to bankruptcy prediction models. In Table 1, an overview of financial distress definitions in existing work is provided. It is clear that the lack of a unified distress definition in the literature makes it difficult to compare the findings and performance of the models proposed in different studies.

3 Methodologies for Financial Distress Prediction

In this section, the core methodologies in FDP, including datasets and their preprocessing, feature engineering, modeling, and evaluation metrics are systematically explored. The goal is to provide readers with a comprehensive technical pipeline for building effective distress prediction systems.

3.1 Datasets

3.1.1 Data Sources and Types

This section introduces the popular data sources and data types used in FDP-related areas.

On the basis of observations in the literature, the majority of data sources come from third-party providers (Table 2). Researchers typically obtain company finan-

Table 2 Four popular financial data sources for FDP

Sources	Explanation
Compustat[a]	Also known as Compustat Financials, provides standardized North American and global financial statements and market data for more than 80,000 public companies, with historical data as far back as 1950
CSMAR[b]	A comprehensive research database based on Chinese financial data. Users can obtain historical data from the Chinese securities market, including stocks, bonds, futures, and other resources, by selecting the date range, company code, and data fields
EDGAR[c]	The electronic data collection, analysis, and retrieval system of the U.S. Securities and Exchange Commission (SEC). It provides documents filed by public companies with the SEC, as well as information provided by mutual funds (including money market funds), exchange-traded funds (ETFs), variable annuities, and individuals
CRSP[d]	Provided by the University of Chicago. It has research data products, including stock data, real estate data, etc. and market indexes

[a] http://www.compustat.com
[b] https://data.csmar.com
[c] https://www.sec.gov/edgar
[d] https://www.crsp.org

cial data, market data, and other relevant economic indicators from these databases. Sometimes, multiple different databases are combined. For example, the Center for Research in Security Prices (CRSP) and Compustat data are frequently combined in financial research. Researchers use CRSP data for historical stock prices and returns, and Compustat data to assess the financial health of companies. Common identifiers (such as PERMNO and GVKEY) are used to link data from both sources for comprehensive analysis.

One study utilized financial data such as total assets, total current assets, and net income from Compustat North America, as well as stock return and volatility data from CRSP, and textual disclosure data from 10 K annual reports in the Electronic Data Gathering, Analysis, and Retrieval system (EDGAR). By applying deep learning models, the accuracy of predicting company conditions is improved [23]. Another study analyzed foreign currency and commodity derivative holdings of over 2,000 non-financial firms via textual data from 10 K filings in the EDGAR database and financial data from the Compustat database, accessed via Wharton Research Data Services (WRDS), covering the fiscal years 1996–1997. It also included stock return and volatility data from the CRSP, and commodity price data from the Bureau of Labor Statistics, such as quarterly changes in the aggregate producer price index for crude oil. By examining the relationship between leverage and hedging, this study highlights its implications for financial distress and corporate risk management [5]. In addition to these two U.S. databases mentioned, there is another database from the U.S., called the Thomson Reuters (TR) Eikon database, which included 6696 firm-year observations of large U.S.-listed firms from 2010 to 2019. The dataset comprised corporate social performance (CSP) measures using Refinitiv ESG scores, along with financial variables such as total assets and total debts. The study employed regression analysis to investigate the effect of CSP on the likelihood of bankruptcy [24]. Another study is based on the Chinese market [25], which uses textual data from annual reports of listed firms in the Chinese market, sourced from the China Stock Market & Accounting Research Database (CSMAR). The study extracted sentiments from the MD&A sections and audit reports using deep learning algorithms, combining these with traditional financial and market variables to predict financial distress.

There are a few smaller databases for countries other than the U.S. and China. For example, the FINSTAT database contains annual financial statements of Slovak companies, including financial information and qualitative indicators for more than 90,000 SMEs from 2015 to 2019. The financial data encompassed various ratios like efficiency, leverage, liquidity, margin, and profitability. Qualitative indicators included sector, region, size, legal form, ownership, and company age. [26] utilized this database to assess the impact of crisis periods on bankruptcy prediction models and found that models perform significantly worse during crisis periods than non-crisis periods [26].

Some financial data can also be obtained from freely available public or personal repositories. For instance, one study constructs and shares a financial dataset derived from 8,262 public companies listed on the US stock market between 1999 and 2018 for experimental purposes. The study evaluates bankruptcy prediction performance on this dataset using multiple machine learning models, providing benchmark results

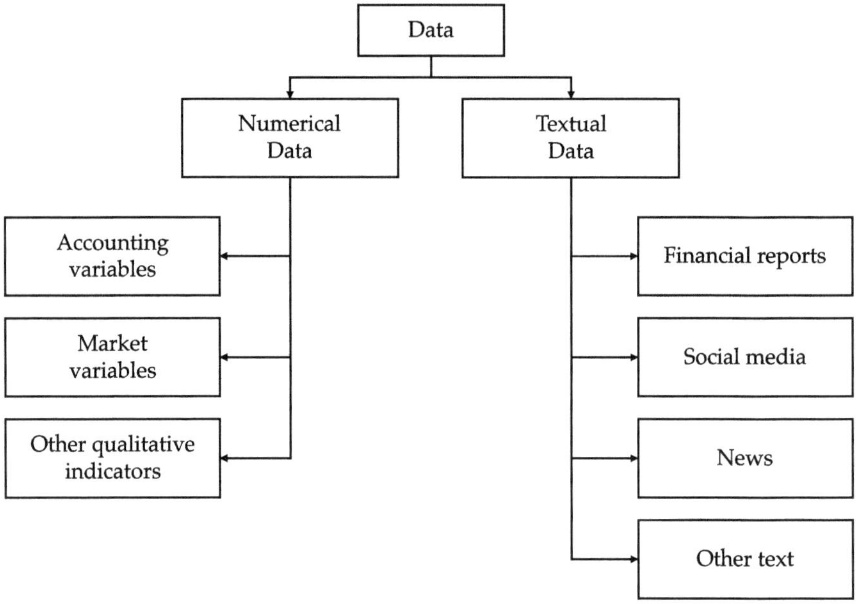

Fig. 1 An illustration of the architecture for the data types in this survey

for future research [27]. Open ML repositories like Kaggle also offer many datasets. These data sources usually come with open-source experimental code, which is a major advantage. The open-source code is particularly useful for reducing the workload from the data preprocessing step. As opposed to data provided by reputable providers, however, it is hard to verify the correctness of these datasets.

In terms of data types, previous research on corporate finance has focused primarily on information from structured numerical data. Recent studies have emphasized the influence of textual data, such as media coverage and news articles, on model's performance [25]. This survey focuses on these two categories, as shown in (Table 2). Numerical data is obtained from accounting variables, market variables, and so on; while textual data is normally taken from financial reports, social media, news, or other texts (Fig. 1).

3.1.2 Challenges in Datasets

As the field of FDP has developed, more data has also been collected, but some challenges remain to be solved.

The nature of the financial industry makes it extremely difficult to obtain valuable data. Financial data often involves sensitive information, and the accuracy and completeness of the data is problematic. Therefore, it is necessary to find a way to use data for effective prediction under the premise of protecting privacy. The trade-off for this premise is often a possibility of affecting the comprehensiveness and accuracy of the model.

The imbalance nature of financial distress data is another important challenge. The number of companies that are in trouble is usually far fewer than that of companies that are operating normally. The majority number of healthy corporate samples are likely to dominate the minority number of distressed companies, making it difficult for the model to identify the characteristics of distressed companies effectively. In other words, the imbalance of sample categories greatly affects the prediction of the model and leads to the lower accuracy of detecting financial distressed samples.

Currently, the training data used in existing research are mainly numerical data such as financial data and company annual reports. The potential of text data, such as social media data, news reports, and financial report texts, has not been deeply studied. Effective integration of data from different data sources and different data types is worthy of further study.

Some other issues are how to achieve early warning for the prediction of financial distress, how to process large amounts of data in the sufficient time, what is the interaction between different data features, and how to select effective data features. If only outdated historical data is considered, the model will likely be unable to reflect new changes and dynamics in the market in a timely manner. In terms of different research purposes, there are differences in the processing of data from cross-country, cross-industry, and different types of companies. It is worthwhile to further study whether a set of data processing methods can be proposed to overcome these challenges.

3.2 Common Preprocessing Techniques

3.2.1 Data Preprocessing

The purpose of data preprocessing is to improve the quality of the initial dataset and transform the input data into a form suitable for the target application [28]. Effective preprocessing not only enhances model performance but also increases the reliability of the data analysis results. Various data types necessitate distinct preprocessing steps. In this section, the common preprocessing steps, including data cleaning, numerical data normalization and standardization, textual data preprocessing, and handling imbalanced data are presented.

Data Cleaning

The first step in data processing is data cleaning, which aims to identify and correct errors in datasets, and eliminate noise [29]. Data cleaning mainly refers to tasks such as dealing with missing values, fixing random errors, and correcting errors, to ensure that the dataset is cleaned to ensure the consistency of the datasets for the next step in preprocessing. Data cleaning addresses several challenges in the FDP problem.

Missing values, as one of the common challenges in data cleaning, might be caused by errors in the data collection process, insufficient disclosure of information, or other reasons. In [30], the authors impute for the missing observations via multiple imputations with chain equations—MICE. In [31], a wide range of methods including simple methods, interpolation methods, statistical modeling, machine learning, deep learning methods, and multiple interpolations are applied. There are several representative methods such as the simple method of mean or median interpolation, which applies to data with a small number of missing values by replacing the missing values with the mean or median of the data. In addition, machine learning methods such as support vector machine (SVM), use regression or classification models to interpolate missing values and are suitable for data with nonlinear relationships.

The handling of random errors (e.g., removing duplicates) and outliers (e.g., removing extreme values) in the dataset can also have a significant effect on the prediction performance of financial problems. The reason for the duplicated occurrence might be the multiple recordings of the same event or the integration of data from different sources. Duplicates can lead to bias and overfitting of the model. Common methods to ensure the uniqueness of each data point include filtering and matching on the basis of unique identifiers, such as company ID and timestamps. Financial data can contain outliers due to data entry errors, extreme market events, or other factors. The main methods for outlier detection are statistical methods, machine learning-based methods, and time series analysis-based methods [32]. Using the mean and standard deviation of financial data is one way to detect and address outliers. In [33], the authors use z-score normalization to minimize the effect of the outliers. Supervised and unsupervised learning models such as using Isolation Forest algorithms (IFs) and SVM, can be utilized to identify and address outliers [32].

Numerical Data Normalization and Standardization

Normalizing data is essential to ensure that features contribute equally to the model's learning process. This process scales data to a range, typically between zero and one, to eliminate the effect of different feature ranges on model training. Normalization is more applicable to bounded data and it changes the range of the data. While the range of the data is changed, normalization maintains the original data distribution. [34] used normalization methods include Min-Max Scaling and Decimal Scaling Normalization. Min-Max Scaling adjusts the data to a specified range, ensuring that all feature values are compared on the same scale, whereas Decimal Scaling Normalization focuses on normalizing the data by shifting the position of the decimal point.

Standardization, on the other hand, transforms the data into a standard normal distribution with zero mean and one standard deviation. Standardization is more applicable to algorithms that are capable of handling unbounded data. Even though the distribution of the data is changed to a Gaussian distribution, the range of the data remains the same. The main methods of standardization include z-score standardization, which is achieved by subtracting the mean and dividing by the standard deviation.

In financial data processing, financial data often contain features of different magnitudes and ranges, such as numerical values in balance sheets and market data. Through normalization and standardization, the model can converge faster in the training process and improve its generalizability and robustness.

Textual Data Preprocessing

Textual data, such as news articles, financial reports, and social media posts, contain a lot of valuable information. However, its preprocessing pipeline is different from numerical data owning to the unstructured nature of textual information. Several common textual preprocessing steps are text cleaning, text normalization, and vector vectorization. Text cleaning includes steps such as removing punctuation, numbers, and stop words, and correcting spelling errors. Text normalization converts text to lowercase form and performs stemming to reduce lexical diversity and improve the generalization ability of the model. Lastly, vectorization is the process of transforming textual data into vector representations, such that words with similar semantic meanings are closer together in the vector space.

An example use case of textual data preprocessing is shown in [30], where the authors described a detailed preprocessing pipeline for financial text data. First, the original documents are extracted from the EDGAR system with the removal of the tables, charts, additional PDFs, and redundant elements. Then, the Management Discussion and Analysis (MDA) section is extracted. Text standardization, such as converting text to lowercase, and removing punctuation and redundant white space characters, was performed. A financial lexicon constructed by [35] was then used to calculate the proportion of positive, negative, and uncertain words in the text. Finally, a dictionary-based sentiment classifier is utilized to classify sentiment according to rules and extract text features.

Handling Imbalanced Data

In financial distress problems, the data collected is often imbalanced, with the number of healthy companies typically far exceeding that of distressed companies. This imbalance can result in biased models that correctly predict the majority class (healthy companies) but fail to accurately identify the minority class (distressed companies). Therefore, after data preprocessing, certain techniques are employed to balance the dataset before the training process. Data augmentation increases the quantity and balance of training data by generating new samples from existing samples, which can address the issue of imbalanced data [28].

Among the techniques for data augmentation, the synthetic minority over-sampling technique (SMOTE)-based methods [36] are commonly applied to the FDP imbalanced data problem [37–39]. For example, the LightSpace-SMOTE and the focal loss were used in [37] to generate minority class samples and adjust sample weights, for FDP in high-dimensional imbalanced scenarios. Reference [38] also used SMOTE to generate new minority class samples to improve classification performance in FDP using machine learning models. In [39], a hybrid sampling method combining

Table 3 Augmentation methods for handling imbalanced data in the literature

Authors	Augmentation methods for handling imbalanced data
Chen et al. [37]	LightSpace-SMOTE and focal loss
Liu et al. [38]	SMOTE
Zelenkov and Volodarskiy [39]	Hybrid sampling (SMOTE and random undersampling)
Aljawazneh et al. [42]	ADASYN (ADaptive SYNthetic sampling approach)
Liang et al. [40]	LR-SMOTE (based on K-means)
Zhou et al. [41]	SVM-SMOTE (Support Vector Machine with SMOTE)

SMOTE and random undersampling was applied to generate hard-to-classify minority class samples in bankruptcy prediction problems. Additionally, there are other methods such as LR-SMOTE (based on K-means) [40] and SVM-SMOTE (support vector machine with SMOTE) [41]. In addition to SMOTE-based methods, [42] compared the performance of deep learning in FDP by using ADASYN (ADaptive SYNthetic sampling approach) to generate minority class samples. Table 3 lists some representative methods for handling imbalanced data.

3.2.2 Feature Engineering

Feature engineering involves developing features that best suit the given data, models, and tasks using techniques such as feature extraction and dimensionality reduction.

Feature Extraction

Feature extraction involves converting raw data into informative features for use in machine learning models. As outlined in Sect. 3.1.1, "Data Sources and Types," this paper broadly categorizes the extracted data into two types: numerical data extraction and textual data extraction.

The numerical data includes accounting variables, market variables, and other quantitative indicators. Based on the characteristics of these data and the financial attributes attached to them, there are time-series data features (such as moving averages, rolling statistics) [43], statistical features (such as the mean, median, standard deviation, and skewness) [44], and domain-specific features (industry-specific indicators) [45].

Textual data includes financial news articles, company financial reports, social media financial posts, and other qualitative metrics. Depending on the characteristics of the data and the semantic financial information it conveys, features can include sentiment analysis (e.g., positive, negative) [30], topic modeling (e.g., potential themes, word clusters) [46], and named entity identification (e.g., mentions of companies, products, or key events).

Dimensionality Reduction

In practice, many features in raw data are irrelevant to the training task. Dimensionality reduction reduces the number of features in the dataset while retaining as much relevant information as possible. These techniques help solve overfitting caused by irrelevant features and decrease the computational burden of processing features with similar information [43]. Among them, feature selection is considered a dimensionality reduction approach that focuses on removing redundant, noisy, and noninformative features from the original feature set [47].

In [45], the authors applied 15 feature engineering methods, including PCA, LDA, an autoencoder, reducing the feature space of the entire dataset from 1.54 to 10.29%, which effectively improved the performance. In [44], the author used an integrated feature selection approach that combines expert knowledge with the wrapper method. Specifically, the authors categorized financial features into seven groups based on expert domain knowledge and then applied the wrapper method to identify the optimal subset of features within each category. Compared with the traditional method, this approach significantly enhanced the model prediction accuracy, achieving an average accuracy of 81.36% and a lower error rate.

3.3 *Machine Learning Models for Financial Distress Prediction*

Over the years, many machine learning models have been proposed and applied to FDP tasks. First, we formulate the problem of FDP system construction. Consider a dataset containing n samples $\mathcal{D} = (X_i, y_i)$ where $i = 1, 2, \ldots, n$. For each sample, we have a dependent variable $y_i = \{0, 1\}$ (which could take the value 0 for healthy firms and 1 for distressed firms) and a set of k independent variables (e.g., features) $X = (x_1, x_2, \ldots, x_k)$ (which can be financial ratios or accounting variables of a firm in a particular year), we aim to construct a model f such that $f(X) = y$. Due to the time-series nature of the problem, X is usually extracted from financial reports of one or more years before the year of interest that y is taken from. In this section, we review the common models to construct f and the literature that used them.

3.3.1 Altman's Z-Score Model

In one of the earliest studies about financial distress, in 1968, Altman presented his findings about five financial ratios with high predictive power. After analyzing 66 firms using multiple discriminant analysis (MDA), the Z-score model is a linear combination of five financial ratios:

$$Z = 1.2x_1 + 1.4x_2 + 3.3x_3 + 0.6x_4 + x_5 \tag{1}$$

where x_{i1} measures liquidity (working capital/total assets), x_{i2} measures cumulative profitability (retained earnings/total assets), x_{i3} measures profitability (earnings before interest and taxes/total assets), x_{i4} is a leverage ratio (market value of equity/total liabilities), x_{i5} is a total asset turnover ratio (sales/total assets). Depending on the value of Z, a company could be classified into one of three zones: "distress" zone ($0 < Z < 1.81$), "grey" zone ($1.81 < Z < 2.99$), and "safe" zone ($2.99 < Z$). Over the years, variations of the Z-score model for different markets were also proposed, as well as modifications of the original Z-score formula by Altman himself [48]. While Altman's Z-score model was proposed quite a long time ago and it is not a learning algorithm, it is still widely used in recent literature as a baseline model (see [30, 49]).

3.3.2 Linear Models

Linear models are among the most popular statistical methods applied to financial distress modeling and are typically used in comparison with MDA [50]. We describe two popular linear models for FDP which are Logistic Regression and Support Vector Machines. These models belong to the class of linear models since they are linear in the parameters (not necessarily linear in the input X).

Logistic Regression

The model Logistic Regression models the probability that y_i belongs to one of the two classes given X_i. The probability of failure (distress) p_i of each sample X_i is defined as:

$$p_i = \frac{exp(\beta_0 + \sum_{i=1}^{n} \beta_i x_i)}{1 + exp(\beta_0 + \sum_{i=1}^{n} \beta_i x_i)} \quad (2)$$

where β_i are the weights of the model.

Typically, a company is classified as distressed if $p_i > 0.5$ and healthy otherwise [50, 51]. The parameters β_i are optimized by minimizing the log-likelihood function:

$$l(\beta) = -\sum_{i=0}^{n} (y_i log(p_i) - (1 - y_i) log(1 - p_i) \quad (3)$$

where n is the number of samples, y_i is the true class for sample i, p_i is the predicted probability for sample i. As it turns out, minimizing the binary cross-entropy can also be done by using Maximum Likelihood Estimation (MLE) since they are equivalent.

In [51], the author found different optimal models for determining the financial failure of Small and Medium-sized Enterprises (SMEs) using the step-wise method of estimating logistic regression. In another subsequent study, the authors found that linear regression outperforms other models with an overall accuracy of 93.33% in a normal Moroccan economy context [52]. The performance of Logistic Regression is also usually used in comparison with other models. In [50], the authors compared the performance of logistic regression, random forest, and neural network models in industrial enterprises operating in Slovak. A similar comparison is conducted by [53] for more than 12 thousand SME companies in the Baltic states. In [54], the authors compared the performance of logistic regression and support vector machines on 172 firms traded in Borsa Istanbul. While both methods were shown to achieve good predictive accuracy, the SVM model (when used in combination with the feature selection phase) shows better performance. We describe the working mechanism of SVMs in the following section.

Support Vector Machines

Support Vector Machine (SVM) is another commonly used method for constructing FDP models. SVM aims to determine the optimal hyperplane for separating two classes. While being a linear model, SVM can find a non-linear way to separate the input data by first mapping the original input X into a high-dimensional feature space $\varphi(X)$ using the (non-linear) function φ. The model (or hyperplane) is then defined as:

$$f(X) = w \times \varphi(X) + b \tag{4}$$

where w and b are parameters that can be estimated by minimizing the regularized risk function:

$$L = min\frac{1}{2}||w||^2 + C\frac{1}{\lambda}\sum_{i=1}^{\lambda} L_\epsilon(y_i, f(x_i)) \tag{5}$$

where:

$$L_\epsilon(y_i, f(x_i)) = \begin{cases} |y_i - f(x_i)| - \epsilon & \text{if } |y_i - f(x_i)| \geq 0 \\ 0 & \text{otherwise} \end{cases} \tag{6}$$

Similar to Logistic Regression, SVM is another commonly used baseline model. The performance of Linear Discriminant Analysis and SVM on Indonesian companies was compared in [55]. In [56], the authors compared the performance of neural networks, decision trees, and SVM on a dataset containing UAE firms' financial data between 2010 and 2017. [57] found that SVM achieves better results than the Naive Bayes model when the data is balanced with the Kmeans-SMOTE method. Several modifications of the original SVM were also proposed for the task of distress prediction. Observing that the variables used for predicting corporate financial distress can be divided into several groups such as market transactions, growth ability, solvency, etc., [58] proposed the Grouping Sparse Principle Component Analysis-SVM (GSPCA-SVM) method. Instead of training the SVM model on all features, GSPCA-SVM first groups the features into k groups, and then extract k sparse features by running sparse principle component analysis on k groups. Finally, k sparse features are used as input for the SVM model. The authors showed that better prediction models can be built with GSPCA-SVM by identifying the most important feature indicators in each category [58]. In more recent work, [59] used Particle Swarm Optimization (PSO) to optimize the parameters of SVM. Their result shows that PSO helped improve the model accuracy by about 3% compared with using only SVM in predicting the financial distress of property and real estate companies listed on the Indonesian Stock Exchange.

3.3.3 Tree and Ensemble Models

Decision Tree

Decision Tree is another commonly used ML algorithm for constructing a FDP model. The main idea of this algorithm is to build a tree containing *decision nodes* and *leaf node*. For each sample, each decision node will make some comparisons and then decide to continue on the left side of the tree or the right side of a tree. The comparison is made by comparing a feature value to some constant. The leaf nodes are where the comparisons are no longer made, and the sample is decided to belong to which class. The question then becomes how can we generate the tree (or decision nodes).

Some of the algorithms for generating decision trees are Classification and Regression Tree (CART), Chi-squared Automatic Interaction Detection (CHAID), or C5.0 algorithms. The CART algorithm is a two-step method, first, creates a maximum binary decision tree, and then prunes to avoid overfitting. The Gini index can be used to measure node impurity in CART. C5.0 is a similar algorithm, which creates a binary decision tree with clean terminal nodes and then uses entropy to measure node impurity. On the other hand, CHAID generates a non-binary tree and prunes as the tree is growing and the chi-square statistic is used to split individual nodes.

One of the key advantages of Decision Trees compared to other ML methods is DT is one of the few self-interpretable models. That is, for each sample, we know exactly why the model makes the prediction. Reference [60] is one work that explores the explainability of decision trees (and other methods) for FDP. The authors of [61] also conducted a comparison study between decision trees and artificial neural networks. In [62], the authors described in detail the creation of decision trees for classifying 2,359,731 enterprises from emerging markets and provided a detailed discussion about the importance and splitting criterion of each variable.

Random Forests

One disadvantage of decision trees is their high variance, which often leads to low accuracy. Random forests [63] reduce the variance of decision trees by aggregating the output of multiple decision trees trained on slightly different versions of the original training set. Different versions of the training sets are created using a technique called bagging. Let B be the number of new training sets. B training sets are created by selecting a random sample with replacement of the original training set B times. Then, B classifiers f are trained on B new training set, and the output of the final ensemble model $\hat{f}$ is decided by aggregating B classifier, for example, taking the average:

$$\hat{f} = \frac{1}{B} \sum_{b=1}^{B} f_b(x') \tag{7}$$

Random Forest employs another type of bagging scheme called "feature bagging". That is, different decision trees are trained using a random subset of the features. Due to the randomized nature of the random forest, we lost the interpretability of the decision tree but the variance of the model is decreased without increasing the bias.

Boosting Trees

Boosting Trees is another ensemble model. Similar to Random Forest, Boosting Trees rely on the output of many classifiers and aggregate their results. The key difference between the two algorithms is that the trees in Boosting Trees are weak learners (e.g., shallow decision trees). Some very popular boosting trees are AdaBoost [64], XGBoost [65], LightGBM [66]. We refer interested readers to the referenced papers to learn more about how these boosting algorithms work in detail.

Boosting trees is another type of model favored by distress prediction researchers. Reference [67] proposed the combination of feature importance and XGBoost (called FS-XGBoost) for the task of bankruptcy prediction. Compared with other methods, FS-XGBoost yields a more accurate prediction for small and medium French firms. In [20], permutation importance (PIMP) is proposed to correct the bias in feature importance before passing to gradient-boosted trees. They also found that their proposed PIMP-XGBoost offers higher prediction accuracy for distress prediction in the context of listed companies in China between 2001 and 2019. Two issues when applying XGBoost to financial risk management were identified by [68]. First, XGBoost does not take into account the class imbalance issue in the problem of FDP. Second, XGBoost has low interpretability due to its complex ensemble structure. The authors proposed a cost-sensitive weighted XGBoost model (XGBoost-W-BT) and constructed an approximate tree of this model to address these two issues [68]. In a similar line of work, [69] is tried to apply interpretable methods to XGBoost.

3.3.4 Neural Networks

In recent years, neural network and deep learning models become extremely popular due to their high performance and ability to be applied to multiple domains. In the context of finance and specifically FDP, multiple works also tried to apply neural network models. Some work used simple perceptron models while others used

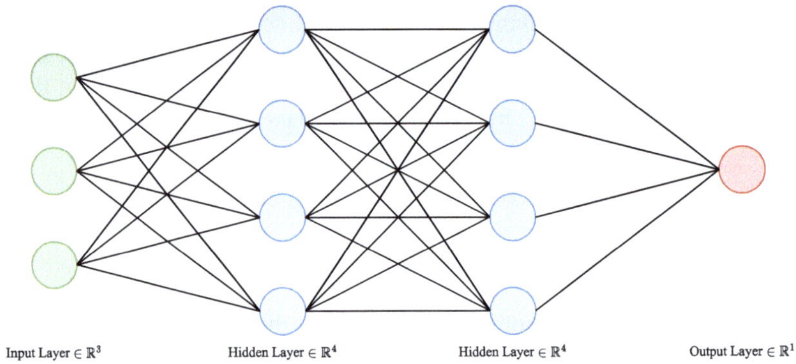

Fig. 2 A 4-layer neural network

more sophisticated deep learning models like Long Short-Term Memory (LSTM). Figure 2 illustrates a 4-layer neural network.

Each neural network has at least one input layer, one output layer, and one or more hidden layers. Each layer contains one or multiple nodes, called "neurons". In Fig. 2, the input layer has 3 neurons, two hidden layers have 4 neurons each, and the output layer has 1 neuron. Except for the input layer and the output layer, each neuron receives inputs from previous layer neurons, calculates the output value, and then passes the result to the next layer's neurons. Let g be the non-linear activation function, x_i be inputs from the previous layer, w_i be the weights corresponding to the inputs, and b be the bias, then the output value is calculated by Eq. 8.

$$x_{out} = g\left(\sum_i w_i x_i + b\right) \tag{8}$$

The non-linear activation function is important since without it, all layers in the network collapse into one linear transformation. Common non-linear activation functions are $tanh$, $sigmoid$, rectified linear unit (ReLU), and max.

In [70], the authors compared the performance of LR and neural network models on Zimbabwe-listed companies. While they found that the performance of LR is superior to neural networks, their neural network architecture is somewhat too simple with only one hidden layer containing three nodes. In [71], a multi-layer perceptron artificial neural network is also proposed for predicting financial distress in Hanoi and Ho Chi Minh stock exchanges between 2007 and 2019. They found that the model is suitable for early forecasting financial distress in Vietnam, with 94.7% accuracy in the manufacturing sector and 94.2% accuracy in the service sector. Reference [72] is another work where the author incorporates a whale optimization algorithm with a neural network to find optimal hyperparameters. Reference [67] combined fuzzy set qualitative comparative analysis and convolutional neural network for bankruptcy prediction.

3.4 Evaluation Methods

In this section, we discuss some common methods for separating training and testing sets, as well as metrics used to select the most suitable model.

3.4.1 Dataset Separation

When training machine learning models, to accurately measure the generalizability of the models and prevent overfitting, separating the dataset into training and test sets is needed. In a typical machine learning setting, 20–30% of the dataset is randomly selected as the test set, and the remaining samples are used for training. The final model is chosen based on their performance on the test set. In an FDP setting, however, there is more than one way to construct the features for each sample and to separate the training and testing sets.

As mentioned, for each pair of sample (X, y), there is more than one way to construct X. Even though the names and the number of features remain the same for every sample, the time dimension gives researchers flexibility in choosing the year(s) in which X is drawn. Two common ways to construct X in the literature are for two different purposes: n-year-ahead prediction and long-horizon prediction. Let t be the year in which we are interested in predicting the distress status. N-year-ahead method uses the financial variables available at the year $t - n$ as features. A concrete example is for the one-year-ahead prediction task, the dataset is prepared so that X contains the features collected in the year $t - 1$ and the label y is the financial status of the firm at the year t. This method of constructing the features X was used in [16, 20, 73].

Contradicting to using data from only a single year to construct X, the long horizon method uses data spans several years. For this method, a looking back window length (WL) is usually defined to specify how many years of data are included in X. For example, a looking back WL equals 5 means that the features from a five-year period ($t - 1$ to $t - 5$) are used as features. The purpose of this type of preparation is to see how much data in the past is needed for accurate predictions. Typically, different values for WL are experimented with to determine the optimal WL. We note that when the WL equals 1, this method is the same as the one-year-ahead method. Reference [27] is one of the works that experimented with the long horizon method.

Once the construction of the individual sample is finalized, the dataset $\mathcal{D} = (X_i, y_i)$ is obtained. Since each label y_i is associated with a particular year, more flexibility is available for researchers in selecting how to separate the dataset into training and testing sets. The typical separation method is to randomly select 20 to 30% of the samples as the test set, and this is called out-of-sample testing. While widely used [16, 68, 73], this separation method does not take into account the nature of the real-life distress prediction problem, that is we aim to predict future events using historical data. Therefore, another method to separate training and test sets is to use the year associated with y_i as the deciding threshold (e.g., data before 2015 is used to train the model, and data after 2015 is used to test the model). The threshold year should be chosen so that the ratio of train and test samples remains around 7/3 still.

3.4.2 Metrics

Some common metrics for evaluating the performance of distress models are Accuracy, Balance Accuracy, Precision, Recall, Specificity, F1-score, Type I error, Type II error, and Area under the ROC Curve.

First, we define True Positive (TP), True Negative (TN), False Positive (FP), and False Negative (FN) as the following:

- TP: Correctly classified distressed firms.
- TN: Correctly classified non-distressed (healthy) firms.
- FP: Incorrectly classified distressed firms.
- FN: Incorrectly classified non-distressed (healthy) firms.

Then the other metrics can be calculated as shown in the following formulas:

$$\text{Accuracy} = \frac{TP + TN}{TP + TN + FP + FN} \tag{9}$$

$$\text{Precision} = \frac{TP}{TP + FP} \tag{10}$$

$$\text{Recall} = \frac{TP}{TP + FN} \tag{11}$$

$$\text{Specificity} = \frac{TN}{TN + FP} \tag{12}$$

$$\text{Balanced Accuracy} = \frac{\text{Recall} + \text{Specificity}}{2} \tag{13}$$

$$\text{F1-score} = \frac{2 \times \text{Precision} \times \text{Recall}}{\text{Precision} + \text{Recall}} \tag{14}$$

$$\text{Type I error} = \frac{FP}{TN + FP} \tag{15}$$

$$\text{Type II error} = \frac{FN}{TP + FN} \tag{16}$$

4 Comparative Study

While many machine learning techniques and models were proposed for the construction of financial distress prediction models, it is interesting to see how their performance differs on the same dataset. For this purpose, we conducted experiments with some of the techniques and models mentioned in the previous section and presented their results.

4.1 Dataset

For reproducibility and transparency, we conducted our experiments on a publicly available dataset [27]. This dataset contains 18 accounting and financial variables of 8262 different companies during the period 1999 and 2018. This feature set was previously referred to by [2, 3, 74]. The original dataset was annotated with bankruptcy and non-bankruptcy labels. For our work, which is to detect distress signals of a corporation before the point of bankruptcy, we denote a company is in distress if it falls into one of the three cases: (1) the net profit of the following year is negative, (2) the company goes bankrupt the following year, (3) the profit is reduced for two consecutive years. This definition is partially similar to the definition of the profit reduction stage introduced by [21]. In Table 4, we present the total number of firms, bankrupt firms, and distressed firms per year.

4.2 Experimental Setup

To evaluate the performance and generality of different machine learning models, the dataset is separated into three sets: a training set, a validation set, and a test set. The training set consists of data from 1999 to 2011, the validation set consisting of data from 2012 to 2014, and the test set consisting of data from 2015 to 2018.

First, we conducted the one-year-ahead experiment, that is, to use the data at year $t-1$ to predict the financial status of a firm at year t. Second, we would like to have an estimate of the number of years of data needed to make more accurate predictions. For this, we conducted experiments with different looking-back window lengths, ranging from 1 to 5 years. Five models including LR, SVM, DT, RF, and XGBoost were run. Hyperparameter searches were conducted to find the optimal hyperparameters for all models. Finally, we use data-augmented methods (under-sampling, over-sampling, and SMOTE) for the best models.

Table 4 Number of bankrupt firms and distressed firms per year from 2000 to 2019

Year	Firms	Bankrupt firms	Distressed firms
2000	5308	3	1881
2001	5226	7	2100
2002	4897	10	2336
2003	4651	17	2238
2004	4417	29	1801
2005	4348	46	1623
2006	4205	40	1486
2007	4128	51	1528
2008	4009	59	1567
2009	3857	58	1667
2010	3743	23	1617
2011	3625	35	1351
2012	3513	25	1213
2013	3485	26	1349
2014	3484	28	1420
2015	3504	33	1529
2016	3354	33	1531
2017	3191	29	1415
2018	3014	21	1123
2019	2723	36	300
Total	78682	609	31075

4.3 *Results*

We conducted experiments with five different machine learning models to have a holistic view of their performance under different scenarios. In the first scenario, financial variables from only the previous year are used to predict the distress status in the target year. This setting is also known as the "one-year ahead" or "long horizon method with a window-length of 1". Table 5 presents the results of five models according to five evaluation metrics. RF is the best model for this scenario, with an AUC score of 0.7490 and a macro-F1 score of 0.7455. Linear SVM is the weakest model with an AUC score of 0.7094 and a macro-F1 score of 0.7151.

In the second scenario, we aimed to understand the amount of historical data needed for the models to achieve their best performance. Too little historical data could hinder the models performance, while too much historical data could introduce noise and distract the models from useful signals. To incorporate historical data, we ran long horizon experiments with different window lengths, with the maximum window length being five years preceding the target year. The results of these experiments are shown in Tables 6, 7, 8 and 9.

Table 5 One year long horizon experiment result for five models with five evaluation metrics (WL = 1)

Model	AUC score	Macro F1	Accuracy	Precision	Recall
LR	0.7245	0.7277	0.7542	0.6655	0.6216
SVM	0.7094	0.7151	0.7473	0.6671	0.5779
DT	0.7458	0.7423	0.7606	0.6540	0.6944
RF	**0.7490**	**0.7455**	**0.7636**	**0.6581**	0.6983
XGBoost	**0.7490**	0.7454	0.7635	0.6577	**0.6988**

Table 6 Two years long horizon experiment result for five models with five evaluation metrics (WL = 2)

Model	AUC score	Macro F1	Accuracy	Precision	Recall
LR	0.7271	0.7283	0.7522	0.6553	0.6402
SVM	0.7160	0.7200	0.7485	**0.6602**	0.6036
DT	0.7460	0.7405	0.7574	0.6451	0.7066
RF	0.7492	0.7452	0.7629	0.6559	0.7015
XGBoost	**0.7524**	**0.7475**	**0.7643**	0.6554	**0.7114**

Table 7 Three years long horizon experiment result for five models with five evaluation metrics (WL = 3)

Model	AUC score	Macro F1	Accuracy	Precision	Recall
LR	0.7347	0.7337	0.7548	0.6525	0.6649
SVM	0.7236	0.7257	0.7510	0.6567	0.6287
DT	0.7570	0.7482	0.7623	0.6449	0.7386
RF	0.7566	0.7513	0.7677	0.6592	0.7182
XGBoost	**0.7681**	**0.7608**	**0.7754**	**0.6649**	**0.7430**

Table 8 Four years long horizon experiment result for five models with five evaluation metrics (WL = 4)

Model	AUC score	Macro F1	Accuracy	Precision	Recall
LR	0.7351	0.7338	0.7546	0.6513	0.6674
SVM	0.7216	0.7233	0.7484	0.6517	0.6287
DT	0.7543	0.7465	0.7614	0.6458	0.7295
RF	0.7558	0.7509	0.7677	0.6604	0.7144
XGBoost	**0.7679**	**0.7600**	**0.7742**	**0.6621**	**0.7462**

Table 9 Five years long horizon experiment result for five models with five evaluation metrics (WL = 5)

Model	AUC score	Macro F1	Accuracy	Precision	Recall
LR	0.7358	0.7347	0.7557	0.6533	0.6672
SVM	0.7217	0.7234	0.7484	0.6516	0.6290
DT	0.7566	0.7485	0.7631	0.6472	0.7340
RF	0.7524	0.7481	0.7653	0.6583	0.7077
XGBoost	**0.7671**	**0.7598**	**0.7744**	**0.6635**	**0.7420**

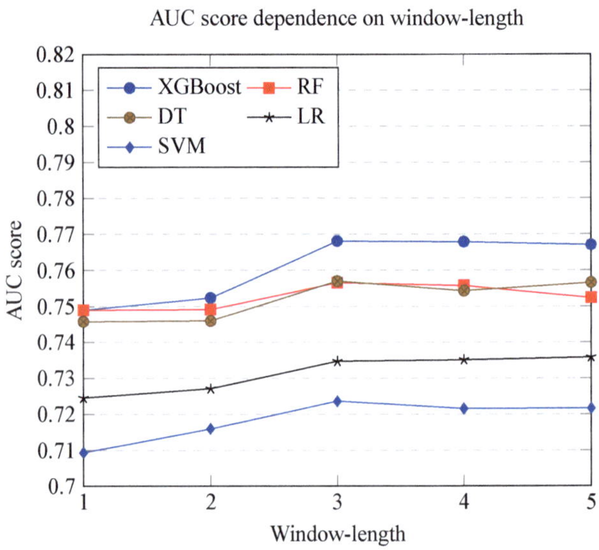

Fig. 3 The performance of the five models by AUC score, concerning different amounts of historical data used

In Fig. 3, we illustrate the performance of the models in terms of AUC score concerning different window lengths. It can be seen from the figure that all models benefit from incorporating up to three years of historical data, and the performance of the models is slightly decreased when four or five years of historical data are used.

In the last set of experiments, we combined the findings from Fig. 3 and different augmentation methods. That is, we conducted two addition sampling techniques, namely SMOTE, and Random Under Sampling (RUS), with three years of historical data as input. The sampling methods were applied to the training set only, making this set a balanced set while testing and validation were conducted on the original imbalanced sets. Table 10 shows the performance of the models when different sampling methods are applied. From these results, we produced Figs. 4 and 5 to visualize the effect of sampling techniques in terms of the two metrics AUC and Recall scores.

Table 10 Three years long horizon experiment results (WL = 3) with different augmentation methods

Model	Augmentation method	AUC score	F1 score	Accuracy	Precision	Recall
LR	No sampling	0.7347	0.7337	0.7548	0.6525	0.6649
SVM	No sampling	0.7236	0.7257	0.7510	0.6567	0.6287
DT	No sampling	0.7570	0.7482	0.7623	0.6449	0.7386
RF	No sampling	0.7566	0.7513	0.7677	0.6592	0.7182
XGBoost	No sampling	**0.7681**	**0.7608**	**0.7754**	**0.6649**	**0.7430**
LR	SMOTE	0.7457	0.7360	0.7500	0.6277	0.7306
SVM	SMOTE	0.7400	0.7332	0.7494	0.6320	0.7075
DT	SMOTE	0.7613	0.7442	0.7539	0.6217	**0.7869**
RF	SMOTE	0.7585	0.7490	0.7627	0.6441	0.7439
XGBoost	SMOTE	**0.7691**	**0.7565**	**0.7682**	**0.6456**	0.7723
LR	RUS	0.7417	0.7326	0.7471	0.6250	0.7228
SVM	RUS	0.7378	0.7316	0.7482	0.6315	0.7018
DT	RUS	0.7640	0.7462	0.7555	0.6227	**0.7935**
RF	RUS	0.7638	0.7518	0.7639	**0.6412**	0.7636
XGBoost	RUS	**0.7677**	**0.7540**	**0.7652**	0.6401	0.7764

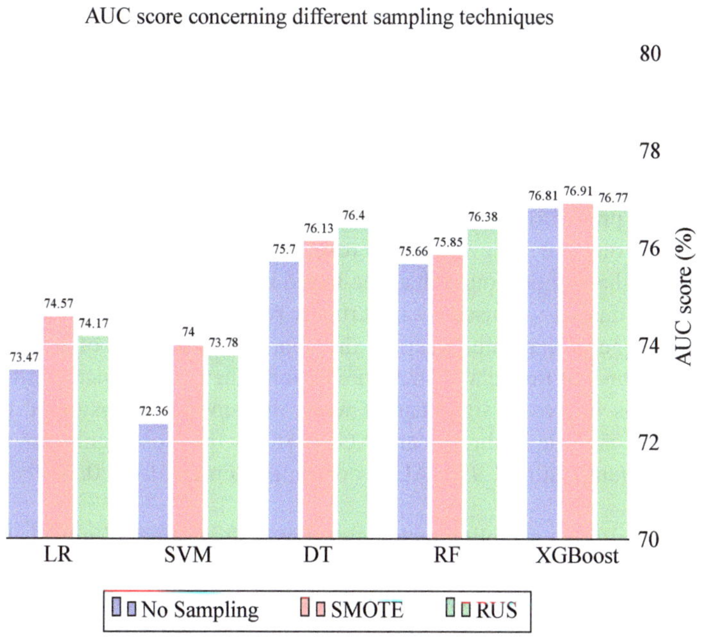

Fig. 4 The comparison of AUC scores for three years long horizon experiment results (WL = 3) with three different augmentation methods

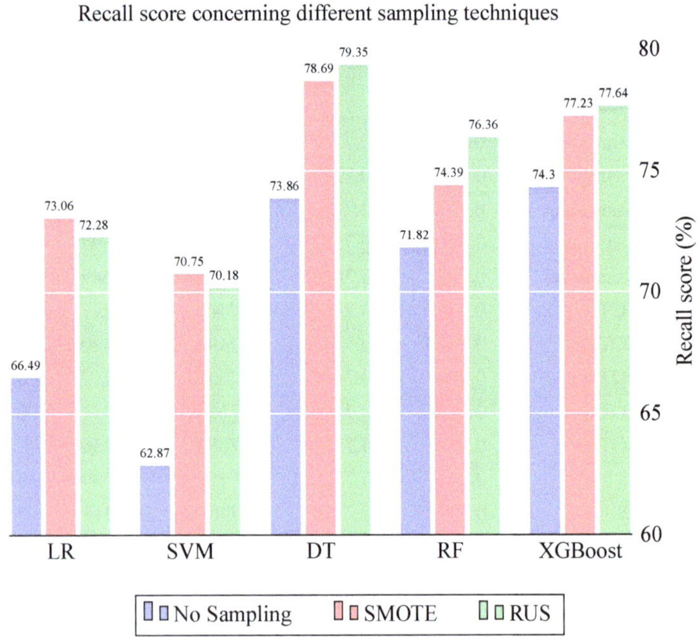

Fig. 5 The comparison of recall scores for three years long horizon experiment results (WL = 3) with three different augmentation methods

4.4 *Discussions*

The results from Table 5 show that generally, tree-based models are the best models for distress prediction from financial variables. Among the tree-based models, XGBoost is the one that achieves the best results. These results align with the finding in [27], where the authors found that tree-based models outperform other methods when the same dataset was used for bankruptcy prediction.

Figure 3 suggests that three years is the appropriate amount of historical data needed for the task of financial distress prediction. The performance of all machine learning models improved significantly when the amount of historical data was increased from one year to three years. XGBoost saw the most significant improvement improvement, with a 1.91% AUC score improvement, from 0.7490 (WL = 1) to 0.7681 (WL = 3).

Regarding different sampling techniques, Fig. 4 shows that all models benefited from resolving the imbalanced issue in the training set. Weaker models such as LR and SVM show greater improvement with SMOTE sampling algorithms, while tree-based models show greater improvement when the simple RUS is applied (Fig. 4). In particular, by addressing the imbalanced issue with RUS, DT and RT can achieve a similar performance to the complex XGBoost model. In Fig. 5, we illustrated another interesting finding. That is, the Recall scores are significantly increased, up to nearly

8% in the case of SVM, when sampling methods are applied. After sampling, DT is the model with the highest Recall score (79.35%), with an increase of 5.49% compared to pre-sampling. This finding suggests that in the business cases where Recall scores are of interest, augmentation is a very effective method to improve the models.

5 Research Directions

The study of the FDP has been explored and has achieved valuable stages. Future direction should pay more attention to several key areas such as data integration, emerging machine learning models and their applications in sentiment analysis, real-time prediction, and model interpretability.

First, with the growth of data volume and the complexity of data, how to extract and integrate multiple data sources effectively becomes a problem worth exploring [75]. Structured data and unstructured data can be considered for effectively integration. Structured data such as financial statements and market data, and unstructured data such as text data (news articles and social media posts), audio data (recordings of company conference calls), image data (charts and graphs in a company's annual report), and video data (news videos and company promotional videos) were used. For example, there is an FDP model that integrates the emotional characteristics and financial indicators of corporate management during earnings calls. The results show that the predictive performance of the model significantly improves with the addition of emotional features, especially in predicting financial distress trends [76]. Diverse data types offer models a more comprehensive perspective, improving the accuracy of predicting financial distress. Moreover, improving data cleansing techniques to ensure high-quality integration of data from multiple sources is also a key task.

Additionally, the application of emerging machine learning models and integrated models in FDP has yet to be explored [12, 19, 75]. For example, graph neural networks (GNNs) can be utilized to analyze the risk chain and extract implicit information from disclosed documents [77]. Deep reinforcement learning can improve model adaptability under changing market conditions [78]. Generative adversarial networks (GANs) can generate synthetic data to solve data scarcity and imbalanced problems [79]. Large language models (LLMs) such as the GPT-family models can deeply analyze textual data in detail, capturing features and emotions that traditional models were not able to capture [80]. With these emerging models, future research can achieve greater accuracy and reliability in FDP. For example, by applying sentiment analysis methods, models can comprehensively handle market sentiment regarding financial status more accurate [30]. GNNs can reveal complex networks of relationships between companies and provide deeper risk assessment. Deep reinforcement learning can optimize prediction strategies through data engineering in dynamic market environments.

Currently, fast fluctuations in the market environment demand real-time prediction of financial distress [81]. The situation of the market and the company's finances could change significantly in a short period, and it has been proven that negative future stock crashes positively impact short-term changes in FDP [82]. Additionally, integrating dynamic and static models can be considered in the future to balance long-term trends and short-term fluctuations, and construct a real-time analysis and decision-making system. Therefore, real-time prediction offers timely evaluation for profit-related groups that act quickly to avoid potential financial loss from financial distress. This is critical feature to maintain the competitiveness and stability of a company.

Finally, with the complexity of data and models, the demand for explanations of methodologies is growing [83]. Particularly in the study of FDP, stakeholders and other profit-related groups need to clearly understand the logic behind the financial distress derived from the model to take appropriate actions. Future research aims to develop more methodologies that can explain model decisions and keep the decision process transparent and logical. So that, decision-makers can understand how the model can draw conclusions and identify the key factors that influence the decision. For example, with tree-based gradient boosting models, the specific financial variables and features are examined with a significant influence on prediction. Moreover, it offers financial distress an effective methodology and explanation [73]. Methods such as SHAP [84] or LIME [85] are also worth exploring in the context of FDP.

By exploring these future research directions, researchers and practitioners can develop more advanced, reliable, and efficient tools for predicting financial distress, thus advancing the fields of FDP and financial risk management.

6 Conclusions

In this review, we systematically examine the key concepts, data sources, preprocessing techniques, machine learning models used in FDP, and future directions. We survey recent papers to clarify the content of the most commonly used predictor variables, datasets, and methodologies for constructing early warning systems for FDP. Through a comprehensive review, we summarized the current trend in this field and compared the commonly used methodologies, discussing their strength and weaknesses. And some valuable future research directions have also been proposed.

Owing to the limitations in the breadth and depth of this survey, it is not easy for us to present all the FDP-related techniques. Therefore, our comparative study applies to a single dataset and a selection of generally representative data processing techniques and machine learning models. Our results show that XGBoost yielded the best performance in terms of AUC score, while DT gave the best performance in terms of recall. Sampling techniques such as SMOTE and RUS further enhanced these results.

In future work, empirical analyses should be extended to other datasets and machine learning models, with a focus on the development of novel integration methods that combine structured and unstructured data as well as the application of emerging machine learning models to the FDP. The interpretability and transparency of the models can also be improved to ensure the reliability of the prediction results and the transparency of the decision-making process.

References

1. Doumpos, M., Zopounidis, C.: A multicriteria discrimination method for the prediction of financial distress: the case of Greece. Multinatl. Financ. J. **3**, 71–101 (1999)
2. Altman, E.I.: Financial ratios, discriminant analysis and the prediction of corporate bankruptcy. J. Financ. **23**(4), 589–609 (1968)
3. Beaver, W.H.: Financial ratios as predictors of failure. J. Account. Res. **4**, 71–111 (1966)
4. Beaver, W.H., Correia, M., McNichols, M.F., et al.: Financial statement analysis and the prediction of financial distress. Found. Trends® Account. **5**(2), 99–173 (2011)
5. Purnanandam, A.: Financial distress and corporate risk management: theory and evidence. J. Financ. Econ. **87**(3), 706–739 (2008)
6. Zhang, Y., Wang, S., Ji, G.: A rule-based model for bankruptcy prediction based on an improved genetic ant colony algorithm. Math. Probl. Eng. **2013**(1), 753251 (2013)
7. Ohlson, J.A.: Financial ratios and the probabilistic prediction of bankruptcy. J. Account. Res. 109–131 (1980)
8. Zmijewski, M.E.: Methodological issues related to the estimation of financial distress prediction models. J. Account. Res. 59–82 (1984)
9. Alaka, H.A., Oyedele, L.O., Owolabi, H.A., Kumar, V., Ajayi, S.O., Akinade, O.O., Bilal, M.: Systematic review of bankruptcy prediction models: towards a framework for tool selection. Expert Syst. Appl. **94**, 164–184 (2018)
10. Matenda, F.R., Sibanda, M., Chikodza, E., Gumbo, V.: Bankruptcy prediction for private firms in developing economies: a scoping review and guidance for future research. Manag. Rev. Q. 1–40 (2021)
11. Kuizinienė, R.D.D., Krilavičius, T., Maskeliūnas, R.: Systematic review of financial distress identification using artificial intelligence methods. Appl. Artif. Intell. **36**(1), 2138124 (2022)
12. Md Jahidur Rahman and Hongtao Zhu: Predicting financial distress using machine learning approaches: evidence China. J. Contemp. Account. Econ. **20**(1), 100403 (2024)
13. Beaver, W.H., Correia, M., McNichols, M.F.: Financial statement analysis and the prediction of financial distress. Found. Trends Account. **5**(2), 99–173 (2011)
14. Danilov, K.: Corporate bankruptcy: assessment, analysis and prediction of financial distress, insolvency, and failure. SSRN Electron. J. (2014)
15. du Jardin, P.: A two-stage classification technique for bankruptcy prediction. Eur. J. Oper. Res. **254**(1), 236–252 (2016)
16. Mai, F., Tian, S., Lee, C., Ma, L.: Deep learning models for bankruptcy prediction using textual disclosures. Eur. J. Oper. Res. **274**(2), 743–758 (2019)
17. Moscatelli, M., Parlapiano, F., Narizzano, S., Viggiano, G.: Corporate default forecasting with machine learning. Expert Syst. Appl. **161**, 113567 (2020)
18. Gandhi, P., Loughran, T., McDonald, B.: Using annual report sentiment as a proxy for financial distress in us banks. J. Behav. Financ. **20**(4), 424–436 (2019)
19. Huang, B., Yao, X., Luo, Y., Li, J.: Improving financial distress prediction using textual sentiment of annual reports. Ann. Oper. Res. **330**(1), 457–484 (2023)
20. Qian, H., Wang, B., Yuan, M., Gao, S., Song, Y.: Financial distress prediction using a corrected feature selection measure and gradient boosted decision tree. Expert. Syst. Appl.: Int. J. **190**(C) (2022)
21. Farooq, U., Qamar, M.A.J., Haque, A.: A three-stage dynamic model of financial distress. Manag. Financ. **44**(9), 1101–1116 (2018)
22. Farooq, U., Qamar, M.A.J.: Predicting multistage financial distress: reflections on sampling, feature and model selection criteria. J. Forecast. **38**(7), 632–648 (2019)

23. Mai, F., Tian, S., Lee, C., Ma, L.: Deep learning models for bankruptcy prediction using textual disclosures. Eur. J. Oper. Res. **274**(2), 743–758 (2019)
24. Florian Habermann and Felix Bernhard Fischer: Corporate social performance and the likelihood of bankruptcy: evidence from a period of economic upswing. J. Bus. Ethics **182**(1), 243–259 (2023)
25. Huang, B., Yao, X., Luo, Y., Li, J.: Improving financial distress prediction using textual sentiment of annual reports. Ann. Oper. Res. **330**(1), 457–484 (2023)
26. Papík, M., Papíková, L.: Impacts of crisis on SME bankruptcy prediction models' performance. Expert Syst. Appl. **214**, 119072 (2023)
27. Lombardo, G., Pellegrino, M., Adosoglou, G., Cagnoni, S., Pardalos, P.M., Poggi, A.: Machine learning for bankruptcy prediction in the American stock market: dataset and benchmarks. Futur. Internet **14**(8) (2022)
28. Mumuni, A., Mumuni, F.: Automated data processing and feature engineering for deep learning and big data applications: a survey. J. Inf. Intell. (2024)
29. Li, P., Rao, X., Blase, J., Zhang, Y., Chu, X., Zhang, C.: CleanML: a study for evaluating the impact of data cleaning on ML classification tasks. In: 2021 IEEE 37th International Conference on Data Engineering (ICDE), pp. 13–24. IEEE (2021)
30. Nguyen, B.-H., Huynh, V.-N.: Textual analysis and corporate bankruptcy: a financial dictionary-based sentiment approach. J. Oper. Res. Soc. **73**(1), 102–121 (2022)
31. Tawakuli, A., Havers, B., Gulisano, V., Kaiser, D., Engel, T.: Survey: time-series data preprocessing: a survey and an empirical analysis. J. Eng. Res. (2024)
32. Blázquez-García, A., Conde, A., Mori, U., Lozano, J.A.: A review on outlier/anomaly detection in time series data. ACM Comput. Surv. (CSUR) **54**(3), 1–33 (2021)
33. Imron, M.A., Prasetyo, B.: Improving algorithm accuracy k-nearest neighbor using z-score normalization and particle swarm optimization to predict customer churn. J. Soft Comput. Explor. **1**(1), 56–62 (2020)
34. Saranya, C., Manikandan, G.: A study on normalization techniques for privacy preserving data mining. Int. J. Eng. Technol. (IJET) **5**(3), 2701–2704 (2013)
35. Loughran, T., McDonald, B.: When is a liability not a liability? textual analysis, dictionaries, and 10-Ks. J. Financ. **66**(1), 35–65 (2011)
36. Chawla, N.V., Bowyer, K.W., Hall, L.O., Kegelmeyer, W.P.: SMOTE: synthetic minority oversampling technique. J. Artif. Intell. Res. **16**, 321–357 (2002)
37. Chen, Y., Kuang, X., Guo, J.: LiFoL: an efficient framework for financial distress prediction in high-dimensional unbalanced scenario. J. Artif. Intell. Res. **11**(2), 2784–2795 (2023)
38. Liu, Y., Zeng, Q., Li, B., Ma, L., Ordieres-Meré, J.: Anticipating financial distress of high-tech startups in the European union: a machine learning approach for imbalanced samples. J. Forecast. **41**(6), 1131–1155 (2022)
39. Zelenkov, Y., Volodarskiy, N.: Bankruptcy prediction on the base of the unbalanced data using multi-objective selection of classifiers. Expert Syst. Appl. **185**, 115559 (2021)
40. Liang, X.W., Jiang, A.P., Li, T., Xue, Y.Y., Wang, G.T.: LR-smote—an improved unbalanced data set oversampling based on k-means and SVM. Knowl.-Based Syst. **196**, 105845 (2020)
41. Zhou, B., Li, Z., Zhang, S., Zhang, X., Liu, X., Ma, Q.: Analysis of factors affecting hit-and-run and non-hit-and-run in vehicle-bicycle crashes: a non-parametric approach incorporating data imbalance treatment. Sustainability **11**(5), 1327 (2019)
42. Aljawazneh, H., Mora, A.M., García-Sánchez, P., Castillo-Valdivieso, P.A.: Comparing the performance of deep learning methods to predict companies' financial failure. IEEE Access **9**, 97010–97038 (2021)
43. Aghakhani, S., Alhajj, R., Rokne, J., Chang, P.: An effective LmRMR for financial variable selection and its applications. In: 2017 IEEE International Conference on Information Reuse and Integration (IRI), pp. 535–543. IEEE (2017)
44. Lin, F., Liang, D., Yeh, C.-C., Huang, J.-C.: Novel feature selection methods to financial distress prediction. Expert Syst. Appl. **41**(5), 2472–2483 (2014)
45. Kuizinienė, D., Savickas, P., Kunickaitė, R., Juozaitienė, R., Damaševičius, R., Maskeliūnas, R., Krilavičius, T.: A comparative study of feature selection and feature extraction methods for financial distress identification. PeerJ Comput. Sci. **10**, e1956 (2024)

46. Aziz, S., Dowling, M., Hammami, H., Piepenbrink, A.: Machine learning in finance: a topic modeling approach. Eur. Financ. Manag. **28**(3), 744–770 (2022)
47. Al-Tashi, Q., Abdulkadir, S.J., Rais, H.M., Mirjalili, S., Alhussian, H.: Approaches to multi-objective feature selection: a systematic literature review. IEEE Access **8**, 125076–125096 (2020)
48. Altman, E.I., Iwanicz-Drozdowska, M., Laitinen, E.K., Suvas, A.: Distressed firm and bankruptcy prediction in an international context: a review and empirical analysis of Altman's z-score model (2014)
49. Desheng, W., Ma, X., Olson, D.L.: Financial distress prediction using integrated z-score and multilayer perceptron neural networks. Decis. Support Syst. **159**, 113814 (2022)
50. Gregova, E., Valášková, K., Adamko, P., Tumpach, M., Jaros, J.: Predicting financial distress of Slovak enterprises: comparison of selected traditional and learning algorithms methods. Sustainability **12**, 3954 (2020)
51. Zizi, Y., Oudgou, M., El Moudden, A.: Determinants and predictors of SMEs' financial failure: a logistic regression approach. Risks **8**(4), 107 (2020)
52. Zizi, Y., Jamali-Alaoui, A., El Goumi, B., Oudgou, M., El Moudden, A.: An optimal model of financial distress prediction: a comparative study between neural networks and logistic regression. Risks **9**(11), 200 (2021)
53. Malakauskas, A., Lakštutienė, A.: Financial distress prediction for small and medium enterprises using machine learning techniques. Eng. Econ. **32**(1), 4–14 (2021)
54. Doğan, S., Koçak, D. and Atan, M.: Financial distress prediction using support vector machines and logistic regression. In: Kenan Terzioğlu, M. (ed.) Advances in Econometrics, Operational Research, Data Science and Actuarial Studies: Techniques and Theories, pp. 429–452. Springer International Publishing, Cham (2022)
55. Santoso, N., Wibowo, W.: Financial distress prediction using linear discriminant analysis and support vector machine. J. Phys: Conf. Ser. **979**(1), 012089 (2018)
56. Sreedharan, M., Khedr, A.M., El Bannany, M.: A comparative analysis of machine learning classifiers and ensemble techniques in financial distress prediction. In: 2020 17th International Multi-Conference on Systems, Signals & Devices (SSD), pp. 653–657 (2020). ISSN: 2474-0446
57. Maulana, D.J., Saadah, S., Yunanto, P.E.: Kmeans-SMOTE integration for handling imbalance data in classifying financial distress companies using SVM and Naïve bayes. J. RESTI (Rekayasa Sistem dan Teknologi Informasi) **8**(1), 54–61 (2024)
58. Zeng, S., Li, Y., Yang, W., Li, Y.: A financial distress prediction model based on sparse algorithm and support vector machine. Math. Probl. Eng. **2020**(1), 5625271 (2020)
59. Ayuni, N.W.D., Lasmini, N.N., Dewi, K.C.: Predicting financial distress of property and real estate companies using optimized support vector machine - particle swarm optimization (SVM-PSO). Bull. Soc. Inform. Theory Appl. **8**(1), 97–106 (2024)
60. Tran, K.L., Le, H.A., Nguyen, T.H., Nguyen, D.T.: Explainable machine learning for financial distress prediction: evidence from Vietnam. Data **7**(11), 160 (2022)
61. Aydin, N., Sahin, N., Deveci, M., Pamucar, D.: Prediction of financial distress of companies with artificial neural networks and decision trees models. Mach. Learn. Appl. **10**, 100432 (2022)
62. Podhorska, I., Vrbka, J., Lazaroiu, G., Kovacova, M.: Innovations in financial management: recursive prediction model based on decision trees. Mark. Manag. Innov. 276–292 (2020)
63. Breiman, L.: Random forests. Mach. Learn. **45**, 5–32 (2001)
64. Schapire, R.E.: A brief introduction to boosting. In: Proceedings of the 16th International Joint Conference on Artificial Intelligence, vol. 2, IJCAI'99, pp. 1401–1406. Morgan Kaufmann Publishers Inc., San Francisco, CA, USA (1999)
65. Chen, T., Guestrin, C.: Xgboost: a scalable tree boosting system. In: Proceedings of the 22nd ACM SIGKDD International Conference on Knowledge Discovery and Data Mining, KDD '16, pp. 785–794. Association for Computing Machinery, New York, NY, USA (2016)
66. Ke, G., Meng, Q., Finley, T., Wang, T., Chen, W., Ma, W., Ye, Q., Liu, T.Y.: LightGBM: a highly efficient gradient boosting decision tree. In: Advances in Neural Information Processing Systems vol. 30. Curran Associates, Inc. (2017)

67. Ben Jabeur, S., Stef, N., Carmona, P.: Bankruptcy prediction using the XGBoost algorithm and variable importance feature engineering. Comput. Econ. **61**(2), 715–741 (2023)
68. Liu, W., Fan, H., Xia, M., Pang, C.: Predicting and interpreting financial distress using a weighted boosted tree-based tree. Eng. Appl. Artif. Intell. **116**, 105466 (2022)
69. Carmona, P., Dwekat, A., Mardawi, Z.: No more black boxes! explaining the predictions of a machine learning XGBoost classifier algorithm in business failure. Res. Int. Bus. Financ. **61**, 101649 (2022)
70. Muparuri, L., Gumbo, V.: On logit and artificial neural networks in corporate distress modelling for Zimbabwe listed corporates. Sustain. Anal. Model. **2**, 100006 (2022)
71. Nguyen, O.T., Nguyen, D.V., Doan, T.C., Nguyen, A.H., Ta, A.T.: Prediction of financial distress in Vietnam using multi-layer perceptron artificial neural network. In: Proceedings of the 11th International Conference on Emerging Challenges: Smart Business and Digital Economy 2023 (ICECH 2023), pp. 264–290. Atlantis Press (2024)
72. Elhoseny, M., Metawa, N., Sztano, G., El-Hasnony, I.M.: Deep learning-based model for financial distress prediction. Ann. Oper. Res. (2022)
73. Liu, J., Li, C., Ouyang, P., Liu, J., Chong, W.: Interpreting the prediction results of the tree-based gradient boosting models for financial distress prediction with an explainable machine learning approach. J. Forecast. **42**(5), 1112–1137 (2023)
74. Tian, S., Yan, Yu., Guo, H.: Variable selection and corporate bankruptcy forecasts. J. Bank. Financ. **52**, 89–100 (2015)
75. Chaves, R.M., Rossi, A.L.D. and Garcia, L.P.F.: A financial distress prediction using a non-stationary dataset. In: Anais do XX Encontro Nacional de Inteligência Artificial e Computacional, pp. 300–314. SBC (2023)
76. Hajek, P.: Speech emotion recognition from earnings conference calls in predicting corporate financial distress. In: IFIP International Conference on Artificial Intelligence Applications and Innovations, pp. 216–228. Springer, Berlin (2022)
77. Sasaki, H., Fujii, M., Sakaji, H., Masuyama, S.: Enhancing risk analysis with GNN: edge classification in risk causality from securities reports. Int. J. Inf. Manag. Data Insights **4**(1), 100217 (2024)
78. Avramelou, L., Nousi, P., Passalis, N., Tefas, A.: Deep reinforcement learning for financial trading using multi-modal features. Expert Syst. Appl. **238**, 121849 (2024)
79. Tanaka, F.H.K.D.S., Aranha, C.: Data augmentation using GANs (2019). arXiv:1904.09135
80. Kheiri, K., Karimi, H.: Sentimentgpt: exploiting GPT for advanced sentiment analysis and its departure from current machine learning (2023). arXiv:2307.10234
81. Gajamannage, K., Park, Y., Jayathilake, D.I.: Real-time forecasting of time series in financial markets using sequentially trained dual-LSTMs. Expert. Syst. Appl. **223**, 119879 (2023)
82. Andreou, C.K., Andreou, P.C., Lambertides, N.: Financial distress risk and stock price crashes. J. Corp. Financ. **67**, 101870 (2021)
83. Soo Hyun Cho and Kyung-shik Shin: Feature-weighted counterfactual-based explanation for bankruptcy prediction. Expert Syst. Appl. **216**, 119390 (2023)
84. Lundberg, S.M., Lee, S.I.: A unified approach to interpreting model predictions. In: Proceedings of the 31st International Conference on Neural Information Processing Systems, NIPS'17, pp. 4768–4777. Curran Associates Inc., Red Hook, NY, USA (2017)
85. Ribeiro, M.T., Singh, S., Guestrin, C.: Why should I trust you?: explaining the predictions of any classifier. In: Proceedings of the 22nd ACM SIGKDD International Conference on Knowledge Discovery and Data Mining, San Francisco, CA, USA, 13–17 Aug 2016, pp. 1135–1144 (2016)

SPredict: Stock Market Prediction with Social Media Sentiment Analysis and Machine Learning

Safad Ismail and Harsha Vasudev

Abstract Stock market is a dynamic financial marketplace where individuals and organizations engage in the buying, selling, and trading of publicly traded company shares. Stock market forecasting involves estimating future changes in stock values through various analytical methods such as statistical analysis, historical data trends, and economic indicators. Traditional methods such as technical and fundamental analysis are widely used. However, sentiment analysis particularly using social media data has emerged as a promising area for enhancing prediction accuracy. This work integrates sentiment analysis from X data with advanced Machine Learning (ML) models to examine the effect of market perception on stock prices. Techniques such as Auto Regressive Integrated Moving Average (ARIMA), Moving Average (MA), and Long Short Term Memory (LSTM) networks were used to predict stock prices. Sentiment analysis were performed on X data to examine the effect of market perception on stock prices using the VADER and Text Blob lexicon sets. The result showed that adding sentiment analysis can improve the stock price prediction models accuracy. Statistical Evaluation Metrics showed that LSTM model has achieved higher correlation values, indicating superior predictive performance over the ARIMA and MA Models. Additionally, VADER outperformed Text Blob, demonstrating higher accuracy and more reliable results. The findings have significant implications for investors, financial analysts, and researchers interested in stock market prediction. They highlight the potential of sentiment analysis as a supplementary tool for understanding market behavior.

S. Ismail (✉)
Department of Cyber Security, Muthoot Institute of Technology and Science, Kochi, Kerala, India
e-mail: safadismail@mgits.ac.in

H. Vasudev
Indian Institute of Information Technology, Design & Manufacturing, Kurnool, India
e-mail: harsha@iiitk.ac.in; harsha.vasudevanpillai@unipd.it

HIT Research Centre, Department of Mathematics, University of Padua, Padua, Italy

H. T. T. Nguyen et al. (eds.), *Data Science in Finance and Accounting*, Studies in Big Data 181, https://doi.org/10.1007/978-3-032-06179-9_5

1 Introduction

The Stock Market is an essential part of the financial system that acts as a marketplace where stocks representing ownership in businesses are traded. It enables companies to raise capital by selling shares to the public and serves as a venue for investors to buy and sell stocks. Regulated exchanges, such as the New York Stock Exchange (NYSE) and National Association of Securities Dealers Automatic Quotation System (NASDAQ) ensure transparency and fair trading. Investors employ methods like technical and fundamental analysis to predict market behavior [1]. The recent developments demonstrate how algorithmic trading, powered by computer models, combines aspects of technical and fundamental analysis to improve liquidity and lower volatility [2, 3]. Sentiment Analysis, particularly on platforms like X formerly known as Twitter [4], provides real-time insights into public opinion helping traders in making informed decisions. The investors use technical analysis and fundamental analysis to determine the stock movement [5–7]. Technical analysts interpret a stock's movement by looking at its historical trends and use graphical analysis to examine the stock's historical graph and forecast its future direction. Fundamental analysts examine a company's financial statements, public perception, and overall business plan to forecast the direction of the stock [8]. The intrinsic value of the company is computed using this data by discounting the anticipated future cash flows to their net present value. In the dynamic landscape of financial markets, the proposed work seeks to advance predictive modeling by integrating technical analysis and using sentiment analysis on platforms like X. The accurate prediction of stock prices remains a complex challenge [9]. Although the conventional techniques such as technical analysis give meaningful information, they often fail to capture the entire range of factors impacting market movements [10]. The proposed interdisciplinary approach aims to enhance understanding and decision making within the evolving intersection of capital markets and cutting-edge computing techniques such as algorithmic trading. The research findings help the investors and analysts to identify potential dangers and make well informed decisions about their plans by raising awareness of market mood and its possible impact on stock prices. By incorporating X sentiment analysis and advanced ML for both technical and sentiment analysis, this research addresses current trends in the field and pushes the boundaries of existing research.

The goal of this study aims to explore the potential benefits of combining ML models with sentiment analysis of social media data, particularly X to improve the accuracy of stock price prediction. The integration of ML and sentiment analysis addressed a gap in existing research by using real time social media data to enhance predictive modeling in stock market forecasting. This approach acknowledges the dynamic nature of financial markets and seeks to provide a more comprehensive understanding of market sentiment and its impact on stock prices.

The rest of the chapter is structured as follows. In Sect. 1.2, prior research on sentiment analysis, and stock market forecasting are reviewed. The research technique and methodologies used in this study are described in Sect. 1.3. Section 1.4 presents the results of the stock price forecasting models, sentiment analysis, and the financial market's reaction to the sentiment study. Section 1.5 concludes the chapter and offers suggestions for future research.

2 Related Works

In recent years many studies have focused on the usage of ML and Artificial Intelligence (AI) for the prediction of stock price [11–13]. These models employ techniques like Neural Networks (NN), Support Vector Machine (SVM), decision trees, and ensemble approaches to extract important insights from historical price data and market indicators. While the above techniques are good in stock price prediction, several challenges affect their accuracy such as high market volatility, noisy stock data, the complexity of model tuning, and difficulties in result interpretation. Moreover, the unpredictable nature of external events and the limitations of data further affect the accurate predictions. Banerjee et al. proposed a stock market prediction system combining ARIMA and LSTM for determining the sector-wise preferences [14]. Their study focused on stocks from three key sectors such as banking, IT, and manufacturing. From the banking sectors stocks of South Indian Bank, Punjab National Bank, and Yes Bank were analyzed. In the IT sector, Wipro, Infosys, and TCS and from the manufacturing sector, Reliance, ITC, and Hindustan Petroleum. According to their research LSTM model is best for long-term stock prediction when compared to the ARIMA model and the LSTM model's prediction for TCS, Wipro, and South Indian Bank stocks can be improved by effectively changing the model's parameters and training it with additional datasets. Similarly, Xiao et al. [15] conducted a research on ARIMA-LSTM hybrid model for stock market prediction. This model combines linear ARIMA and nonlinear LSTM approaches to forecast the stock index correlation coefficients. Although the model performed well in capturing index correlations, it had difficulty in handling financial anomalies and high-frequency noise, which made it less dependable for real-time forecasting. Dines et al. proposed a regression-based ML model using Moving Average (MA) crossovers [16]. While it provided useful insights into trend reversals, its dependence on historical data limited its effectiveness in rapidly changing market conditions. Ho et al. applied ARIMA, NN, and LSTM to forecast Bursa Malaysia stock prices during the COVID-19 pandemic [17]. Their study showed high accuracy, but the lack of sentiment analysis and economic indicators limited the model's overall understanding of market behavior. Deepak et al. examined ML and AI models for stock price prediction but the work did not incorporate other factors such as consumer's attitude, government policies, and amendments [18]. This omission reduced the model's ability to reflect real-world market influences. In contrast, Nasir et al. developed a novel hybrid prediction technique using local mean decomposition with ARIMA and LSTM models to improve crude oil price prediction accuracy [19]. Their approach decreased volatility and resolved the neural network over fitting issue. The model was tested using publicly available West Texas Intermediate data, and its forecast accuracy was evaluated using standard accuracy measures. The study concluded that ARIMA can be effective for short-term prediction when combined with other techniques. However, the model's domain specificity limits its applicability across diverse stock markets. Demirel et al. compared the performance of MLP, SVM, and LSTM across 42 Istanbul Stock Exchange firms over nine years [20]. Using statistical metrics such as R^2, MSE, and RMSE their results showed that LSTM and MLP outperformed the other models. However, the absence of external market indicators again limited predictive robustness. Ayala et al. [21] highlighted the importance of optimizing technical indicators in hybrid ML models to improve trading strategies. However, their work mainly focused on parameter tuning, with less emphasis on how the model adapts to volatile market conditions. Hazlina et al. explored ARIMA, NN, and LSTM models for predicting Malaysian stock prices during the pandemic, with LSTM achieving over 90% accuracy. However, the study was limited by its short evaluation period, few models, and the exclusion of external predictors and sentiment analysis. These limitations reduce the model's generalizability and its ability to respond to market sentiment [22]. Recent studies have recognized the growing role of social media sentiment in financial forecasting. Platforms like Twitter, Reddit, and Facebook provide real-time insights into public sentiment and investor mood [23]. Sentiment analysis tools such as VADER and NLTK have been used to classify financial sentiment from online text. One such study focused on the usage of VADER and NLTK to conduct sentiment analysis of X data [24]. In the proposed work, the VADER tool performed well in classifying the sentiments of X data. However, the overall accuracy was low due to the limited dataset size and the absence of domain-specific vocabulary. In 2020, Sagala et al. addressed the prediction of stock price movements through a combined approach of technical and sentiment analysis [25]. While this combined approach showed potential, its performance varied based on sentiment data quality and required better preprocessing techniques.

Recent studies have shown that AI and ML can be effective for predicting stock prices. Methods such as NNs, SVM, DTs, and combinations such as ARIMA-LSTM are often used to analyze past price trends and market signals. Sentiment analysis especially from social media has also become popular for improving prediction results. However, many models still miss key factors like economic conditions and policy changes, which makes it harder for them to handle market changes and work well across different sectors.

In the past couple of years, the field of stock price forecasting has seen the emergence of several refined approaches that address the limitations of earlier ML models. Among these the work by Mamillapalli et al. work has gained attention for fusing sentiment analysis using the VADER tool with Gated Recurrent Units (GRUs), improving temporal prediction capabilities [26]. However, its performance tends to decline in scenarios where sentiment activity is minimal. Thomas et al. explored the effectiveness of FinBERT and GPT-4 for financial sentiment tasks. FinBERT demonstrates strong results in handling domain-specific language, while GPT-4 shows advantages in understanding nuanced and complex text [27]. Both models, however requires significant computational resources and extensive fine-tuning for optimal results. Talazadeh et al. introduced a Sentiment-Augmented Random Forest Model (SARF), a hybrid method combining sentiment inputs with a random forest model [28]. Although effective in directional stock prediction, the model's reliability diminishes in highly volatile market conditions. Zhu et al. proposed a BERTopic-LSTM framework that uses topic modeling from financial text to enhance LSTM-based forecasting. Yet, its performance suffers when dealing with sparse textual data [29]. Wang et al. developed an attention-driven sentiment mechanism. It has shown superior performance compared to traditional LSTM in unpredictable markets, although it comes with a high computational cost [30]. Similarly, Liu et al. introduced HiDFORMER, a hierarchical transformer designed for time series forecasting that captures both long- and short-term patterns effectively. However, its performance tends to drop when trained on limited data or confronted with sudden external shocks [31]. These recent advancements indicate a clear trend toward integrating sentiment signals and attention-based architectures to build more resilient and context-aware prediction systems.

The proposed framework aims to overcome these issues by

- Combining technical indicators with sentiment analysis gathered from social media and news sources.
- Implementing a hybrid LSTM-based deep learning model tailored for both short- and long-term market predictions.
- Integrating macroeconomic data and key policy developments to improve contextual awareness.
- Applying effective data preprocessing techniques and ensemble methods to reduce overfitting and better manage data noise.

3 Methodology

In order to predict the stock price two sets of datasets were employed as model inputs in this study. The market data that displays daily prices per share, as well as the market social network datasets is utilized for sentiment analysis. The Table 1 provides descriptive statistics and tweet counts for selected stocks, including the mean and variance of stock prices as well as the volume of related tweets. The proposed model framework as shown in Fig. 1 uses social media sentiment analysis to predict the stock price. Three predictive models were employed in this study namely ARIMA, Moving Average, and LSTM. ARIMA is a statistical method suitable for modeling

Table 1 Descriptive statistics and tweet counts of selected stocks

Company	Stock price mean	Stock price variance	Tweet count
Amazon	132.76	953.46	40890.00
Tesla	181.01	10307.85	37422.00
Microsoft	254.05	5769.75	39890.00

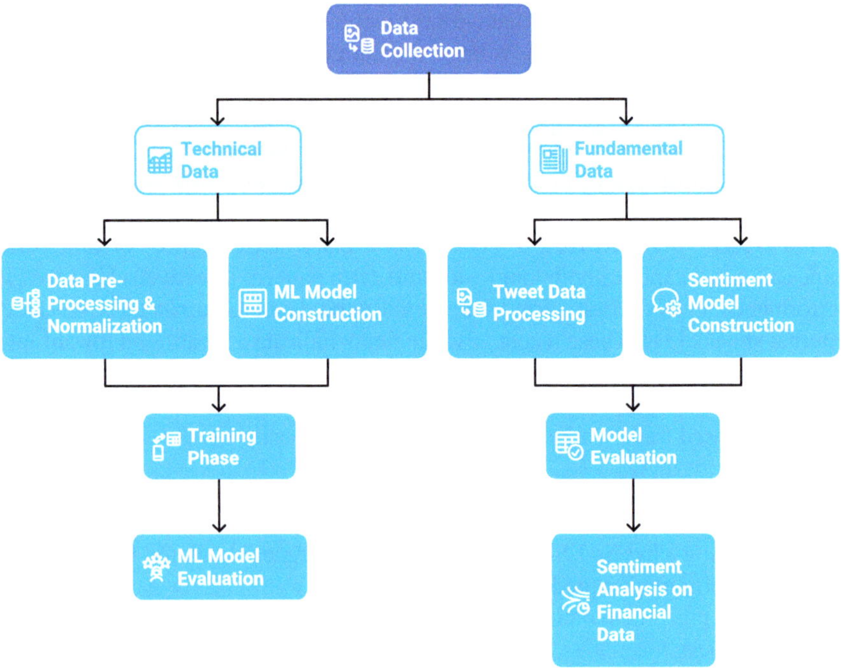

Fig. 1 Proposed research framework

and forecasting time series data with trends. Moving Average provides a smoothed representation of short-term fluctuations, making it ideal for initial trend detection. LSTM is a type of RNN that can learn long-term dependencies, capturing both non-linear patterns and temporal sequences in stock data. These models were selected to compare traditional time series forecasting approaches with deep learning methods. The sentiment scores from TextBlob, and VADER were averaged to obtain a unified sentiment index for each day. These indices were aligned with corresponding daily stock prices and used as an additional feature input in the ARIMA, MA, and LSTM models to examine their impact on price forecasting. Model performance was evaluated using metrics such as RMSE, MAPE, R, and MAE.

3.1 System Flow

The research framework outlines a systematic approach to financial forecasting by integrating both technical and fundamental data through ML. It starts with collecting data, which is then split into two main parts. The first part deals with technical data, where the data is cleaned and prepared through preprocessing and normalization. After that, ML models are built and trained using this data. The model is then tested to check how well it performs. The second part focuses on fundamental data, especially information from tweets. These tweets are processed to understand the emotions or opinions behind them. A separate sentiment analysis model is created and tested for accuracy. Finally, this sentiment information is used alongside financial data to better understand market trends. By using both data types, numbers and public opinions this approach helps to make financial predictions more reliable and insightful.

3.2 System Model

The system diagram provides a structured approach to analyze the relationship between stock market trends and public sentiment on social media as shown in Fig. 2. It begins by collecting tweets related to specific stocks using a stock tweets API and obtaining historical stock price data from reliable sources. The tweets are then processed through two sentiment analysis tools namely TextBlob and VADER, which assess the emotional tone and subjectivity of the content. These sentiment scores are combined with stock price data to form a comprehensive dataset. A correlation analyzer examines this data to identify possible links between market behavior and public sentiment. Based on these insights, the predictive models are used to estimate future stock prices. The system is supported by a user interface for easy access and interpretation of results, with internet connectivity ensuring real-time data updates. This integrated flow supports informed decision-making by combining financial data with sentiment analysis.

3.3 Dataset

The system begins by downloading data from two sources, stock tweets downloader module downloads the stock tweets data from a social media platform like X. The stock price data module downloads historical stock price data in Indian Rupees from yahoo finance website [32]. The collected data are preprocessed and the following features are extracted such as 'low', 'open','high', 'close', 'adjusted close price', and 'volume' are extracted from the financial data. Out of the extracted features adjusted close price is used in this study. For the tweets extracted from X database, they are preprocessed to remove the punctuation's, stop words, etc. The preprocessed tweets are sent to the sentiment analyzer tool VADER and Text Blob designed to assess sentiment in social media posts. The data from both sentiment analysis tools is then integrated into a single CSV file labeled 'Stock Tweets Data'. The Correlation Analyzer analyzes the correlation between the sentiment data in the 'Stock Tweets Data' file and the historical stock price data. The purpose is to find out if there is a correlation between social media sentiment and stock prices. The Stock Price Predictor module uses the results of the correlation analysis to predict the future stock prices using ML models. The system also has a user interface, which allows

Fig. 2 Architecture diagram

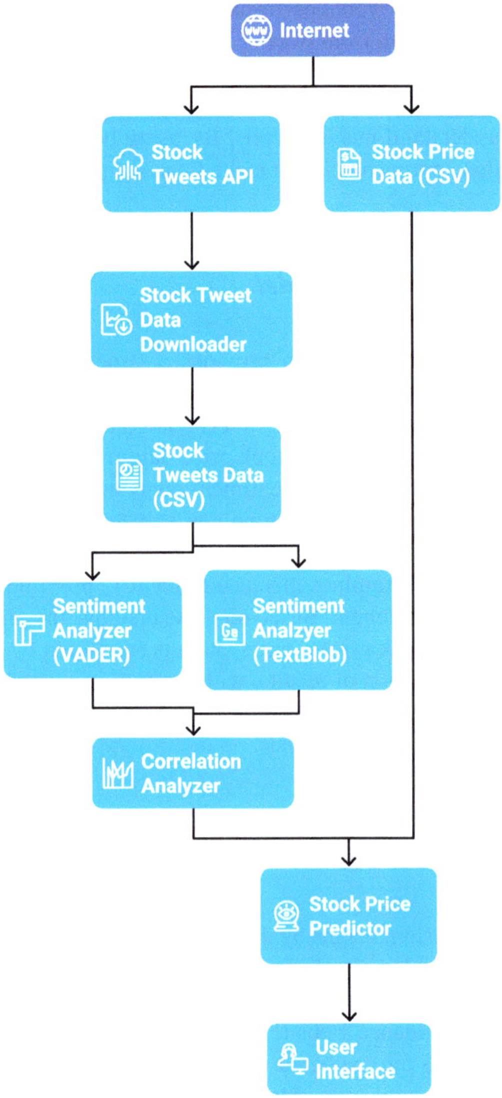

the users to interact with the system, view the results of the sentiment analysis and stock price prediction, and potentially input their own data. The stock price over time is shown in Fig. 3.

The dataset was split chronologically with 70% allocated for training (from *2019-02-01* to *2021-12-30*) and 30% for testing (from *2022-01-01* to *2024-03-19*). Table 1 presents the descriptive statistics and tweet counts of selected stocks. For the LSTM model, hyper-parameters such as the number of epochs, batch size, and number of LSTM units were optimized using grid search and early stopping based on validation

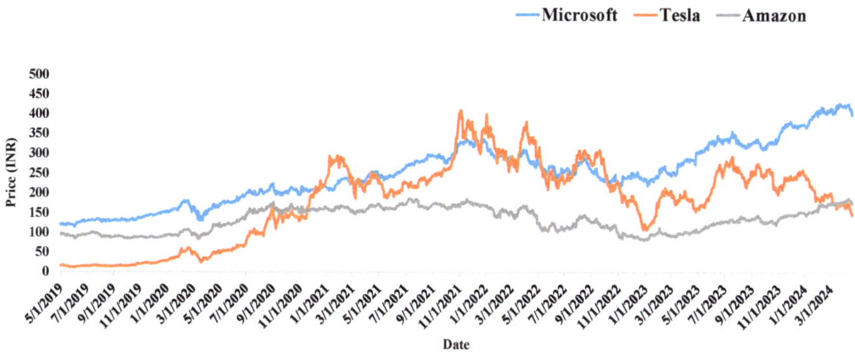

Fig. 3 Stock price over time

loss. The tweets were crawled from the X database and the data collection period for the X dataset ranges from *'2021-10-31'* to *'2022-08-07'*. There are a total of '118,208' tweets related to Amazon, Microsoft, and Tesla in the dataset.

3.4 Data Preprocessing

To ensure the datasets were suitable for modeling and analysis, preprocessing was conducted in two distinct phases. These phases addressed the structural characteristics of numerical stock data and the unstructured nature of textual social media content, referred to respectively as Technical Data Processing and Fundamental Data Processing.

A. Technical Data Processing

The preprocessing of the financial data involved several key steps before feeding to the ML models. First, the Date column was converted to a date-time format and set as the index for time-series analysis. The target variable, 'Adj Close' was then selected for forecasting. Any missing or null values in the dataset were removed to maintain data integrity. To assess the stationary of the series, the Augmented Dickey-Fuller (ADF) test was applied, and if the series was found to be non-stationary, first-order differencing was applied. The data was then scaled using Min-Max scaling to normalize it to a range of [0, 1], particularly for the LSTM model. Finally, time sequences with a 10-step window were created to form the input (X) and next-step output (y) for LSTM based forecasting. These preprocessing steps ensured the data was properly formatted and ready for accurate predictions using ARIMA, MA, and LSTM models. A sample of preprocessed financial data is shown in Table 2.

Table 2 Technical data preprocessing for time series forecasting

Step	Preprocessing step	Description	Input columns	Output	Sample output
1	Parse dates & set index	Convert 'Date' to datetime and use it as index	Date	Time-indexed DataFrame	2019-04-22 → 94.36
2	Select target variable	Choose 'Adj Close' for forecasting	Adj close	1-column DataFrame	94.36, 96.18, 95.08
3	Handle missing values	Drop missing/null values	Adj close	Cleaned series (no NaNs)	No missing values
4	Stationarity check (ADF test)	Use augmented dickey-fuller test to check stationarity	Adj Close	ADF stat, p-value	ADF = −1.78, p = 0.39
5	Differencing (if needed)	Apply first differencing to make data stationary	Adj close	Stationary series	+1.82, −1.10, +0.02
6	Scaling (MinMax for LSTM)	Normalize data to [0, 1] range	Adj close	Scaled array	0.428, 0.605, 0.472
7	Create sequences for LSTM	Create X (input) and y (label) using 10-step windows	Scaled Adj close	X: time sequences, y: next step	X = [0.42..0.60], y = 0.47

B. Fundamental Data Processing

The collected tweets has to be preprocessed to reduce the noise and to normalize the data. For example, special characters, URLs (Uniform Resource Locator), hashtags, and mentions are some of the unnecessary information that frequently appears in tweets. For instance same words may appear in tweets in different forms (example: 'happy', 'happyy', 'happyyyy'), by normalization the data becomes more consistent. The extracted tweets are pre-processed to remove the punctuation's, special characters, URLs, words, digits, gibberish words, and stop words. The processed tweets are lemmatized and tokenized, and POS Tagging is performed on the tokenized words and chunking is performed over it. The time series analysis of tweet count is shown in Fig. 4 and Table 3 provides a summary of the preprocessed tweet data.

3.5 Correlation Matrix

A correlation matrix was generated by performing a correlation analysis on a dataset. This statistical technique measures the strength and direction of the relationship

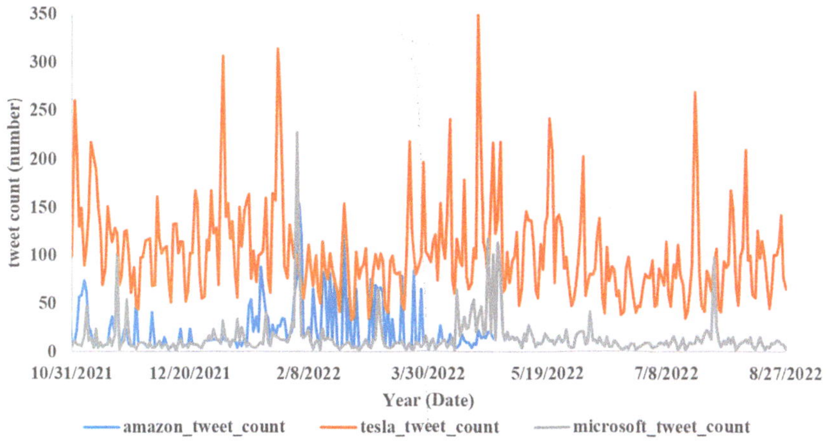

Fig. 4 Time series analysis of tweet count

Table 3 Sample preprocessed tweet features with sentiment scores

Date	Original tweet	Tokens	Lemmatized tokens	POS tags	Noun phrases/Bigrams/Trigrams	Sentiment score
2022-02-04	Jeff bezoz looking at people who bought amzn MSFT	[jeff, looking, people, bought, today]	[jeff, look, people, buy, today]	(jeff, PROPN), (look, VERB), (people, NOUN)	[people], [today] (jeff, look), (look, people), (people, buy) (jeff, look, people), (look, people, buy)	neg: 0.0, neu: 1.0, pos: 0.0
2022-01-21	Bought a little more of amzn amp avlr here im MSFT	[bought, little, gon, na, go, bottom]	[buy, little, go, to, go, bottom]	(buy, VERB), (little, ADJ), (go, VERB)	(buy, little), (little, go) (buy, little, go), (little, go, to)	neg: 0.0, neu: 1.0, pos: 0.0
2022-08-25	Amazon amzn and plug power plug announced toda. MSFT	[plug, power, plug, today, hydrogen, supply, d.]	[plug, power, plug, today, hydrogen, supply, d.]	(plug, VERB), (power, NOUN), (plug, NOUN)	[power, plug, today, hydrogen, supply, deal] (plug, power), (power, plug), (plug, today) (plug, power, plug), (power, plug, today)	neg: 0.0, neu: 0.876, pos: 0.124

between two variables [33]. In this case the dataset contains multiple variables and we aim to understand their inter relationships. To create the correlation matrix the correlation coefficient for each pair of variables is calculated [34]. The correlation coefficient is a number between –1 and 1, indicates the strength and direction of the relationship. When the variables have a positive value, it indicates that they are moving in the same direction while a negative value means they move in opposite direction [35]. The closer the coefficient is to 1 or –1, the stronger the relationship.

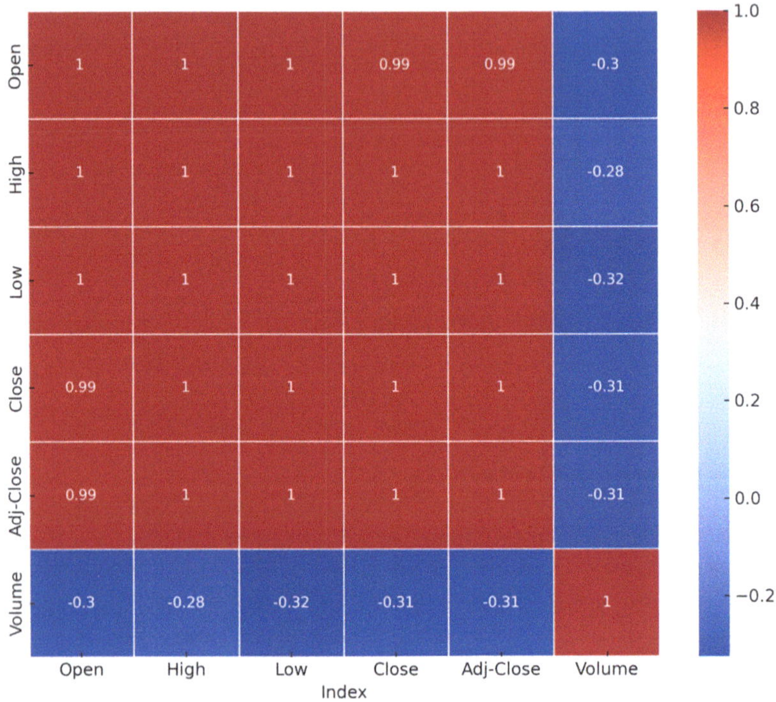

Fig. 5 Correlation analysis of the dataset

After calculating the correlation coefficients for all pairs of variables they are visualized in a matrix format. Each cell in the matrix represents the correlation coefficient between two variables indicating the strength of the correlation. By examining the correlation matrix, one can identify the patterns and relationships between the variables providing valuable insights for understanding the dataset and making informed decisions as shown in Fig. 5.

3.6 Techniques for Predicting Stock Price

The stock values of Amazon, Tesla, and Microsoft are predicted in this study using three different ML techniques namely ARIMA, MA, and LSTM. The forecasting period spans five years, from '2020' to '2024'. This study compares the prediction abilities of these three ML techniques in order to determine which model is most suited for predicting the stock values.

3.6.1 Auto Regressive Integrated Moving Average (ARIMA)

ARIMA is a statistical method used for forecasting time series data, such as stock prices [36]. It does this by considering past values, trends, and random fluctuations. The basic ARIMA (1, 1, 1) model can be expressed as

$$Y_t = \phi_1 Y_{t-1} + \theta_1 \epsilon_{t-1} + \epsilon_t \tag{1}$$

where: Y_t is the value of the time series at time 't'. ϕ_1 represents the auto regressive coefficient. θ_1 is the moving average coefficient to be estimated. ϵ_t is the error term at time 't'. For the prediction of stock price ARIMA Model uses the actual close value of the stock price.

3.6.2 Moving Average (MA)

MA models are a fundamental tool in time series analysis [37]. This model operates on the principle of smoothing out fluctuations within a dataset by averaging consecutive subsets of observations over a specified window or period.

$$Y_t = c + \varepsilon_t + \theta_1 \varepsilon_{t-1} + \theta_2 \varepsilon_{t-2} + \ldots + \theta_q \varepsilon_{t-q} \tag{2}$$

where:

Y_t: is the value of the time series at time t,
ε_t: is the white noise error term at time t,
$\theta_1, \theta_2, \ldots, \theta_q$: are the weights for the previous q error terms.

3.6.3 Long Short Term Memory (LSTM)

Recurrent Neural Network (RNN) with LSTM networks are specifically engineered to identify long-term dependencies in sequential data [38]. In contrast to conventional RNN, LSTM solve the vanishing gradient issue, allowing them to extract information from distant past elements in a sequence. LSTMs use a special structure called a memory cell as shown in Fig. 6, which helps to manage information using three gates [39]. The forget gate removes the old information, the input gate adds new important data, and the output gate controls what information is sent to the next step. These gates help the LSTMs to remember useful information for a longer time, making them better at tasks like predicting sequences. This feature is particularly useful in finance, where stock prices are often influenced by past events like policy changes or economic trends. LSTMs can learn and remember these long-term patterns, making them better at predicting stock prices. Even though financial markets are unpredictable, LSTMs can capture complex patterns, giving them an advantage

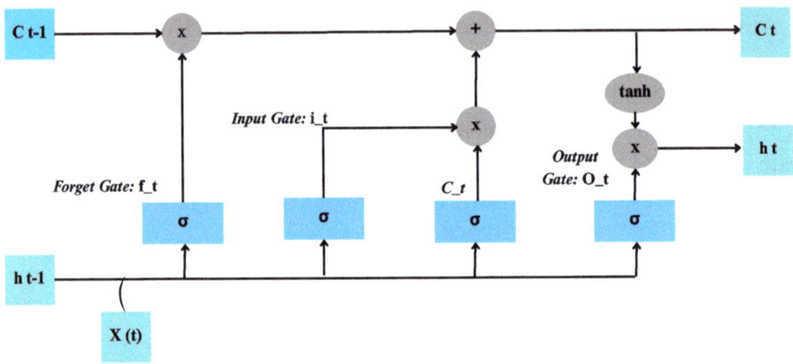

Fig. 6 LSTM model architecture

over traditional methods and regular RNNs. The following equations describe the operations within an LSTM model.

- Forget Gate (f_t): The forget gate decides which information from the previous memory cell (c_{t-1}) should be discarded. It uses the previous hidden state (h_{t-1}) and the current input (x_t) to generate a value between 0 and 1 (through the sigmoid function σ), where 0 means 'forget' and 1 means 'retain'.

$$f_t = \sigma(W_f \times [h_{(t-1)}, x_t] + b_f) \tag{3}$$

- Input Gate (i_t): The input gate determines which new information should be stored in the memory cell. Like the forget gate, it uses the previous hidden state (h_{t-1}) and the current input (x_t) to decide what data is important to keep. The input gate also generates a value between 0 and 1 through the sigmoid function σ.

$$i_t = \sigma(W_i \times [h_{(t-1)}, x_t] + b_i) \tag{4}$$

- Memory Cell Update (c_t): The memory cell c_t is updated by combining the previous memory cell (c_{t-1}) and the new candidate cell state ($\tilde{c}_t$). The forget gate's output decides how much of the old memory to keep, and the input gate's output determines how much of the new memory to add.

$$c_t = f_t \odot c_{(t-1)} + i_t \odot \tilde{c}_t \tag{5}$$

- Output Gate (o_t): The output gate controls what part of the memory cell should be passed to the next hidden state. It uses the current input and the previous hidden state to decide how much of the updated memory cell should influence the next time step. The output is passed through a sigmoid function σ.

$$o_t = \sigma(W_o \times [h_{(t-1)}, x_t] + b_o) \tag{6}$$

Fig. 7 ARIMA model prediction for stock price **a** Amazon **b** Tesla **c** Microsoft

4 Results and Discussions

In this study, three distinct predictive models are employed namely ARIMA, LSTM, and MA to forecast the stock prices of Amazon, Tesla, and Microsoft over a five year period spanning from '2020' to '2024' and to evaluate the impact of sentiment analysis on stock price. The effectiveness of these models in capturing the dynamic nature of the stock market is evident through a clear contrast between the predicted and actual stock values. With the help of time series analysis, ARIMA showed strong performance in identifying short to medium term trends, which was especially clear in Microsoft's stock price forecasts. LSTM demonstrated exceptional flexibility in handling intricate patterns and outperformed other models in forecasting Tesla's erratic stock prices. MA provided stable and consistent predictions, particularly suitable for Amazon's relatively stable stock performance. Our analysis compares the model forecasts with actual stock prices, offering valuable insights for both investors and researchers. This promotes informed decision making and contributes to the ongoing discussion on stock market prediction methods. The prediction results of the ARIMA model, MA Model, and LSTM Models for different stocks are shown in Figs. 7, 8 and 9 respectively.

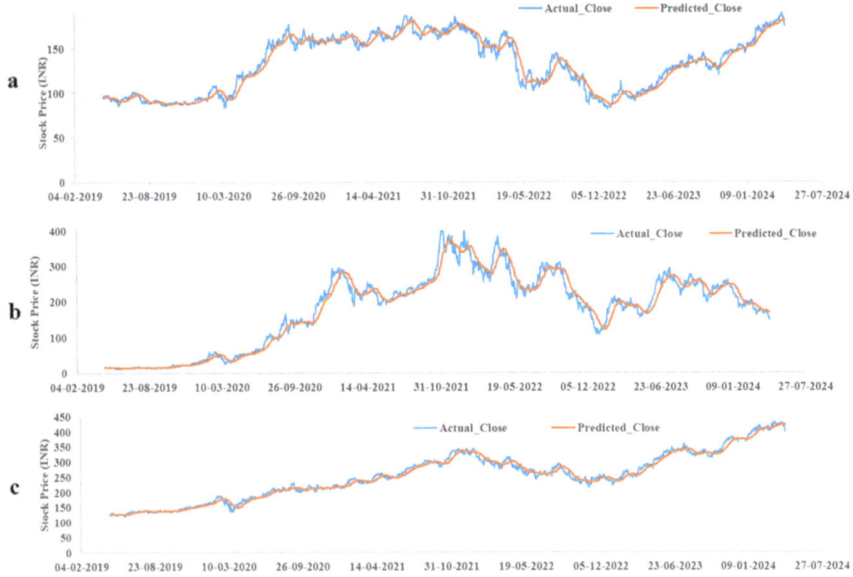

Fig. 8 MA model prediction for stock price **a** Amazon **b** Tesla **c** Microsoft

Fig. 9 LSTM model prediction for stock price **a** Amazon **b** Tesla **c** Microsoft

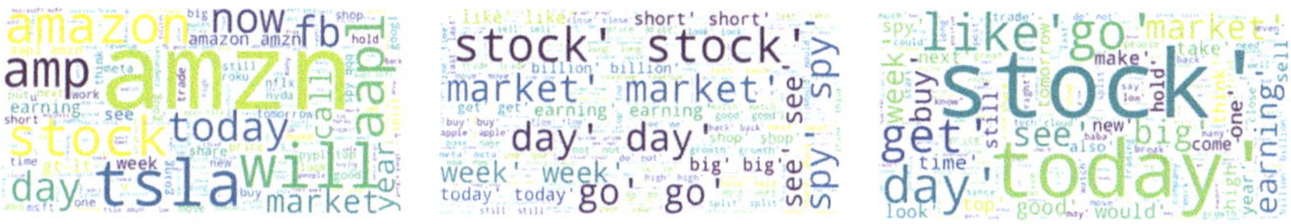

(a) (b) (c)

Fig. 10 Representation of stock market sentiment **a** Unigram **b** Bigram **c** Trigram

4.1 Discussion of Sentiment Analysis

After gathering the relevant data the next stage involves understanding individual's views on Amazon, Tesla, and Microsoft and their offerings. We have used two sentiment analysis tool namely VADER and Text Blob for the tweets related to the companies.

4.1.1 n-gram

n-gram is a continuous series of n elements from a particular text or speech sample used in sentiment analysis [40]. It is an effective sentiment analysis tool for capturing the meaning and context of a statement. These elements can be characters, syllables, or words depending on the level of analysis. A one-gram, or unigram, analyze each word separately [41]. For example consider a statement Amazon stock is soaring today, the unigrams are ['Amazon', 'stock', 'is', 'soaring', 'today']. This approach captures individual word sentiments but may miss context provided by the word combinations. On the other hand bi-gram examines the consecutive words such as ['Amazon stock', 'stock is', 'is soaring', 'soaring today']. This can provide more information than just unigrams by capturing basic contextual interactions between the words. Trigram examine the sequence of three words [42] ['Amazon stock is', 'stock is soaring', 'is soaring today'], trigrams can be used to differentiate between 'not good' and 'good'. The incorporation of unigrams, bigrams, and trigrams in sentiment analysis enhances the model's ability to understand and predict as shown in Fig. 10.

4.1.2 Sentiment Analyzer Using VADER

VADER (Valence Aware Dictionary and Sentiment Reasoner) is a lexicon-based sentiment analysis tool that uses a pre-labeled dictionary of words to determine the sentiment of a given text [43]. In this context, sentiment refers to the overall emotion or opinion expressed in a tweet—categorized as positive, negative, or neutral. To generate the sentiment distribution charts, we used VADER to compute a sentiment score for each tweet, ranging from –1 to 1. Scores near 1 indicate strongly positive

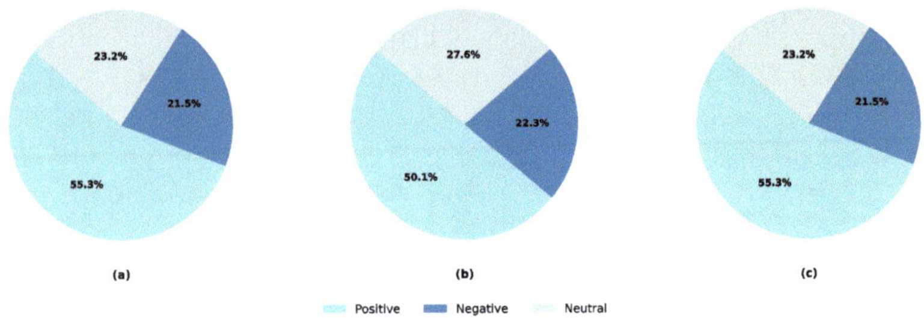

Fig. 11 Sentiment distribution of tweets using VADER **a** Amazon **b** Tesla **c** Microsoft

sentiment, scores near –1 reflect strongly negative sentiment, and scores around 0 represent neutral sentiment. Tweets were classified into three sentiment categories based on their scores. Scores below -0.01 were labeled as negative, those between –0.01 and 0.01 as neutral, and scores above 0.01 as positive. Each tweet was then assigned to a sentiment category based on its score. The final distribution was computed by counting the number of tweets in each category and visualizing the results in the corresponding figures.

- Amazon: A total of 22,612 tweets (55.3%) were positive, 8,791 (23.2%) were negative, and 9,486 (21.5%) were neutral, as illustrated in Fig. 11a.
- Tesla: Out of the analyzed tweets, 18,748 (50.1%) had a positive sentiment, 8,345 (22.3%) were negative, and 10,328 (27.6%) were neutral, as shown in Fig. 11b.
- Microsoft: The sentiment analysis showed 22,059 tweets (55.3%) with positive sentiment, 8,576 (21.5%) with negative sentiment, and 9,254 (23.2%) with neutral sentiment, see Fig. 11c.

4.1.3 Sentiment Analyzer Using Text Blob

To analyze the sentiment of stock-related tweets, we employed the Python module TextBlob. TextBlob supports various Natural Language Processing (NLP) tasks such as sentiment analysis, noun phrase extraction, and Part-Of-Speech (POS) tagging [45]. It determines the sentiment of a given text using a predefined set of words associated with polarity scores. Based on the computed sentiment score, each tweet is categorized as positive, neutral, or negative. Using this approach, we analyzed tweets for Amazon, Tesla, and Microsoft stocks. The sentiment distribution for each company is summarized below.

- Amazon: Out of the analyzed tweets, 19,627 (48%) expressed a positive sentiment, 7,728 (18.9%) were negative, and 13,534 (33.1%) were neutral. These results are illustrated in Fig. 12a.
- Tesla: A total of 18,748 tweets (50.1%) showed a positive sentiment, while 6,810 (18.2%) were negative and 11,862 (31.7%) were neutral, as shown in Fig. 12b.

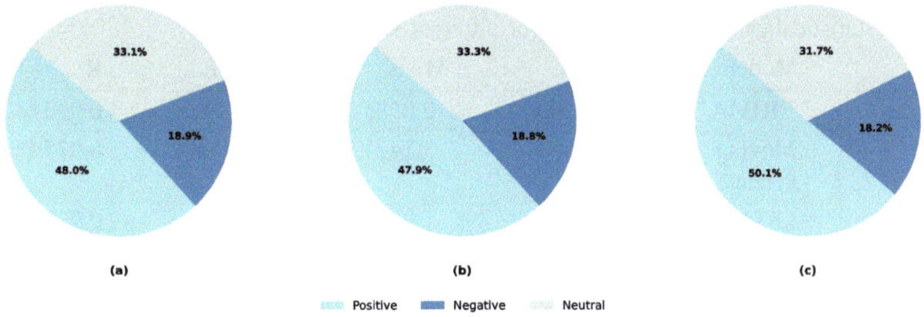

Fig. 12 Sentiment distribution of tweets using text blob **a** Amazon **b** Tesla **c** Microsoft

- Microsoft: The sentiment analysis revealed that 19,107 tweets (47.9%) were positive, 7,499 (18.8%) were negative, and 13,283 (33.3%) were neutral. These proportions are displayed in Fig. 12c.

4.2 Evaluation of Models

To evaluate the performance of each forecasting model in predicting the stock closing price, several metrics have been used. They include RMSE, MAE, MAPE, and the Correlation Coefficient (R) [46–48]. The indicators are expressed mathematically as given below.

$$\mathrm{RMSE} = \sqrt{\frac{1}{n}\sum_{i=1}^{n}(Y_i - \hat{Y}_i)^2} \tag{7}$$

$$\mathrm{MAPE} = \frac{1}{n}\sum_{i=1}^{n}\left|\frac{Y_i - \hat{Y}_i}{Y_i}\right| \times 100\% \tag{8}$$

$$\mathrm{R} = 1 - \frac{\sum_{i=1}^{n}(y_i - \hat{y}_i)^2}{\sum_{i=1}^{n}(y_i - \bar{y})^2} \tag{9}$$

Y_i is the actual observation, $\hat{Y}_i$ is the predicted value obtained from the proposed forecasting model, and N is the number of forecast values.

From the statistical evaluation of the model as depicted from Table 4, the RMSE Value is low for LSTM Model which infers that LSTM is good for predictive performance of stock price when compared to ARIMA and MA. The R function of the LSTM Model for the Amazon, Tesla, and Microsoft stock prices are '0.9955',

Table 4 Statistical evaluation of models for different stocks

Stock	Model	RMSE	MAE	MAPE	R
Amazon	ARIMA	3.9200	2.0585	1.5840	0.9931
	Moving average	5.8670	5.1421	3.9800	0.9754
	LSTM	2.6306	1.9753	1.5607	0.9955
Tesla	ARIMA	7.7796	5.0655	2.8985	0.9832
	Moving average	20.1255	13.8706	8.3533	0.9802
	LSTM	7.6506	5.8211	2.8105	0.9970
Microsoft	ARIMA	5.7027	3.2934	1.3425	0.9952
	Moving average	10.9440	7.7190	3.0712	0.9920
	LSTM	5.3273	4.2080	1.2971	0.9984

'0.9970', and '0.9984' respectively. The R function measures the correlation between predicted values and actual values. A value close to 1 indicates a strong positive correlation, while a value close to –1 indicates a strong negative correlation. In this case, the R value suggests a strong positive correlation between the predicted and actual values generated by the LSTM model.

4.3 Sentiment Analysis on Financial Markets

Understanding how public sentiment influences financial markets has gained significant attention, especially with the rise of social media and NLP tools. This study uses sentiment analysis using VADER and TextBlob models to examine their correlation with stock price movements.

Figure 13 displays the relationship between Amazon's stock closing price and sentiment scores from VADER and TextBlob over five days. The Y-axis shows the stock prices, while the sentiment scores are scaled to fit the same range. VADER scores exhibit more rapid changes that closely align with stock movements, suggesting its responsiveness to daily sentiment. In contrast, TextBlob provides smoother trends, still showing correlation with price changes.

$$\text{VADER\_Score\_(scaled)}(t) = 50 \times \text{VADER}(t) + 121.25 \tag{10}$$

$$\text{TextBlob\_Score\_(scaled)}(t) = 50 \times \text{TextBlob}(t) + 120.75 \tag{11}$$

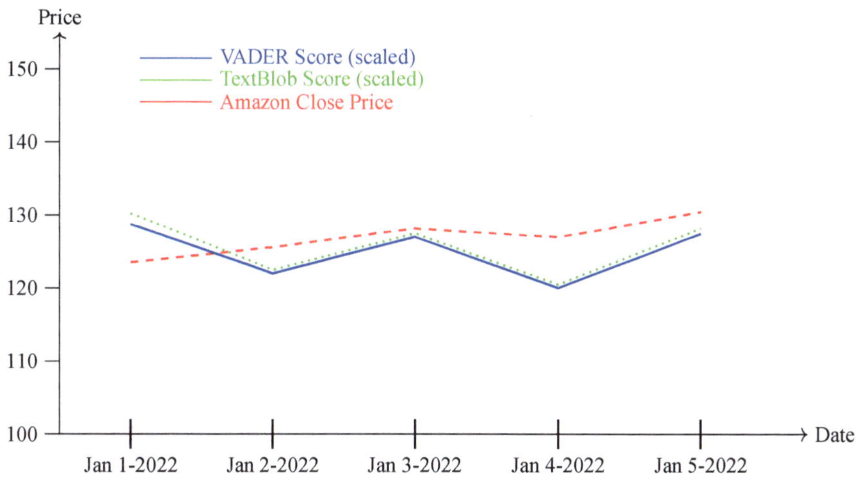

Fig. 13 Temporal plot of sentiment scores and stock price

Table 5 Sample of time-aligned dataset combining sentiment and financial features

Date	VADER score	TextBlob score	Close price	MA(5)	Volatility	Return
2022-01-01	0.15	0.21	123.5	121.3	0.032	0.014
2022-01-02	−0.12	−0.10	125.6	122.1	0.028	0.017
2022-01-03	0.08	0.10	128.2	124.0	0.030	0.021
2022-01-04	−0.20	−0.18	127.0	125.2	0.033	−0.009
2022-01-05	0.10	0.13	130.5	126.8	0.027	0.028

Table 5 highlights a sample dataset aligning sentiment scores with financial metrics such as moving averages, volatility, and daily returns. This integration provides a foundation for assessing the predictive power of public sentiment on financial markets.

Figures 14 and 15 further examine the sentiment-stock relationship across Amazon, Tesla, and Microsoft. Subplot 14a reveals that positive sentiment spikes in early 2022 correspond with an increase in Amazon's stock price. In Subplot 14b, Tesla shows volatile sentiment, yet periods of positivity align with price upticks. Subplot 14c indicates a consistent drop in both sentiment and stock value for Microsoft starting mid-2022. Similarly, Fig. 15 presents comparable trends using the TextBlob sentiment model. Although the sentiment variations are smoother, they still align with stock price movements. Across all three companies, sentiment trends appear to either precede or closely align stock price changes. This supports the hypothesis that social media sentiment can serve as a leading indicator in financial forecasting. A

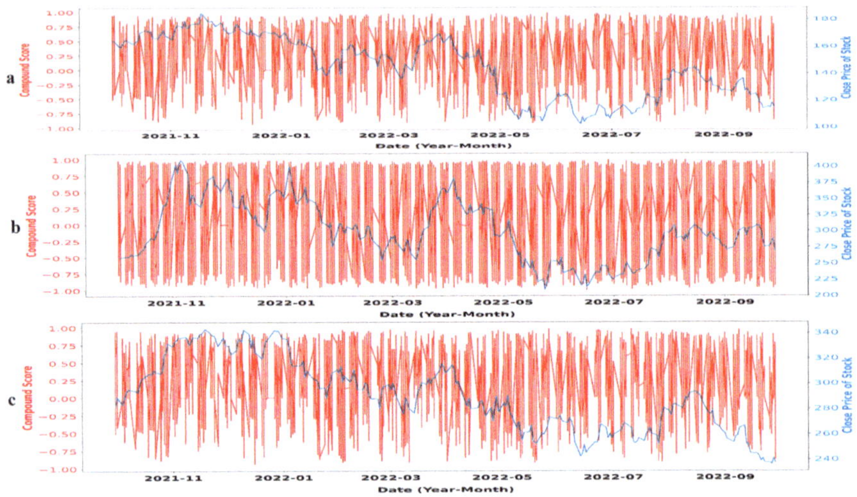

Fig. 14 Compound score versus close price using VADER: **a** Amazon **b** Tesla **c** Microsoft

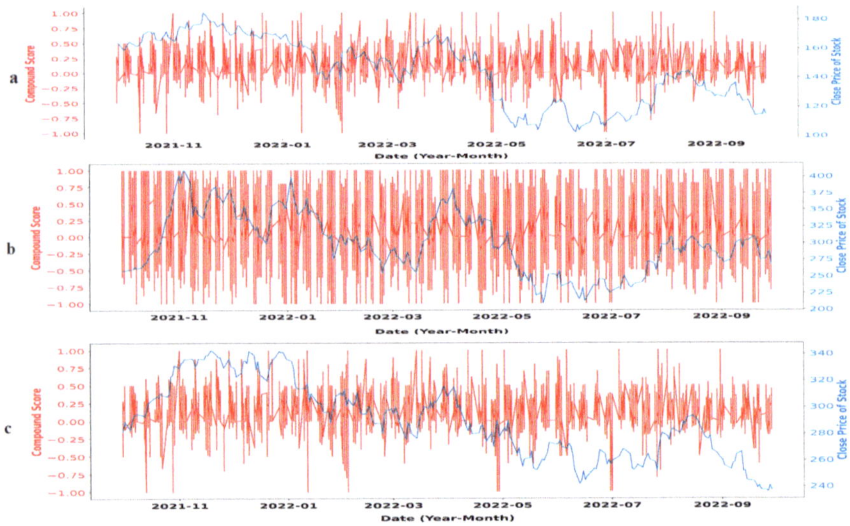

Fig. 15 Compound score versus close price using TextBlob: **a** Amazon **b** Tesla **c** Microsoft

strong positive correlation (Pearson correlation coefficient of 0.6887) was observed between VADER sentiment and stock prices, underscoring sentiment analysis as a valuable tool in market prediction.

4.4 Statistical Analysis for Sentiment Analysis

To understand how the sentiment analysis reflect the stock market behavior, we compared the sentiment models with actual stock close prices. The Table 6 presents the performance of both models across three key metrics, Pearson Correlation, Spearman Correlation, and RMSE. Pearson Correlation measures how closely two things move together (linear relationship between sentiment scores and stock prices) and Spearman Correlation measures how well the trends match (measure the strength and direction of a monotonic relationship between two ranked variables, sentiment scores and stock closing prices).

$$r = \frac{\sum(S_i - \bar{S})(P_i - \bar{P})}{\sqrt{\sum(S_i - \bar{S})^2 \sum(P_i - \bar{P})^2}} \quad (12)$$

where:

- r is the Pearson correlation coefficient,
- S_i is the sentiment score on day i,
- P_i is the stock closing price on day i,
- $\bar{S}$ is the mean of all sentiment scores,
- $\bar{P}$ is the mean of all stock closing prices.

A value of $r = 1$ implies a perfect positive linear correlation, $r = -1$ implies a perfect negative linear correlation, and $r = 0$ indicates no linear relationship.

It is useful for understanding whether higher sentiment scores tend to be associated with higher (or lower) stock prices.

$$\rho = 1 - \frac{6\sum d_i^2}{n(n^2 - 1)} \quad (13)$$

where:

- ρ is the Spearman's rank correlation coefficient,
- d_i is the difference between the ranks of sentiment score and stock closing price on day i,
- n is the total number of observations (e.g., days),
- $\sum d_i^2$ is the sum of squared differences in ranks.

The constant 6 is derived from the mathematical formulation and helps to normalize the correlation value within the range of -1 to $+1$. A value of $\rho = 1$ indicates a perfect positive correlation in ranks, $\rho = -1$ a perfect negative correlation, and $\rho = 0$ implies no correlation.

VADER shows stronger correlations with stock prices, with a Pearson value of 0.68 and a Spearman value of 0.71, compared to TextBlob's 0.41 and 0.39 respectively. Additionally, VADER exhibits a lower RMSE (28.43) than TextBlob (44.87), indicating that its sentiment predictions are more accurate in approximating the stock

Table 6 Statistical comparison of sentiment models with stock closing prices

Metric	VADER sentiment	TextBlob sentiment
Pearson correlation	0.68	0.41
Spearman correlation	0.71	0.39
Root mean square error (RMSE)	28.43	44.87

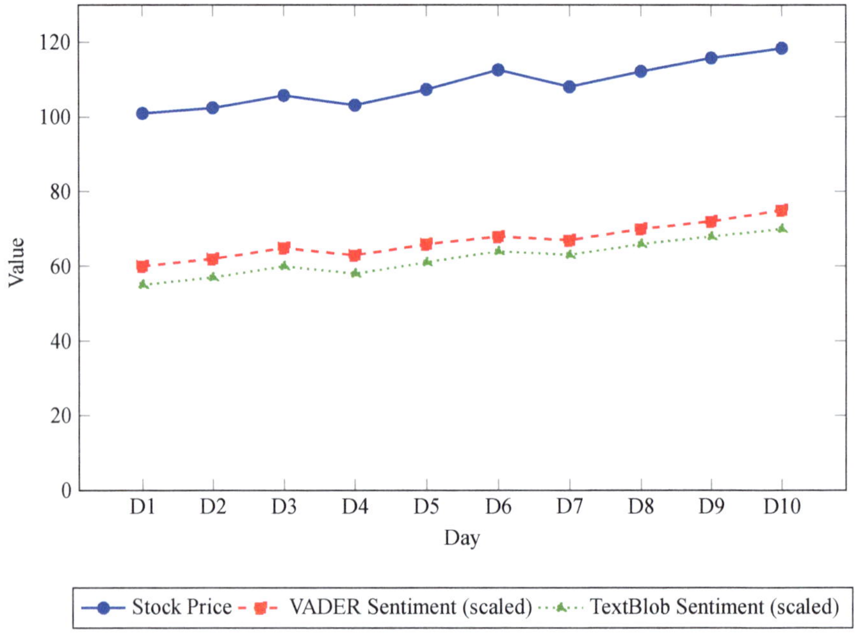

Fig. 16 Comparison of stock closing prices with scaled sentiment scores

price behavior. A time-series plot is provided showing stock closing prices alongside the scaled sentiment scores from both models over a 10 d period. The VADER sentiment line more closely follows the movement of the actual stock prices compared to TextBlob (Fig. 16).

4.5 User Interface

A web page is also created for the users to predict the stock price in which the stock companies are categorized into five domains namely banking, digital marketing, automobile, technology, and finance. Few companies are listed under each domain. Once the user chooses a date and the ML model to predict the stock price for a

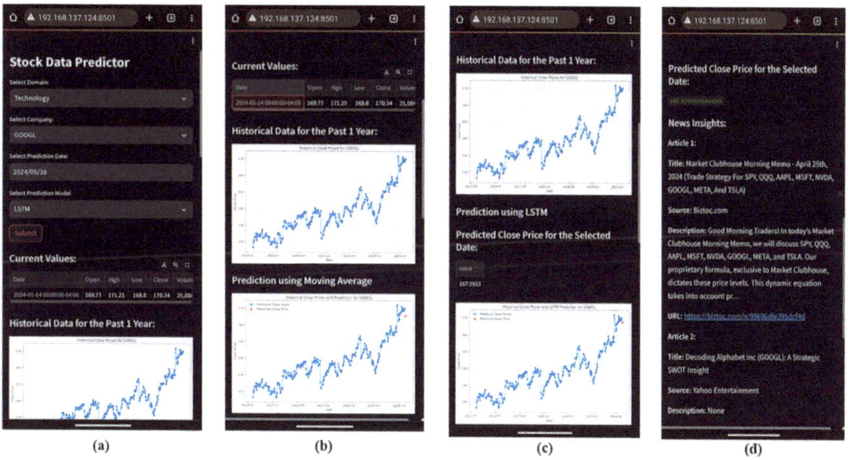

Fig. 17 User interface of the stock prediction system **a** Main interface **b** Historical data and predicted price using MA **c** Predicted prices using LSTM **d** News insights about the company

particular company the library crawls the stock data from the X and feed to the selected pre-trained model and the subsequent predictions were made accordingly. To facilitate the users, a graph of the data is also plotted with the following parameters 'open' ,'high', 'low', 'close price', and 'volume' and the plot of predicted price of the stock will be shown under the predicted tab along with a plot of historical data of previous one year. The details of news insights of the selected company stock will also be displayed along with the predicted price. The user interface is shown in Fig. 17a, b, c, d.

5 Conclusion

This study evaluated the usage of ML approach and sentiment analysis for the accurate prediction of stock price. The tweets of three companies namely Amazon, Tesla, and Microsoft and their financial data are used. Two sentiment analysis tools VADER and Text Blob were used for analyzing the sentiments. The novelty of the work is the integration of ML approach and sentiment analysis. After the preprocessing of data the performance of ML Model is assessed. Out of the three ML approaches LSTM outperformed well in terms of MAE, MAPE, and R Value. In order to summarize the utilization of X data for sentiment analysis the VADER Model outperforms well in the classification of the tweets as compared to Text Blob. Several factors could undermine the accuracy of the results. Firstly, Python libraries like Text Blob and VADER may generate inaccurate sentiment scores for numerous tweets, specifically when faced with sarcasm they often misinterpret negative tweets as positive and vice versa, leading to an erroneous training dataset. The another threat to the precision of the findings is the collection of tweets from the spam accounts. Additionally, on certain occasions despite a positive sentiment, stock performance might be trending downwards for a particular day. Some aspects of this study could be enhanced in future iterations. For instance, collection of tweets from accounts with a substantial number of followers known to have a significant impact on market trends, may enhance the model's accuracy. Furthermore, training and evaluating X data directly could result in higher accuracy than using a prepackaged libraries like VADER and Text Blob.

References

1. Khan, W., Ghazanfar, M.A., Azam, M.A., Karami, A., Alyoubi, K.H., Alfakeeh, A.S.: Stock market prediction using machine learning classifiers and social media, news. J. Ambient. Intell. Humanized Comput. 1–24 (2022)
2. El Mahjouby, M., Bennani, M.T., Lamrini, M., El Far, M., Bossoufi, B., Alghamdi, T.A.: Predicting market performance using machine and deep learning techniques. IEEE Access (2024)
3. Banerjee, P., Nayak, R.: Recommendations on financial models for stock market prediction using Arima and LSTM. In: 2023 OITS International Conference on Information Technology (OCIT), pp. 260–265. IEEE (2023)
4. https://x.com/amazon
5. Gunturu, P.A., Joseph, R., Revant, E.S., Khapre, S.: Survey of stock market price prediction trends using machine learning techniques. In: 2023 International Conference on Artificial Intelligence and Applications (ICAIA) Alliance Technology Conference (ATCON-1), pp. 1–5. IEEE (2023)
6. Kumar, S., Pal, N., Tripathi, A.M.: Improving long short-term memory (LSTM)-based stock market price predictions in the machine learning Era. In: 2024 IEEE International Conference on Computing, Power and Communication Technologies (IC2PCT), vol. 5, pp. 923–928. IEEE (2024)
7. Kumar, S., Aggarwal, U., Gautam, P., Tuteja, A., Matta, P., Maurya, S.: LSTM approach for efficient stock market prediction. In: 2023 3rd International Conference on Intelligent Technologies (CONIT), pp. 1–5. IEEE (2023)
8. Nabipour, M., Nayyeri, P., Jabani, H., Shahab, S., Mosavi, A.: Predicting stock market trends using machine learning and deep learning algorithms via continuous and binary data; a comparative analysis. IEEE Access **8**, 150199–150212 (2020)
9. Kim, S., Ku, S., Chang, W., Song, J.W.: Predicting the direction of US stock prices using effective transfer entropy and machine learning techniques. IEEE Access **8**, 111660–111682 (2020)
10. Pinto, N., da Silva Figueiredo, L., Garcia, A.C.: Automatic prediction of stock market behavior based on time series, text mining and sentiment analysis: a systematic review. In: 2021 IEEE 24th International Conference on Computer Supported Cooperative Work in Design (CSCWD), pp. 1203–1208. IEEE (2021)
11. Vijh, M., et al.: Stock closing price prediction using machine learning techniques. Procedia Comput. Sci. **167**, 599–606 (2020)
12. Patel, J., et al.: Predicting stock and stock price index movement using trend deterministic data preparation and machine learning techniques. Expert. Syst. Appl. **42**(1), 259–268 (2015)
13. Jiang, W.: Applications of deep learning in stock market prediction: recent progress. Expert Syst. Appl. **184**, 115537 (2021)

14. Banerjee, P., Nayak, R.: Recommendations on financial models for stock market prediction using ARIMA and LSTM. In: 2023 OITS International Conference on Information Technology (OCIT). IEEE (2023)
15. Xiao, D., Su, J.: Research on stock price time series prediction based on deep learning and autoregressive integrated moving average. Sci. Program. (2022)
16. Dinesh, S., et al.: Prediction of trends in stock market using moving averages and machine learning. In: 2021 6th International Conference for Convergence in Technology (I2CT). IEEE (2021)
17. Ho, M., Darman, H., Musa, S.: Stock price prediction using ARIMA, neural network and LSTM models. J. Phys: Conf. Ser. **1988**, 012041 (2021). https://doi.org/10.1088/1742-6596/1988/1/012041
18. Kumar, D., Sarangi, P.K., Verma, R.: A systematic review of stock market prediction using machine learning and statistical techniques. Mater. Today: Proc. **49**, 3187–3191 (2022)
19. Nasir, J., et al.: A new approach for forecasting crude oil prices based on stochastic and deterministic influences of LMD using ARIMA and LSTM models. IEEE Access **11**, 14322–14339 (2023)
20. Demirel, U., Cam, H., Ünlü, R.: Predicting stock prices using machine learning methods and deep learning algorithms: the sample of the Istanbul stock exchange. Gazi Univ. J. Sci. **34**(1), 63–82 (2021)
21. Ayala, J., et al.: Technical analysis strategy optimization using a machine learning approach in stock market indices. Knowl.-Based Syst. **225**, 107119 (2021)
22. Ho, M.K., Darman, H., Musa, S.: Stock price prediction using ARIMA, neural network and LSTM models. J. Phys.: Conf. Ser. **1988**(1) (2021). IOP Publishing
23. Zhang, C., Sjarif, N.N., Ibrahim, R.B.: Decision fusion for stock market prediction: a systematic review. IEEE Access **10**, 81364–81379 (2022)
24. Elbagir, S., Yang, J.: Twitter sentiment analysis using natural language toolkit and VADER sentiment. In: Proceedings of the International Multiconference of Engineers and Computer Scientists, vol. 122. no. 16(2019)
25. Sagala, T.W., et al.: Stock price movement prediction using technical analysis and sentiment analysis. In: Proceedings of the 2020 2nd Asia Pacific Information Technology Conference (2020)
26. Mamillapalli, A., Ogunleye, B., Inacio, S.T., Shobayo, O.: GRUvader: sentiment-informed stock market prediction (2024). arXiv:2412.06836
27. Thomas, G.J.: Enhancing TinyBERT for financial sentiment analysis using GPT-augmented FinBERT distillation (2024). arXiv:2409.18999
28. Talazadeh, S., Perakovic, D.: SARF: Enhancing stock market prediction with sentiment-augmented random forest (2024). arXiv:2410.07143
29. Zhu, E., Yen, J.: BERTopic-driven stock market predictions: unraveling sentiment insights (2024). arXiv:2404.02053
30. Wang, S., Bai, Y., Fu, K., Wang, L., Lu, C.T., Ji, T.: ALERTA-Net: A temporal distance-aware recurrent networks for stock movement and volatility prediction. In: Proceedings of the International Conference on Advances in Social Networks Analysis and Mining, pp. 538–542 (2023)
31. Liu, Z., Cao, Y., Xu, H., Huang, Y., He, Q., Chen, X., Liu, X.: Hidformer: hierarchical dual-tower transformer using multi-scale mergence for long-term time series forecasting. Expert Syst. Appl. **239**, 122412 (2024)
32. Yahoo Finance (n.d.).: Amazon.com, Inc. (AMZN), Tesla, Inc. (TSLA), and Microsoft Corporation (MSFT) stock price & news. https://finance.yahoo.com
33. Montgomery, D.C., Peck, E.A., Vining, G.G.: Introduction to Linear Regression Analysis. Wiley (2021)
34. Pham, H. (ed.): Springer Handbook of Engineering Statistics. Springer Nature (2023)
35. Maulud, D., Abdulazeez, A.M.: A review on linear regression comprehensive in machine learning. J. Appl. Sci. Technol. Trends **1**(2), 140–147 (2020)

36. Mashadihasanli, T.: Stock market price forecasting using the ARIMA model: an application to Istanbul Turkiye. J. Econ. Policy Res. **9**(2), 439–454 (2022)
37. Hanif, R., Mustafa, S., Iqbal, S., Piracha, S.: A study of time series forecasting enrollments using fuzzy interval partitioning method. J. Comput. Cogn. Eng. **2**(2), 143–149 (2023)
38. Al-Selwi, S.M., Hassan, M.F., Abdulkadir, S.J., Muneer, A.: LSTM inefficiency in long-term dependencies regression problems. J. Adv. Res. Appl. Sci. Eng. Technol. **30**(3), 16–31 (2023)
39. Ismail, S., Vasudev, H.: Artificial intelligence for personalized electric vehicle energy management. In: Emerging Multisector Applications of AI and IoT, pp. 221–260. IGI Global Scientific Publishing (2025)
40. Zargari, H., Zahedi, M., Rahimi, M.: GINS: a global intensifier-based N-Gram sentiment dictionary. J. Intell. Fuzzy Syst. **40**(6), 11763–11776 (2021)
41. Euna, N.J., Hossain, S.M.M., Anwar, M.M., Sarker, I.H.: Content-based spam email detection using an n-gram machine learning approach. In: Applied Intelligence for Industry 4.0, pp. 176–187. Chapman and Hall/CRC (2023)
42. Vercellotti, M.L., Juffs, A., Naismith, B.: Multiword sequences in English language learners' speech: the relationship between trigrams and lexical variety across development. System **98**, 102494 (2021)
43. Ismail, S., Vasudev, H.: EIROB: an emotionally intelligent ROBot for empathetic support. In: 2024 IEEE International Conference on Signal Processing, Informatics, Communication and Energy Systems (SPICES), pp. 1–6. IEEE (2024)
44. Jacob, S., Jacob, P.M., Cheriyan, J., Ismail, S.: Legal assistance redefined: transforming legal access with AI-powered LegalLink. In: 2024 IEEE International Conference on Signal Processing, Informatics, Communication and Energy Systems (SPICES), pp. 1–6. IEEE (2024)
45. Nair, R.P., Thushara, M.G.: Investigating natural language techniques for accurate noun and verb extraction. Procedia Comput. Sci. **235**, 2876–2885 (2024)
46. Dessain, J.: Machine learning models predicting returns: Why most popular performance metrics are misleading and proposal for an efficient metric. Expert Syst. Appl. **199**, 116970 (2022)
47. Oyedele, A.A., Ajayi, A.O., Oyedele, L.O., Bello, S.A., Jimoh, K.O.: Performance evaluation of deep learning and boosted trees for cryptocurrency closing price prediction. Expert Syst. Appl. **213**, 119233 (2023)
48. Hoque, K.E., Aljamaan, H.: Impact of hyperparameter tuning on machine learning models in stock price forecasting. IEEE Access **9**, 163815–163830 (2021)

Post-IPO Performance Prediction: A Comparative Study of Logistic Regression and Machine Learning Techniques for Thai IPO Firms

Pornpawee Supsermpol, Van-Nam Huynh, and Navee Chiadamrong

Abstract This chapter investigates the key factors influencing post-IPO financial performance, with a focus on internal capability factors and the amount of funds raised, and evaluates the effectiveness of various predictive models in assessing firm success within the context of the Stock Exchange of Thailand (SET). Given the relatively small dataset, typical of an emerging and smaller market like the SET, the chapter compares traditional statistical methods, such as logistic regression, with machine learning techniques, including Decision Tree (DT), Random Forest (RF), Extreme Gradient Boosting (XGBoost), Light Gradient Boosting Machine (LightGBM), Support Vector Machine (SVM), and Artificial Neural Network (ANN). The analysis reveals that RF outperforms logistic regression and more advanced machine learning techniques regarding predictive power. RF performs particularly well in smaller datasets, where its simpler structure allows it to generalize better, while more complex algorithms may struggle with overfitting. Key factors such as funds raised, retention, and cost of goods sold are identified as the most significant contributors to post-IPO financial success prediction, with their relative importance shifting over time as firms mature. The chapter's findings provide valuable insights for investors, firms, and stakeholders, particularly in emerging markets like Thailand.

P. Supsermpol
School of Knowledge Science, Japan Advanced Institute of Science and Technology, Ishikawa, Japan
e-mail: p.supsermpol@gmail.com; p.supsermpol@jaist.ac.jp

P. Supsermpol · N. Chiadamrong (✉)
School of Manufacturing Systems and Mechanical Engineering, Sirindhorn International Institute of Technology, Thammasat University, Pathum Thani, Thailand
e-mail: navee@siit.tu.ac.th

V.-N. Huynh
Japan Advanced Institute of Science and Technology, Ishikawa, Japan
e-mail: huynh@jaist.ac.jp

School of Finance and Accounting, Industrial University, Ho Chi Minh City, Vietnam

H. T. T. Nguyen et al. (eds.), *Data Science in Finance and Accounting*, Studies in Big Data 181, https://doi.org/10.1007/978-3-032-06179-9_6

1 Introduction

An Initial Public Offering (IPO) is a significant milestone for firms, marking their first sale of shares to the public. This event not only provides the firm the opportunity to tap into new financial resources but also brings heightened visibility, credibility, and opportunities for growth [1]. However, the transition from a private to a public firm comes with substantial challenges. Once listed on a stock exchange, firms face increased scrutiny from regulators, shareholders, and the market. They are expected to meet higher standards of transparency and accountability [2] and must consistently demonstrate financial performance to satisfy investor expectations [3].

The post-IPO period is critical for firms as they strive to manage the influx of capital while navigating the operational and strategic demands of their new status. Decisions regarding the deployment of IPO proceeds, adjustments to corporate governance, and balancing growth with profitability are key elements that influence a firm's success or failure in the years following the IPO [4]. For investors, accurately predicting a firm's financial performance post-IPO is of paramount importance, as it informs their investment decisions and expected returns [5].

In emerging markets such as Thailand, the Stock Exchange of Thailand (SET) has become a focal point for IPO activity in Southeast Asia. The dataset for this chapter comprises 134 companies listed on the SET, spanning the years 2002 to 2021. Predicting post-IPO financial performance in such markets can be particularly challenging due to unique economic, regulatory, and market conditions [6]. In this chapter, internal capability factors are analyzed alongside the amount of funds raised during the IPO for predicting financial performance, as they reflect both a firm's operational strength and its ability to capitalize on raised funds to drive post-IPO success. The ability to forecast financial success in this environment is crucial not only for investors and firms but also for policymakers aiming to support the sustainable growth of the capital market [7].

This chapter seeks to explore and review the methodologies used in predicting post-IPO performance. While traditional statistical models such as logistic regression have been widely employed [5], machine learning techniques are increasingly being adopted because they can effectively handle complex, non-linear interactions among variables [8]. This chapter will assess both techniques, comparing their effectiveness in predicting post-IPO success [9]. It also reviews the key factors that influence the financial performance of firms in the post-IPO period. In particular, it examines internal capability factors—such as liquidity, firm size, and leverage—that have historically been used to predict financial outcomes [3]. Additionally, it considers IPO-specific factors, such as the amount of capital raised, which can have a profound impact on post-IPO performance [2, 10].

In summary, this chapter offers a comprehensive overview of the various factors and models used in post-IPO performance prediction. This foundation will be critical for the subsequent analysis, which will apply these factors and models to real-world data to derive insights into the financial trajectories of IPO firms, particularly those listed on the SET. The contributions of this chapter are as follows:

- This chapter provides an in-depth comparison of logistic regression with machine learning techniques in predicting post-IPO financial performance within the context of an emerging market, specifically Thailand.
- The chapter identifies the most significant factors that influence post-IPO performance prediction in each early years post-IPO.
- By focusing on the SET, this chapter addresses the unique challenges of financial performance prediction in emerging markets, where market size is relatively small and data availability may be limited compared to developed markets. These constraints introduce distinct risks and opportunities, making the insights from this chapter particularly relevant for firms, investors, and other stakeholders in similar regions.

2 Literature Review

Predicting post-IPO financial performance has become a central focus in academic research and financial analysis, as firms undergoing IPOs face unique challenges in their transition from private to public entities. Understanding the factors that influence their post-IPO financial health is crucial for investors, firm executives, and other stakeholders. This review examines the primary factors affecting post-IPO financial performance and the metrics commonly used to assess it. It also explores the evolution of predictive models, from traditional statistical methods to advanced machine learning techniques.

Several studies emphasize the role of internal capability factors in driving post-IPO performance. Post-IPO financial performance is shaped by several key factors, primarily categorized into internal capabilities and IPO-specific factors. Internal capabilities can be further divided into financial strengths and operational efficiency, while IPO-specific factors include variables such as the amount of funds raised during the IPO.

Internal capability factors are critical in predicting a firm's financial performance post-IPO. These are typically divided into two categories: financial strengths and operational efficiency. The focus on these internal factors arises from the fact that they offer more direct control for management, providing more actionable insights into a firm's ability to navigate the post-IPO landscape. As Flamini et al. [11] and Jain and Kini [3] have demonstrated, financial strengths, such as firm size, liquidity, leverage, and tangibility, are key to understanding a firm's capacity to withstand market pressures and capitalize on growth opportunities.

Firm size is commonly used as a measure of financial strength, often represented by the logarithm of total assets. Larger firms typically benefit from economies of scale and more extensive resources, allowing them to weather market fluctuations better than smaller firms. Flamini et al. [11] showed that larger firms are generally more stable and perform better post-IPO, as they can leverage their size to access financing and expand their market share. However, Majmudar [12] cautioned that larger firms might face inefficiencies due to bureaucratic complexity, leading to suboptimal asset utilization, which could offset the benefits of size.

Liquidity, which refers to the proportion of current assets relative to current liabilities, is another important financial strength. Firms with higher liquidity can manage their day-to-day operations more effectively, ensuring that short-term obligations are met without financial strain. Jain and Kini [3] found that firms with better liquidity prior to an IPO generally maintain stronger performance post-IPO, as they can better withstand financial volatility, especially for firms in their early post-IPO stages, where immediate financial demands are high. Conversely, Hvide and Møen [13] argued that excess liquidity might indicate inefficiencies, as firms with large cash holdings may not allocate resources optimally, leading to lower long-term returns.

Leverage, typically measured by comparing total liabilities to total assets, plays a dual role. On the one hand, moderate leverage can enhance a firm's financial performance by reducing the cost of capital and increasing returns to shareholders. Robb and Robinson [14] highlighted that firms with balanced leverage strategies tend to perform well post-IPO, as debt allows them to fund expansion while retaining equity. On the other hand, Jensen and Meckling [15] warned that excessive leverage could lead to financial difficulties, particularly if the firm faces challenges in generating enough cash flow to meet its debt.

Tangibility, the proportion of fixed assets to total assets, is another important factor in determining financial strength. Firms with higher tangibility have more collateral to secure financing, which reduces borrowing costs and enhances financial flexibility. Iltas and Demirguness [16] showed that firms with greater asset tangibility tend to perform better post-IPO, as they can secure favorable loan terms and reduce the cost of capital. However, Almeida and Campello [17] noted that an over-reliance on fixed assets could reduce liquidity, limiting a firm's ability to respond to short-term financial challenges.

The second category of internal capability factors, operational efficiency, encompasses variables such as cost of goods sold, asset utilization, retention, and net profit margin. Cost of goods sold, measured as a percentage of total sales, is an indicator of the effectiveness of a firm's management of production expenses. Firms with lower cost of goods sold are typically more profitable, as they can generate higher margins on their products. Pagano et al. [7] found that firms able to control production costs are more likely to succeed post-IPO, as they can offer competitive pricing while maintaining profitability.

Asset utilization, which is the proportion of total sales to total assets, reflects a firm's ability to convert its assets into revenue. Firms with higher asset utilization are generally more efficient in managing their resources, leading to better financial performance. Kim and Ritter [18] demonstrated that firms with higher asset turnover tend to outperform their peers post-IPO, particularly in industries where asset management is crucial for profitability.

Retention, or the proportion of retained earnings to net income, is a measure of how much profit a firm reinvests in its operations rather than distributing to shareholders. Higher retention rates suggest a focus on growth and long-term sustainability. Gompers and Lerner [19] found that firms with higher retention rates tend to perform better in the years following their IPO, as they have more resources to invest in innovation, expansion, and operational improvements.

Net profit margin, calculated by dividing net income by total sales, is a critical indicator of a firm's ability to convert revenue into profit. Pagano et al. [7] demonstrated that firms with greater net profit margins have a higher likelihood of sustaining financial success post-IPO, as strong margins provide a buffer against market volatility and operational challenges.

IPO-specific factors, particularly the amount of funds raised during the IPO, are also essential in influencing post-IPO performance. The funds raised provide firms with the capital necessary to invest in growth opportunities, repay debt, and stabilize their operations. Ritter and Welch [4] reviewed several studies and concluded that firms that raise larger amounts during their IPOs tend to perform better in the long term due to increased financial flexibility. However, Jain and Kini [3] cautioned that misallocation of IPO capital can result in poor financial outcomes, particularly if the funds are diverted to speculative ventures rather than core business activities.

In assessing financial performance, metrics such as Return on Assets (ROA), Return on Equity (ROE), and Return on Sales (ROS) are commonly used. Return on Assets (ROA) measures a firm's ability to generate profit from its assets. Jain and Kini [3] found that many firms experience a decline in ROA after their IPO, indicating potential inefficiencies in managing newly raised capital. Return on Equity (ROE) evaluates how effectively a firm generates returns for its shareholders. Aggarwal et al. [5] demonstrated that firms with high pre-IPO ROE tend to sustain better financial performance after going public. Return on Sales (ROS), which measures profitability relative to sales, is another key indicator of financial health. Kim and Ritter [18] found that firms with higher ROS ratios are more likely to succeed post-IPO, as they can maintain strong margins while growing their revenues. Another commonly used metric is Tobin's Q, which compares a firm's market value to the cost of replacing its assets. When Tobin's Q exceeds one, it indicates that the market places a higher value on the firm's assets than their cost, indicating future growth potential. However, Pagano et al. [7] cautioned that Tobin's Q can be volatile and may not always reflect a firm's true financial condition, especially during periods of market exuberance.

To provide a more comprehensive view of financial performance, aggregating ROA, ROE, and ROS can offer a more balanced perspective on a firm's overall health. Each metric captures a different aspect of financial performance—ROA focuses on asset efficiency, ROE emphasizes shareholder returns, and ROS highlights profitability. Ittner and Larcker [20] argued that using a composite measure of financial performance improves the robustness of predictions by mitigating the limitations of any single metric. In this chapter, the aggregation of ROA, ROE, and ROS allows for a well-rounded evaluation of post-IPO financial performance, offering insights into how efficiently a firm uses its assets, generates returns for its shareholders, and maintains profitable operations.

Predicting post-IPO financial performance has been a focus of research, with various models that span from conventional statistical methods to modern machine learning techniques. These models differ in how they handle data and complexity, offering distinct advantages depending on the application. Linear regression is another widely used model to predict continuous outcomes like stock prices or returns. Pagano et al. [7] and Ritter and Welch [4] applied linear regression to study long-term stock performance and IPO factors. The strength of linear regression lies in its simplicity and interpretability. However, its assumption of linearity limits its effectiveness in capturing complex relationships, prompting the need for more advanced models.

While linear regression is effective for predicting continuous outcomes, it is not suitable for scenarios where the goal is to predict whether a firm will succeed or fail post-IPO—an inherently categorical outcome. For such cases, classification models are necessary. Jain and Kini [3] and Aggarwal et al. [5] used logistic regression to model post-IPO performance, identifying key financial indicators that help classify firms as likely to succeed or fail. Logistic regression is valued for its ability to handle binary outcomes and for offering a more interpretable model when the focus is on categorical predictions. This model is especially useful for investors or decision-makers seeking to classify firms rather than predict specific continuous financial metrics.

Recently, the adoption of machine learning techniques for post-IPO financial predictions has grown significantly because they can effectively capture intricate, non-linear relationships. These models are capable of handling large datasets with many variables and can uncover patterns that traditional models might miss. Tree-based models, such as Decision Tree (DT), Random Forest (RF), Extreme Gradient Boosting (XGBoost), and Light Gradient Boosting Machine (LightGBM), are particularly popular in financial forecasting.

DT is known for its simplicity and interpretability. Delen [21] and Yousaf and Dey [22] applied DT to predict ROA and ROE, finding them effective for classifying firms based on profitability but prone to overfitting in smaller datasets.

RF enhances DT by reducing overfitting and increasing prediction accuracy through bagging ensemble learning. Ballings et al. [23] showed that RF outperformed DT and Support Vector Machine (SVM) in predicting stock price movements, showcasing its strength in handling non-linear financial data. Anand et al. [24] also applied RF to predict firm profitability and cash flow metrics, concluding that RF outperformed traditional regression models in both accuracy and robustness.

XGBoost emerged as a leading option for structured financial prediction due to its scalability and superior performance. Chen and Guestrin [9] presented XGBoost, showcasing its ability to outperform other machine learning techniques in financial prediction tasks by handling large datasets efficiently. Toochaei et al. [25] performed XGBoost combined with effective feature selection, showing that it significantly improved predictive accuracy for financial outcomes.

LightGBM is recognized for its effectiveness in managing large datasets due to its leaf-wise tree growth algorithm, which enables quicker and better performance relative to traditional gradient boosting techniques [26]. Yang et al. [27] compared XGBoost and LightGBM for predicting stock prices and demonstrated that both techniques perform well, with LightGBM showing competitive efficiency. The model's ability to process high-dimensional data quickly makes it an attractive option for financial predictions, though it requires careful hyperparameter tuning to avoid overfitting.

While tree-based algorithms are adept at handling structured data and provide transparency through feature importance rankings, they may struggle with complex, high-dimensional datasets without careful tuning. By contrast, non-tree-based techniques, including SVM and Artificial Neural Network (ANN), are particularly effective for non-linear relationships and high-dimensional data but often come at the cost of interpretability and computational complexity. These techniques can be particularly useful in financial predictions, where intricate relationships between variables exist that simpler models might not capture.

SVM is well-suited for high-dimensional datasets, making them effective in predicting financial outcomes. Huang et al. [28] applied SVM to stock market predictions, demonstrating that SVM excels in managing non-linear data compared to traditional regression models. However, SVM is computationally expensive, especially when selecting the appropriate kernel function for non-linear problems, which may limit their scalability [29].

ANN excels at capturing complex non-linear patterns, making it well-suited for predicting intricate financial outcomes. Zhong and Enke [30] applied ANN to stock returns prediction, finding that ANN outperforms traditional models; however, a major limitation of ANN is their "black box" nature, which makes them less interpretable than other models. Kayakuş et al. [31] also showed that ANN is highly effective in predicting post-IPO financial performance metrics like ROA and ROE, but emphasized the need for greater transparency in such models. Table 1 provides an overview of the advantages and disadvantages of each technique used in the chapter based on insights drawn from existing literature, focusing on key aspects such as interpretability, complexity, ability to handle non-linearity, and the risk of overfitting.

In conclusion, predicting post-IPO financial performance involves analyzing both internal and IPO-specific factors, with common metrics like ROA, ROE, and ROS used to measure outcomes. Machine learning techniques such as DT, RF, XGBoost, LightGBM, SVM, and ANN are frequently applied, but this chapter will compare various machine learning techniques with logistic regression to identify the most effective models and key factors influencing performance, specifically in the context of firms listed on the SET, focusing on the first few years after the IPO.

3 Methodology

This chapter seeks to evaluate and compare the predictive power of logistic regression and several machine learning techniques in predicting the post-IPO financial performance of firms listed on the SET. The methodology focuses on assessing the effectiveness of different models and identifying the key factors influencing performance during the first few years after an IPO.

Table 1 Pros and cons of techniques used for predicting post-IPO financial performance

Technique	Pros	Cons
Logistic regression	– Highly interpretable, easy to understand results [32] – Computationally efficient, especially for small datasets [16]	– Assumes linearity, limiting performance with non-linear relationships [16] – Struggles with complex interactions [12]
DT	– Clear and easy-to-interpret visual structure [21]. – Handles non-linear relationships between variables well [33]	– Prone to overfitting, especially with deep trees [21] – Instability; small changes in data can cause structural changes in the tree [21]
RF	– Reduces overfitting by averaging multiple trees [34] – Performs well on relatively small to medium datasets [34] – Handles non-linear data effectively [34]	– Less interpretable than DT [24] – Computationally expensive, especially with a large number of trees [24]
XGBoost	– High accuracy, especially for complex datasets [9] – Efficient in handling non-linearity [9]	– Complex to tune with many hyperparameters [9] – Less interpretable due to its complexity [9] – Requires more computational resources during training [9]
LightGBM	– Highly efficient for large datasets and fast, particularly with sparse data [26] – Comparable in performance to XGBoost [26]	– Can overfit without careful tuning [26] – Less interpretable due to its complexity [26]
SVM	– Effective in handling non-linear data and high-dimensional datasets [35] – Suitable for datasets with clear boundaries [28]	– Requires meticulous hyperparameter tuning for optimal results [28] – Computationally intensive [35] – Less interpretable than simpler models [35]
ANN	– Powerful for capturing complex, non-linear relationships [30] – Flexible and scalable for large datasets [30]	– Considered a "black box" model with low interpretability [31] – Requires large datasets and significant computational resources [30] – Prone to overfitting [30]

3.1 Research Design

The research design focuses on applying predictive models to evaluate post-IPO financial performance. The core objective is to determine the factors that most significantly influence a firm's performance and to assess which predictive model provides the highest accuracy. The response variable is the financial performance of firms during the first three years post-IPO, measured using a combination of ROA, ROE, and ROS, as defined in Table 2. These metrics offer a comprehensive assessment of profitability, asset efficiency, and operational performance. For this chapter, financial performance is divided into three groups: low, medium, and high. This categorization is based on the aggregate of ROA, ROE, and ROS across the firms' first three years post-IPO. The aggregated financial performance scores are divided into three classes, where the bottom one-third of firms are classified as having low performance, the middle one-third as medium performance, and the top one-third as high performance. This approach allows for a straightforward classification of firms based on their relative financial success post-IPO.

The factors used in predicting the financial performance are divided into two major categories: internal capability factors and IPO-specific factors, both of which have been widely used in past research to predict post-IPO outcomes [4]. Internal capability factors capture the operational and financial strengths of the firm prior to the IPO, while IPO-specific factors relate to the conditions surrounding the IPO itself.

The financial strengths factors include:

- Firm size: Larger firms benefit from economies of scale and greater access to resources, which can improve operational efficiency and profitability post-IPO [11].
- Liquidity: Higher liquidity provides financial flexibility to meet obligations and invest in growth, which stabilizes performance post-IPO, though excess liquidity may suggest inefficient use of assets [13].
- Leverage: Leverage can enhance returns through debt-financed growth but introduces financial risk [15].
- Tangibility: High tangibility provides collateral for borrowing, supporting financing options post-IPO, but may reduce a firm's operational flexibility [17].

Table 2 Response variable

Response variable	Definition
Aggregation of ROA, ROE, and ROS	ROA + ROE + ROS
ROA	Net income/Total assets
ROE	Net income/Total shareholders' equity
ROS	Net income/Total sales

Table 3 Factors

Category	Factor	Acronym	Definition
Financial strengths	Firm size	SIZ	ln(Total assets)
	Liquidity	LIQ	Total current assets/Total current liability
	Leverage	LEV	Total liabilities/Total assets
	Tangibility	TAN	Total fixed assets/Total assets
Operational efficiency	Cost of goods sold	COG	Cost of goods sold/Total sales
	Asset utilization	ASS	Total sales/Total assets
	Retention	RET	Retained earnings/Net income
	Net profit margin	NPM	Net income/Total sales
IPO-specific factor	Amount of fund raised	FUN	IPO funds raised/Total assets

Operational efficiency measures include:

- Cost of goods sold: Lower cost of goods sold indicates better operational efficiency and higher margins, directly contributing to profitability post-IPO [7].
- Asset Utilization: Higher asset utilization reflects efficient use of resources to generate sales, improving financial performance post-IPO [18].
- Retention: High retention supports reinvestment in growth, contributing to long-term profitability, but may reduce immediate returns to shareholders [19].
- Net profit margin: A higher net profit margin indicates strong cost management and profitability [7].

The IPO-specific factor included in this chapter is:

- Amount of fund raised: It provides an indication of how much capital the firm gained from its IPO, which can significantly influence its post-IPO financial trajectory [18].

These variables are presented in Table 3, and are expected to be key factors of financial outcomes for firms during the post-IPO phase. The inclusion of both internal and IPO-specific factors provides a more detailed understanding of the forces shaping financial performance in the early years following a public offering.

3.2 Data Collection

The dataset for this chapter consists of financial data from 134 firms listed on the SET that completed their initial public offerings (IPOs) between 2003 and 2018, as shown in Figs. 1 and 2. These firms were selected based on their availability of

Fig. 1 Number of firms listed on the SET by year

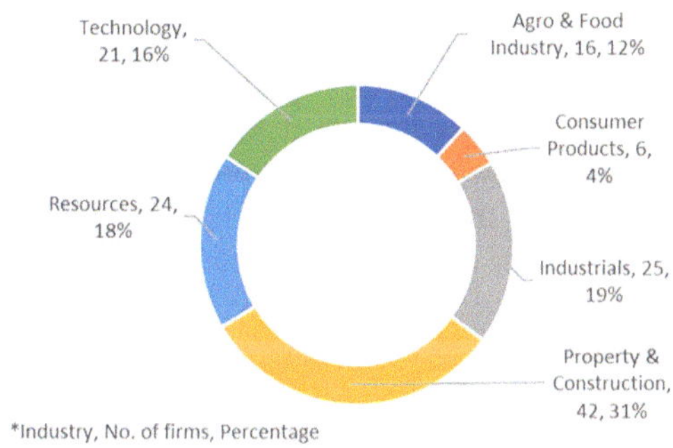

Fig. 2 Percentage of firms listed on the SET by industry

financial data pre- and post-IPO, and their inclusion in publicly accessible financial databases and filings with the Securities and Exchange Commission (SEC). The sample includes firms from various industries, excluding financial and service sectors, to ensure comparability in the analysis. The exclusion of financial firms is necessary due to their distinct regulatory requirements and financial structures, which differ significantly from those of non-financial firms. The number of firms in the dataset is relatively small by nature, reflecting the limited number of IPOs within the Thai market. Efforts were made to include as many firms as possible; however, some were excluded due to incomplete financial data, financial distress, and industry exclusions to maintain uniformity and avoid skewing the analysis.

For each firm, one year of pre-IPO financial data is collected for predicting financial performance in the first three years post-IPO. To standardize the comparison across firms that went public in different years, relative transition years are used instead of calendar years. Year 0 represents the IPO year, and financial performance is tracked for three years post-IPO. This approach allows for consistent comparisons of financial trajectories across firms, regardless of their IPO timing.

The financial performance of each firm is measured using data from their annual financial statements for the three years following the IPO. This period provides sufficient time to assess how effectively firms leverage the capital raised during the IPO and manage their operational strategies in the early years of public trading.

All financial data were extracted from the firms' financial statements, including balance sheets, income statements, and cash flow statements, as well as IPO prospectuses. Data cleaning and preprocessing steps included removing missing or incomplete entries, handling outliers, and standardizing variable formats to ensure consistency across the dataset.

3.3 Model Implementation

In a previous study, numerical prediction models using regression analysis were tested for forecasting post-IPO financial performance, but the results were unsatisfactory, with nearly 50% Mean Absolute Percentage Error (MAPE), indicating low accuracy [36]. Due to these limitations, this chapter adopts a class-based prediction approach, categorizing financial performance into low, medium, and high classes based on aggregated ROA, ROE, and ROS. This approach focuses on classification accuracy rather than the prediction of precise numerical values, making it more suitable for the complexity of financial performance prediction.

Given the complexity and potential non-linearity of relationships between factors and the financial performance, machine learning techniques are particularly well-suited for this task. To evaluate post-IPO financial performance, a traditional statistical method and machine learning techniques are implemented. Logistic regression is chosen as the baseline statistical method, while DT, RF, XGBoost, LightGBM, SVM, and ANN represent a range of powerful machine learning techniques recognized for their capacity to handle non-linear and complex relationships.

3.3.1 Logistic Regression

Logistic regression is a statistical model employed to predict the likelihood of a binary outcome using one or more predictor variables [32]. The model used in this chapter is Ordinal Logistic Regression (OLR), a type of logistic regression designed for predicting outcomes with ordered categories. Unlike binary logistic regression, which predicts dichotomous outcomes, ordinal logistic regression is appropriate when the dependent variable has a natural order but is not continuous. In this case, the model predicts post-IPO financial performance, classified into three ordered categories: low, medium, and high.

Ordinal logistic regression models the likelihood that an observation falls into a given class or a lower one, assuming a proportional odds structure across thresholds. The model is expressed as:

$$\log \frac{P(Y \leq k)}{P(Y > k)} = \beta_0^{(k)} + \sum_{i=1}^{p} \beta_i X_i + \sum_{i=1}^{p-1} \sum_{j=i+1}^{p} \beta_{ij} X_i X_j \tag{1}$$

where $P(Y \leq k)$ is the likelihood of the dependent variable belongs to or is lower than class k, and $X_1, X_2, \ldots, X_p$ are the independent variables. The regression coefficients (β) are constant across the different thresholds, satisfying the proportional odds assumption. In this chapter, the ordinal logistic regression also allows for first-level interaction terms between independent variables (e.g., firm size and leverage). This interaction term $\beta_{12}X_1X_2$ captures the combined effect of these two predictors on the financial performance outcome, improving the model's ability to account for more complex relationships. Logistic regression is favored for being straightforward and easy to interpret, which makes it useful for understanding the relationships between factors and outcomes. However, its reliance on linearity restricts its capacity to model complex, non-linear patterns within the dataset [32].

3.3.2 Decision Tree (DT)

DT is implemented for its simplicity and transparency, making it ideal for understanding how different factors contribute to firm performance classification. It works by systematically dividing the dataset into branches using the most important features, leading to an intuitive visual representation of the classification process. Each node represents a factor, and each leaf represents an outcome, which allows users to easily interpret the paths leading to different performance classes [33]. At each node, the feature that results in the largest reduction in impurity is selected for splitting. Impurity is calculated with the Gini Index, which decreases as the data becomes more uniform. It is calculated as:

$$\text{Gini} = 1 - \sum_{i=1}^{c} p_i^2 \tag{2}$$

where p_i represents the likelihood that a data point is in class i and c is the total number of classes. A feature that minimizes the Gini Index is selected for splitting the data.

One of the main challenges with DT is the risk of overfitting, particularly when the tree becomes deep and overly complex. Overfitting arises when the model is too finely tuned to the training data, capturing noise instead of general patterns. To prevent overfitting, several techniques are used, such as pruning, restricting the depth of the tree, and ensuring a minimum threshold of samples in each leaf [33].

3.3.3 Random Forest (RF)

RF improves upon DT by employing an ensemble learning technique, which mitigates the risk of overfitting. This model involves training several trees on randomly selected portions of the dataset and then combining their predictions to produce a more accurate and stable prediction. This technique is particularly effective in capturing the interactions between variables while reducing the variance associated with single DT [34]. RF operates under two key principles:

- Each DT is built using a bootstrap sample (Bagging), meaning the training data is randomly sampled with replacement to create a subset. This ensures that each tree is exposed to a marginally different dataset, promoting diversity among the trees.
- A randomly chosen set of features is selected at each DT split, and the data is split solely using the best feature from that set. This randomization helps reduce dependency among trees, which increases the robustness of the model.

Once RF is trained, predictions are made by combining the results from all individual trees. In the case of classification, the final output is typically based on the majority vote of the trees.

3.3.4 Extreme Gradient Boosting (XGBoost)

XGBoost is a highly effective ensemble learning technique built on the principles of gradient boosting, which generates decision trees in a step-by-step process, where each tree is designed to reduce the errors of the previous ones [9]. It is renowned for its exceptional performance, particularly with large and intricate datasets. Unlike RF, where trees are built independently, XGBoost builds trees in a sequence, with each subsequent tree focusing more on the misclassified samples from the previous trees. This "boosting" process ensures that the model improves iteratively with each step. XGBoost also includes regularization mechanisms that reduce overfitting, making it more robust than traditional boosting techniques. One of its standout features is the application of a second-order Taylor expansion of the loss function, which incorporates both first and second derivatives (gradients and Hessians) to more accurately guide each update, leading to faster convergence. The training process for XGBoost involves adding trees iteratively. The algorithm calculates the gradients and Hessians of the loss function to guide the growth of the new tree. The goal is to minimize the following regularized objective function:

$$L(\theta) = \sum_{i=1}^{n} l(y_i, \hat{y}_i) + \sum_{k=1}^{K} \Omega(f_k) \tag{3}$$

where $l(y_i, \hat{y}_i)$ represents the loss function for the i-th instance, and $\Omega(f_k)$ is the regularization term controlling the complexity of the model. The model learns by minimizing this objective function using gradient descent, improving over time as new trees are added. While XGBoost offers significant advantages, it is also computationally intensive, especially for large datasets. Additionally, the algorithm requires careful hyperparameter tuning, as improper tuning can lead to suboptimal performance or overfitting [9].

3.3.5 Light Gradient Boosting Machine (LightGBM)

LightGBM is a gradient boosting system designed for efficiency and scalability, particularly suited to large datasets and high-dimensional data [26]. It is designed to build trees with faster computation and lower memory usage compared to traditional boosting algorithms like XGBoost. LightGBM achieves this through innovations in tree construction, making it highly effective in scenarios requiring both speed and performance. Instead of growing trees level-by-level like conventional tree-based techniques, it uses a leaf-wise growth strategy. With this approach, LightGBM grows the tree by splitting the leaf with the highest loss reduction. This results in deeper trees where necessary, capturing more complex patterns and improving model accuracy, particularly in datasets with unbalanced classes or numerous features. The objective function in LightGBM is minimized through gradient boosting by training each tree to address the errors from the preceding ones. Its regularization terms are used to mitigate overfitting by penalizing model complexity, similar to XGBoost. The optimization problem in LightGBM can be expressed as:

$$L(\theta) = \sum_{i=1}^{n} l(y_i, \hat{y}_i) + \lambda \sum_{k=1}^{K} \| f_k \|^2 \tag{4}$$

where $l(y_i, \hat{y}_i)$ represents the loss function measuring the difference between the actual and predicted outcomes, and $\lambda \sum_{k=1}^{K} \| f_k \|^2$ is the regularization term. The framework optimizes this objective function using gradient-based techniques. Compared to traditional boosting algorithms like XGBoost, LightGBM is designed to handle large datasets more efficiently. It achieves this not only by growing trees leaf-wise but also by selectively sampling data and features, making it ideal for high-dimensional data. One limitation is that its leaf-wise strategy can cause overfitting, especially when applied to small datasets [26].

3.3.6 Support Vector Machine (SVM)

SVM is applied for its capability in handling high-dimensional datasets and classifying data with non-linear boundaries. It operates by identifying the best hyperplane that maximizes the margin between different classes. For data that is not linearly distinct, kernel functions are introduced to map the data into a higher-dimensional space, enabling better classification accuracy [35]. While SVM is highly effective in dealing with complex datasets, its performance needs to be meticulously modified by adjusting hyperparameters like regularization and kernel selection. In the context of this chapter, SVM with the Radial Basis Function (RBF) kernel is applied to categorize post-IPO financial performance. The RBF kernel, also referred to as the Gaussian kernel, is particularly useful when dealing with non-linear correlations between variables and target outcomes [37]. This makes SVM with the RBF kernel highly effective for modeling complex datasets, such as those involving financial performance metrics. The RBF kernel converts data into a higher-dimensional space where linear distinction is feasible when the data is not capable of linear separation in its initial structure. The kernel function for RBF is defined as:

$$K(x, x') = \exp(-\gamma \|x - x'\|^2) \tag{5}$$

where x and x' are two data points, and γ is a parameter that controls how much influence each training point has. A smaller γ makes the decision boundary smoother, while a larger γ allows the model to detect more intricate patterns. SVM with the RBF kernel offers several key advantages in classification tasks. However, SVM can be resource-intensive, particularly for large datasets, because it requires calculating the distance between every pair of data points in the transformed space. This issue is mitigated to some extent by the use of efficient algorithms for kernel computations, but it remains a consideration when applying SVM to large-scale problems [37].

3.3.7 Artificial Neural Network (ANN)

ANN mimics the structure and functioning of the brain. It consists of layers of interconnected nodes, where each node receives input data through weighted connections, processes it, and passes the result to subsequent layers. This enables the network to use several processing layers to discover intricate patterns and connections in the data [38]. An input layer, multiple hidden layers, and an output layer make up the ANN's architecture. In this chapter, the ReLU (Rectified Linear Unit) activation function is used for the hidden layers to prevent the vanishing gradient problem [39], while the output layer uses the softmax function to convert the raw predictions into probabilities for each class. The cross-entropy loss function, which assesses the disparity between expected and actual labels, is minimized during the learning process. The optimization is performed using stochastic gradient descent, which iteratively updates the network's weights to improve its classification accuracy. During backpropagation, the network adjusts the weights to reduce the prediction error.

ANN is particularly powerful when dealing with large datasets with high dimensionality, as it can automatically learn which features are most important for making accurate predictions. The use of multiple hidden layers allows ANN to capture higher-order interactions between the input variables, which can be difficult for simpler models like logistic regression or DT. However, ANN requires substantial computational resources, especially when the network architecture becomes complex. They can also be susceptible to overfitting if the model is too large or if the data is insufficiently diverse [40].

Each model chosen for this chapter brings unique strengths to predicting post-IPO financial performance. Ordinal logistic regression offers a simple, interpretable baseline but is limited by its assumption of linearity. DT provides transparency but can overfit without careful management. RF addresses overfitting and improves accuracy by using an ensemble technique. XGBoost and LightGBM enhance performance further with gradient boosting and regularization, making them powerful on complex datasets. SVM handles non-linear classification well, though it requires careful tuning, while ANN captures intricate non-linear relationships but demands substantial computational resources. These diverse techniques allow us to comprehensively assess post-IPO performance prediction, offering insights into the strengths and weaknesses of each model in classifying firms into different performance classes.

3.4 *Cross-Validation and Hyperparameter Tuning*

Cross-validation and hyperparameter tuning are critical steps in ensuring that machine learning techniques perform well and generalize to unseen data. In this chapter, k-fold (5-fold) cross-validation is employed to assess the models' ability to predict post-IPO financial performance. Cross-validation is essential in machine learning, especially with smaller datasets, as it reduces the risk of overfitting and ensures that the model adapts well to unseen data [41]. This technique divides the dataset into five equal parts, using four parts for training and one for validation, repeating this process five times so that each part serves as the validation set once. The final performance metrics are averaged across all folds to provide a reliable evaluation, mitigating overfitting and ensuring that the models are not excessively tailored to the training data. Given that the dataset consists of 134 firms, 5-fold cross-validation is chosen as an optimal balance between maintaining sufficient training data and ensuring robust testing. Dividing the dataset into too many folds would result in each fold containing very few data points, which can lead to unreliable evaluations and increased variance in the results [42].

For each machine learning technique, hyperparameter tuning is performed using grid search. Grid search is a thorough search process that determines which set of predefined hyperparameters yields the optimal model performance by analyzing every possible combination. The grid search is applied within the cross-validation framework, ensuring that hyperparameters are optimized based on their performance across multiple validation sets, reducing the risk of overfitting [43]. Each model has a unique set of hyperparameters that are tuned during the grid search. The best hyperparameters found for each model for each of the three post-IPO years are shown in Tables 4, 5 and 6. These parameters are selected based on model performance (accuracy) on validation sets during the 5-fold cross-validation process.

3.5 *Process of Model Implementation*

The model implementation process begins with acquiring the raw data, as illustrated in Fig. 3. This step involves collecting financial and IPO-related data from firms listed on the SET, spanning from one year pre-IPO to three years post-IPO. After obtaining the raw data, a data preparation step is undertaken, which includes cleaning the dataset, handling missing values, and standardizing numerical factors to ensure

Table 4 Final hyperparameters for first year post-IPO models

Model	Hyperparameters
DT	max_depth: 7, min_samples_leaf: 1, min_samples_split: 2
RF	max_depth: 5, min_samples_leaf: 2, min_samples_split: 2, n_estimators: 100
XGBoost	learning_rate: 0.1, max_depth: 7, n_estimators: 100
LightGBM	learning_rate: 0.1, n_estimators: 150, num_leaves: 15
SVM (RBF)	C: 10, gamma: 'scale'
ANN	activation: 'ReLU', batch_size: 16, epochs: 50, neurons: 64, num_layers: 2, optimizer: 'adam'

Note For DT, the max_depth controls how deep the tree grows, while min_samples_leaf and min_samples_split ensure that each leaf and split has enough samples to avoid overly specific splits. In RF, n_estimators refers to the number of trees, and other parameters are similar to DT. XGBoost and LightGBM use learning_rate to control the contribution of each tree. For SVM (RBF), C determines the trade-off between maximizing the margin and minimizing classification error, while gamma defines how far the influence of a single training example reaches, which is critical for handling non-linear relationships. In ANN, activation specifies the activation function, influencing neuron output, batch_size and epochs regulate the training process; while neurons and num_layers define the network's structure, and optimizer affects how the weights are updated during backpropagation

Table 5 Final hyperparameters for second year post-IPO models

Model	Hyperparameters
DT	max_depth: 5, min_samples_leaf: 1, min_samples_split: 2
RF	max_depth: 5, min_samples_leaf: 2, min_samples_split: 5, n_estimators: 100
XGBoost	learning_rate: 0.3, max_depth: 10, n_estimators: 50
LightGBM	learning_rate: 0.1, n_estimators: 150, num_leaves: 15
SVM (RBF)	C: 10, gamma: 'scale'
ANN	activation: 'ReLU', batch_size: 16, epochs: 100, neurons: 32, num_layers: 2, optimizer: 'adam'

Note The hyperparameter descriptions are the same as those explained under Table 4

Table 6 Final hyperparameters for third year post-IPO models

Model	Hyperparameters
DT	max_depth: 7, min_samples_leaf: 2, min_samples_split: 2
RF	max_depth: 7, min_samples_leaf: 1, min_samples_split: 2, n_estimators: 100
XGBoost	learning_rate: 0.3, max_depth: 7, n_estimators: 50
LightGBM	learning_rate: 0.1, n_estimators: 150, num_leaves: 15
SVM (RBF)	C: 10, gamma: 'scale'
ANN	activation: 'ReLU', batch_size: 16, epochs: 50, neurons: 64, num_layers: 2, optimizer: 'adam'

Note The hyperparameter descriptions are the same as those explained under Table 4

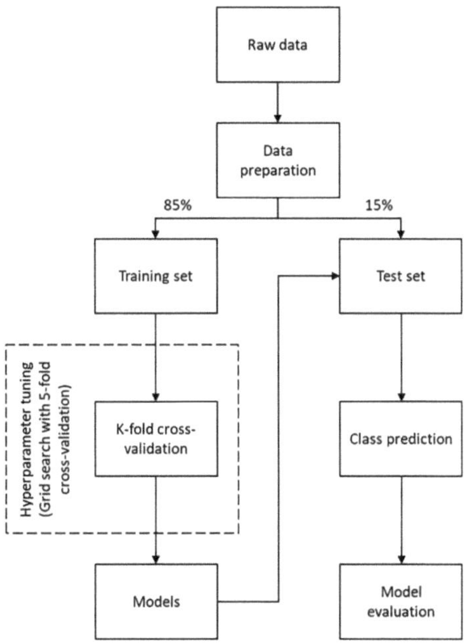

Fig. 3 Flow of model implementation process

consistency. Additionally, the financial performance response variable (an aggregate of ROA, ROE, and ROS) is classified into three performance classes, based on 1/3 and 2/3 quantile cut-offs.

The cleaned and preprocessed data is then split into training and test sets, with 85% allocated to training and 15% reserved for testing. The training set is utilized to develop and optimize the models, while the test set is used exclusively for evaluating final model performance on unseen data, ensuring robust generalization capabilities.

Within the training set, a 5-fold cross-validation with a grid-search strategy is performed to fine-tune hyperparameters and validate model stability.

The final step involves evaluating the models on the test set, using performance metrics such as accuracy, AUC, F1-score, and Cohen's Kappa. These metrics offer a thorough evaluation of the models' ability to predict post-IPO financial performance accurately. This flow ensures that the models are trained and tested in a structured, robust manner, with minimal risk of overfitting and a high degree of generalizability.

3.6 *Model Evaluation*

The evaluation of the models employed to predict post-IPO financial performance is based on several key metrics, each offering insights into different aspects of the models' effectiveness. These metrics—accuracy, AUC (Area Under the Curve), F1-score, and Kappa—provide a thorough evaluation of classification performance.

3.6.1 Accuracy

Accuracy is the most basic metric, which calculates the ratio of correct predictions to the total predictions. The formula for accuracy is:

$$\text{Accuracy} = \frac{\text{Number of correct predictions}}{\text{Total number of predictions}} = \frac{TP + TN}{TP + TN + FP + FN} \tag{6}$$

where:

- TP (True Positive): Firms correctly classified into a specific performance class.
- TN (True Negatives): Firms correctly identified as not belonging to the specified performance category.
- FP (False Positives): Firms incorrectly classified into a performance category when they do not belong to it.
- FN (False Negatives): Firms that belong to a performance category but were incorrectly classified into a different category.

While accuracy is simple, it might be deceptive in imbalanced datasets, resulting in bias in favor of the dominant class [44].

3.6.2 Area Under the Curve (AUC)

The Area Under the Receiver Operating Characteristic Curve (AUC-ROC) is a key metric for evaluating a model's capability to discriminate across classes, particularly when predicting binary or multiclass outcomes. An AUC score of 1 denotes flawless classification, whereas a score of 0.5 suggests no discriminative power [45]. The AUC curve provides a graphical depiction of the model's capacity to strike a balance between sensitivity and specificity by plotting the true positive rate (recall) versus the false positive rate at different threshold values. Generally, an AUC of 0.7 or above is considered acceptable, reflecting moderate to strong ability to differentiate between classes [46].

3.6.3 F1-Score

The F1-score provides a single statistic that balances precision and recall by taking the harmonic average of the two. Precision focuses on the positive class and the proportion of expected positive outcomes that were really realized [47]. It is especially important when false positives carry a significant cost, such as incorrectly predicting a firm's strong post-IPO performance. Precision is defined as:

$$\text{Precision} = \frac{TP}{TP + FP} \tag{7}$$

A model with high precision minimizes the number of false positives, ensuring that firms classified as high performers post-IPO are truly high performers. Recall, sometimes referred to as sensitivity or true positive rate, gauges how well the model can detect every real positive. It is essential in situations when it would be expensive to miss a real positive [48]. Recall is calculated as:

$$\text{Recall} = \frac{TP}{TP + FN} \tag{8}$$

A high F1-score shows that the model well balances recall and precision, which makes it useful in situations where both false positives and false negatives might be expensive [49]. The F1-score is calculated as follows:

$$\text{F1} = 2 \times \frac{\text{Precision} \times \text{Recall}}{\text{Precision} + \text{Recall}} \tag{9}$$

This allows for a more balanced evaluation, as it takes into account not only how many instances were predicted correctly but also the quality of those predictions in terms of both precision and recall.

3.6.4 Kappa

Cohen's Kappa is a widely used statistic for evaluating the level of concordance between two raters or two classification methods beyond what would be expected by chance. It provides an estimate of the proportion of agreement after accounting for the agreement that would occur by random chance alone [50]. The Kappa statistic is particularly valuable in scenarios where data may be imbalanced, such as multiclass classification tasks, making it more informative than simple accuracy.

$$\text{Kappa} = \frac{p_o - p_e}{1 - p_e} \tag{10}$$

where:

- p_o = the observed agreement between a classifier and the true labels.
- p_e = the expected agreement by chance.

A Kappa value of 1 represents perfect agreement, meaning the classifier's predictions align exactly with the actual outcomes. A Kappa value of 0 signifies that the level of agreement is no better than random chance. When the values are negative, the agreement is less than random. Cohen's Kappa values are commonly interpreted as follows: values above 0.21 indicate fair agreement, suggesting some correlation between predicted and actual results. A Kappa greater than 0.41 represents moderate agreement, showing a reasonable match between predicted and actual classifications, though some misclassifications remain. Values higher than 0.61 reflect substantial

Table 7 Descriptive statistics of the dataset

Descriptive statistics	LIQ	LEV	COG	ASS	TAN	SIZ	RET	NPM	FUN
Mean	1.5961	0.5976	0.7385	1.3419	0.4853	21.6052	1.2698	0.1846	0.4847
Std	1.8196	0.1946	0.1705	1.2273	0.2569	1.2950	1.9838	0.3682	0.5369
Min	0.3606	0.1300	0.1936	0.004	0.0136	19.5037	−3.7951	−0.2797	0.0549
25%	0.8806	0.4693	0.6499	0.699	0.2869	20.7341	0.3212	0.0481	0.2127
50%	1.1664	0.6309	0.7533	1.0755	0.4922	21.3596	1.0236	0.0927	0.3498
75%	1.6340	0.7151	0.8514	1.5659	0.6965	22.1829	1.7745	0.1696	0.5409
Max	17.5974	1.0561	1.0345	10.656	0.9582	25.3476	11.4276	2.9552	5.1944

Note The Std (standard deviation) measures the dispersion of data points around the mean. 25, 50, and 75% indicate the 25th, 50th (median), and 75th percentiles of the data distribution, respectively

agreement, meaning the model performs well with minimal errors, demonstrating strong alignment with actual outcomes [51].

The metrics—accuracy, AUC, F1-score, and Kappa—provide a comprehensive framework for evaluating model performance in predicting post-IPO financial outcomes. By combining these evaluation metrics, the chapter offers a nuanced overview of the effectiveness of both statistical and machine learning techniques in predicting financial performance post-IPO.

4 Results and Discussion

Before diving into the performance of the models, it is crucial to understand the key characteristics and relationships among the factors. Table 7 provides the descriptive statistics for the nine factors used in this chapter. From the table, it can be observed that there is significant variation across firms, particularly in firm size, which shows a wide range from 9.29 to 25.55. The standard deviations for many factors such as liquidity and firm size indicate considerable heterogeneity in the sample.

The correlation heatmap shown in Fig. 4 highlights the relationships between the factors. Most notably, the amount of funds raised and cost of goods sold show the highest negative correlations at –0.4417. In general, a correlation coefficient above 0.7 (or below –0.7) between two factors is considered a strong indicator of multicollinearity [52]. Therefore, based on these results, there is no strong evidence of multicollinearity in this dataset.

4.1 Model Performance

The analysis is conducted using Minitab 21 for the statistical modeling, specifically the logistic regression models, and Python 3.8 with scikit-learn for machine learning

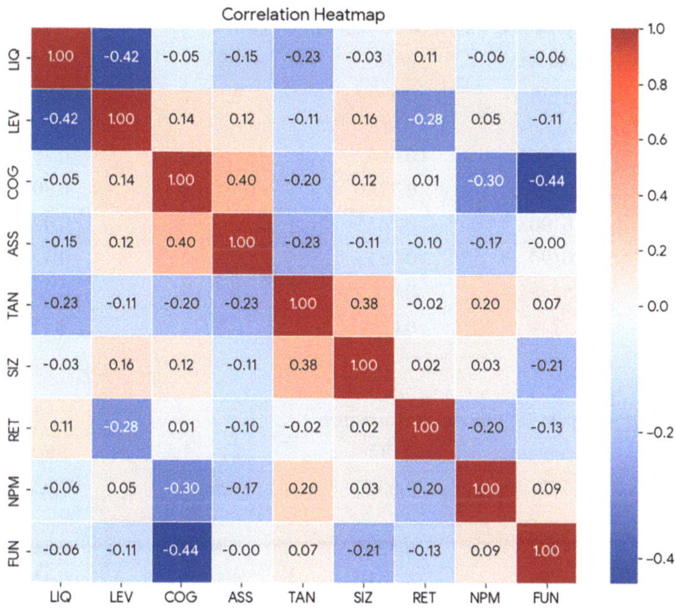

Fig. 4 Correlation heatmap of factors

Table 8 First year post-IPO model performance

Model	Accuracy	AUC	F1	Kappa
Logistic regression	0.5714	0.6774	0.5310	0.3571
DT	0.6190	0.7143	0.6124	0.4286
RF	**0.7143**	**0.8027**	**0.7111**	**0.5714**
XGBoost	0.6667	0.7449	0.6626	0.5000
LightGBM	0.6667	0.7449	0.6626	0.5000
SVM (RBF)	0.5714	0.7517	0.5356	0.3571
ANN	0.6667	0.7925	0.6568	0.5000

techniques. The performance of the machine learning techniques in predicting post-IPO financial performance is evaluated across three years post-IPO, as well as by averaging the metrics over the three-year period. The metrics used include accuracy, AUC, F1-score, and Kappa. The models compared include Logistic regression, DT, RF, XGBoost, LightGBM, SVM (RBF), and ANN. The results are presented for each of the three years post-IPO and as an average across the three years as presented in Tables 8, 9, 10 and 11. The machine learning techniques are evaluated using their best-tuned hyperparameters, as outlined in the Sect. 3.4.

The results demonstrate that the performance of the models tends to decline as the number of years post-IPO increases. This decrease in predictive performance is to be

Table 9 Second year post-IPO model performance

Model	Accuracy	AUC	F1	Kappa
Logistic regression	0.5238	0.5889	0.5221	0.2857
DT	0.5714	0.6667	0.4556	0.3571
RF	**0.6667**	**0.7789**	**0.6468**	**0.5000**
XGBoost	0.6190	0.7415	0.6111	0.4286
LightGBM	0.6190	0.7313	0.6178	0.4286
SVM (RBF)	0.5238	0.4320	0.5189	0.2857
ANN	0.5714	0.6633	0.5370	0.3571

Table 10 Third year post-IPO model performance

Model	Accuracy	AUC	F1	Kappa
Logistic regression	0.5714	0.6369	0.5348	0.3571
DT	0.4762	0.5204	0.4630	0.2143
RF	**0.6190**	**0.7619**	**0.6071**	**0.4286**
XGBoost	0.5238	0.7041	0.5058	0.2857
LightGBM	0.5238	0.7347	0.5111	0.2857
SVM (RBF)	0.5714	0.7279	0.5673	0.3571
ANN	0.5714	0.6131	0.5597	0.3505

Table 11 Average post-IPO model performance across three years

Model	Accuracy	AUC	F1	Kappa
Logistic regression	0.5556	0.6344	0.5310	0.3333
DT	0.5555	0.6338	0.5103	0.3333
RF	**0.6667**	**0.7812**	**0.6550**	**0.5000**
XGBoost	0.6032	0.7302	0.5932	0.4048
LightGBM	0.6032	0.7370	0.5972	0.4048
SVM (RBF)	0.5556	0.6372	0.5406	0.3333
ANN	0.6032	0.6896	0.5845	0.4026

expected. Since the models are trained using pre-IPO data, their ability to accurately predict financial performance diminishes over time. This may result from shifts in management, operational adjustments, or evolving market conditions, which make it harder for the models to predict accurately.

In the first year post-IPO, RF demonstrates the best overall performance, achieving 71.43% accuracy, an AUC of 80.27%, an F1-score of 71.11%, and a Kappa value of 0.5714. The AUC score exceeding 0.8 indicates very strong discriminative power, as an AUC above 0.7 is generally considered indicative of a strong classifier [46]. RF's Kappa value, which falls between 0.41 and 0.60, indicates moderate agreement, further solidifying its reliability [51]. Other models, including XGBoost, LightGBM, and ANN, also perform well, each achieving 66.67% accuracy. However, their AUC values—74.49, 74.49, and 79.25% respectively—fall below RF at 80.27%. ANN notably has a high AUC, suggesting good classification ability, but its Kappa value of 0.5000 suggests it has slightly lower reliability than RF, though still demonstrating moderate agreement.

In the second year, RF continues to outperform other models, maintaining 66.67% accuracy, an AUC of 77.89%, and an F1-score of 64.68%. Its Kappa value of 0.5000 remains within the moderate agreement range, reinforcing RF's consistent and reliable performance over time. XGBoost and LightGBM follow closely behind, with XGBoost achieving an AUC of 74.15% and LightGBM 73.13%, though their lower Kappa values of 0.4286 indicate less consistent classifications, only reaching fair agreement. ANN displays good performance in accuracy (57.14%), but with an AUC of 66.33%, its classification ability is lower than that of RF and XGBoost.

In the third year, RF again achieves strong performance, with 61.90% accuracy, an AUC of 76.19%, and an F1-score of 60.71%, while maintaining a Kappa value of 0.4286, which suggests fair agreement. SVM with the RBF kernel performs similarly well, with an AUC of 72.79% and Kappa of 0.3571; however, its lower Kappa value and accuracy suggest weaker classification reliability compared to RF. ANN demonstrates a respectable AUC of 61.31% but falls behind RF in both accuracy and reliability.

Overall, RF consistently outperforms all other models across all years in terms of accuracy, AUC, and F1-score, demonstrating the most reliable classification of post-IPO financial performance. The model's AUC values, consistently above 0.7, affirm its strong ability to distinguish between different financial performance categories. Its Kappa value of 0.5000 indicates moderate agreement, suggesting that RF provides the most balanced and reliable classifications compared to other models in this analysis. Despite strong individual performances from models like ANN, XGBoost, and LightGBM in specific years, RF's consistency and superior average performance make it the most suitable model for predicting post-IPO financial performance. Notably, across all three years, the traditional statistical method, logistic regression, consistently underperforms compared to machine learning techniques. The relatively poor performance of logistic regression, in contrast to machine learning techniques, reinforces the advantages of more sophisticated, non-linear models in capturing complex relationships between variables, particularly for the multi-class prediction of post-IPO financial performance.

The results indicate that RF performs the best among the models tested, particularly in terms of accuracy, AUC, F1-score, and Kappa. Despite being considered less advanced than boosting algorithms like XGBoost and LightGBM, RF demonstrates better overall performance, which could be attributed to several factors.

- While XGBoost and LightGBM handle non-linear relationships well, RF benefits from its ensemble technique, introducing diversity through different data subsets, resulting in better generalization. This is evidenced by AUC values consistently above 0.78, showing RF's effectiveness in distinguishing financial performance classes. RF's built-in averaging mechanism helps prevent overfitting, particularly in small datasets like the 134 firms used here. In contrast, XGBoost and LightGBM, which rely on boosting techniques that iteratively correct errors, tend to overfit more easily. This happens because boosting aggressively focuses on correcting previous mistakes, making the models sensitive to noise in the data, especially when the dataset is small or has fewer features [34].

- Boosting algorithms excel in capturing complex interactions between features. However, in this dataset, which contains nine factors, the interactions between features may not be as complex, and thus, RF's simpler mechanism of averaging multiple trees works better without introducing unnecessary complexity [53].
- Models like ANN and SVM tend to perform better on larger datasets, where their capability to capture intricate and non-linear patterns becomes more apparent. Given the size of this dataset, RF is able to capture important patterns without overfitting, while more complex models may struggle to generalize as effectively. Research supports that more complex models like ANN may not outperform simpler models in small-to-medium-sized datasets [54].

In summary, RF's success in this chapter can be ascribed to its capacity to manage non-linear correlations, resilience to overfitting, and simplicity in terms of tuning. While other advanced models like XGBoost and LightGBM perform well in specific metrics, RF's overall consistency in accuracy, AUC, F1-score, and Kappa values makes it the most reliable model for predicting post-IPO financial performance.

4.2 Key Factors

The analysis of feature importance from the RF offers significant insights into which factors most significantly impact a firm's financial performance prediction in the first three years post-IPO. This chapter assesses feature importance using SHAP (SHapley Additive exPlanations) values, which provide a robust framework for interpreting the impact of individual factors on the model's outputs. SHAP values are grounded in cooperative game theory and guarantee that each feature's importance is fairly attributed by considering all possible feature combinations, making them particularly useful in understanding the non-linear interactions often present in financial data [55]. The feature importance for predicting post-IPO financial performance, as derived from SHAP values using the RF results, is presented in Figs. 5, 6 and 7.

In the initial year after an IPO, the focus tends to be on the capital raised and how it is deployed to fuel growth. Funds raised stands out as the most important factor. This is expected, as the immediate availability of IPO proceeds is critical for firms seeking to invest in growth, operational expansion, and strategic initiatives. The amount of capital raised often reflects the firm's ability to secure necessary funding, which can significantly influence early financial performance [7]. Firms that effectively

Fig. 5 Feature importance for the first year post-IPO

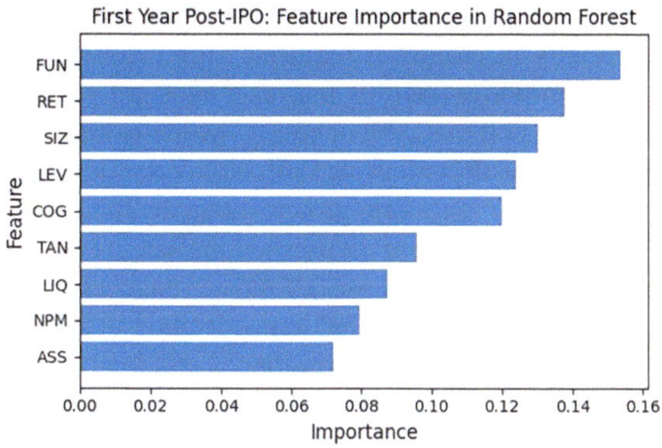

Fig. 6 Feature importance for the second year post-IPO

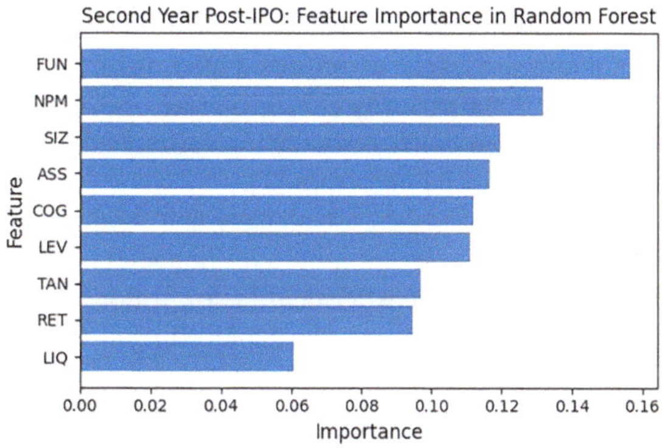

Fig. 7 Feature importance for the third year post-IPO

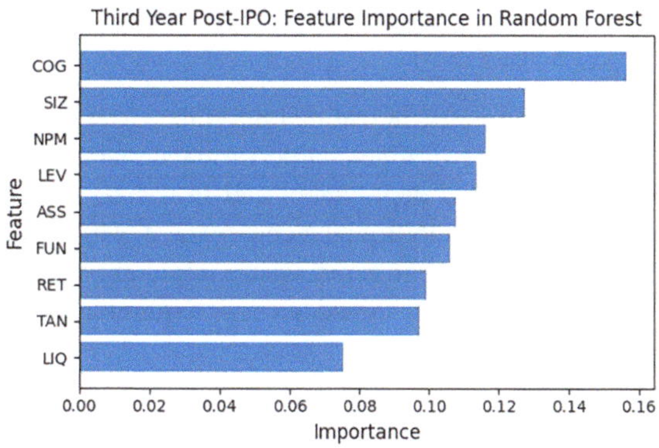

deploy IPO proceeds can achieve rapid growth and gain competitive advantages, especially in industries requiring significant capital investments. Retention is also highly important as this internal financing source is often more flexible and less risky than external debt or equity, which is especially valuable in the early stages of a firm's post-IPO lifecycle [56]. High retention rates signal to investors that the firm is focused on long-term growth, contributing to stronger performance. Larger firms typically benefit from established market presence, stronger brand recognition, and economies of scale. These advantages enable larger firms to operate more efficiently, and secure better financing terms, leading to better post-IPO outcomes [7].

By the second year, operational efficiency and profitability start to take precedence, as firms transition from using capital to generating sustainable financial returns. While the capital raised continues to be a key driver of performance prediction, net profit margin rises in importance, reflecting investors' increased focus on profitability. Firms that can control their costs and generate profits from their operations are likely to perform better as they move beyond the growth phase and into operational maturity [57]. Firm size continues to be relevant, as larger firms tend to benefit from better resource allocation and market leverage, which contribute to improved financial outcomes [7].

By the third year, cost management and operational stability become critical, reflecting a shift toward long-term financial health and performance. Cost of goods sold becomes the most critical factor, followed by firm size and net profit margin. As firms settle into the market, controlling costs becomes essential to maintain profitability. Firms that manage their production and operational costs effectively can better sustain financial performance in the long term [57]. Firm size remains important, as larger companies often have the infrastructure and market position to better manage costs and withstand economic fluctuations. Net profit margin continues to be a significant factor, as sustained profitability is crucial for long-term financial stability.

The progression of feature importance over the three years post-IPO highlights the dynamic nature of financial performance factors. Early post-IPO success is largely driven by capital deployment and firm size, but as firms mature, operational efficiency and profitability take center stage. This transition is consistent with the lifecycle of newly public firms, where initial growth is fueled by investment and market expansion, while sustained success requires a focus on cost management and profitability [7, 56].

4.3 Implications

The findings from this chapter also have important implications for various stakeholders, particularly firms and investors. Understanding the key factors that influence post-IPO financial performance and evaluating the effectiveness of different predictive models can enhance decision-making across these groups.

For firms, the results underscore the importance of strategic capital deployment and operational efficiency. In the early post-IPO phase, the ability to effectively use raised funds is crucial for growth. As firms progress, attention shifts toward maintaining high profitability through cost control and efficient operations. Firms that focus on improving their net profit margin and controlling cost of goods sold are more likely to sustain long-term financial success. These insights can help management teams develop targeted strategies that align with the evolving priorities of their investors and market demands.

For investors, the insights from this chapter highlight the critical factors that influence firm performance in the years following an IPO. The prominence of funds raised in the first year suggests that investors should closely monitor how firms deploy the capital raised during their IPOs. As the firm matures, factors like net profit margin and cost of goods sold become more significant, indicating a shift toward the importance of operational efficiency and profitability. Investors can use this information to adjust their strategies, focusing on these changing priorities over time. Additionally, the models developed in this chapter can be applied to predict the financial performance of firms planning to list on the SET, aiding investors in making more informed decisions before investing. The strong performance of the RF models suggest that integrating machine learning techniques into financial analysis could help investors better understand complex relationships within financial data and improve decision-making accuracy.

From a methodological perspective, this chapter contributes to the growing use of machine learning techniques in financial prediction. The superior performance of RF, compared to more complex models like XGBoost and LightGBM, suggests that while advanced algorithms can be powerful, simpler ensemble techniques may provide better accuracy and interpretability in certain contexts. The use of SHAP values also demonstrates the value of explainable AI, offering greater transparency in understanding which factors most influence the predictions. Future studies may expand on these findings by applying similar models in different financial contexts or exploring new techniques to improve post-IPO performance predictions.

In summary, this chapter offers valuable insights for improving post-IPO financial performance analysis and decision-making, emphasizing the need for strategies that adapt to changing priorities as firms progress beyond their IPOs. The use of machine learning techniques like RF provides a strong foundation for predicting financial outcomes, and the findings can guide firms and investors in making informed decisions.

5 Conclusions

This chapter has explored the predictive factors and models for post-IPO financial performance, focusing on firms listed on the SET. By applying both traditional statistical methods like logistic regression and machine learning techniques such as DT, RF, XGBoost, and LightGBM, SVM, and ANN, the chapter has provided a comprehensive analysis of the factors of firm success in the first three years following an IPO. The findings showed that RF consistently outperformed not only traditional models like logistic regression but also more advanced machine learning techniques.

RF's superior performance could be credited to its ensemble technique, which minimized overfitting by combining the predictions of several trees. This robustness allowed it to manage non-linearity in the financial data more effectively than logistic regression, which assumed linearity. Additionally, RF worked particularly well with this relatively small dataset, as it was less prone to overfitting compared to more complex models, making it highly suitable for datasets of limited size. RF outperformed more advanced techniques like XGBoost and LightGBM due to its simplicity and better suitability for the specific dataset, where interpretability and handling of smaller datasets with relatively fewer hyperparameters were crucial. In comparison, SVM and ANN, while powerful in handling complex and high-dimensional data, tended to require more intensive tuning and computational resources. RF's balance between performance, ease of tuning, and lower risk of overfitting made it the most reliable model for this chapter.

Key factors such as funds raised, firm size, retention, and cost of goods sold emerged as significant predictors of financial performance prediction across the different post-IPO periods. These factors highlighted the importance of capital allocation and operational efficiency as firms transitioned from private to public status. The evolution of factor importance over time suggested that firms needed to adapt their strategies as their priorities shifted from growth in the early years to cost management and profitability in later years.

Despite its contributions, this chapter had some limitations. A notable limitation was the relatively small sample size of 134 firms, which could have affected the ability to generalize the results to other markets or contexts. Additionally, the chapter focused exclusively on internal firm-specific factors and did not account for external market conditions or macroeconomic variables, which could have also played a role in post-IPO performance. The models used in this chapter, while effective, required careful hyperparameter tuning, and the optimal settings may have varied across different datasets or time periods.

For future research, expanding the sample size and incorporating external factors such as market volatility, economic conditions, or industry-specific risks could provide a more holistic understanding of what drives post-IPO success. Additionally, further exploration of hybrid models that combine the strengths of statistical and machine learning techniques may offer new insights into improving prediction accuracy and interpretability. Finally, the inclusion of real-time data and more advanced ensemble techniques, such as deep learning, could enhance the predictive capabilities of future studies, especially in rapidly changing financial markets.

In conclusion, this chapter underscored the importance of adapting predictive models and strategies as firms evolved post-IPO. Machine learning techniques, particularly Random Forest, offered strong predictive power, and by focusing on the most influential factors, stakeholders could better navigate the complexities of post-IPO financial performance. However, continued refinement of models and a broader consideration of external factors will be crucial for future research in this area.

Competing Interests The authors have no conflicts of interest to declare that are relevant to the content of this chapter.

References

1. Ritter, J.R.: The long-run performance of initial public offerings. J. Financ. **46**(1), 3–27 (1991)
2. Loughran, T., Ritter, J.: Why has IPO underpricing changed over time? Financ. Manage. **33**(3), 5–37 (2004)
3. Jain, B.A., Kini, O.: The post-issue operating performance of IPO firms. J. Financ. **49**(5), 1699–1726 (1994)
4. Ritter, J.R., Welch, I.: A review of IPO activity, pricing, and allocations. J. Financ. **57**(4), 1795–1828 (2002)
5. Aggarwal, R., Bhagat, S., Rangan, S.: The impact of fundamentals on IPO valuation. Financ. Manage. **38**(2), 253–284 (2009)
6. Derrien, F.: IPO pricing in "hot" market conditions: Who leaves money on the table? J. Financ. **60**(1), 487–521 (2005)
7. Pagano, M., Panetta, F., Zingales, L.: Why do companies go public? an empirical analysis. J. Financ. **53**(1), 27–64 (1998)
8. Patel, J., Shah, S., Thakkar, P., Kotecha, K.: Predicting stock and stock price index movement using trend deterministic data preparation and machine learning techniques. Expert Syst. Appl. **42**(1), 259–268 (2015)
9. Chen, T., Guestrin, C.: Xgboost: a scalable tree boosting system. In: Proceedings of the 22nd ACM SIGKDD International Conference on Knowledge Discovery and Data Mining, pp. 785–794 (2016)
10. Field, L.C., Hanka, G.: The expiration of IPO share lockups. J. Financ. **56**(2), 471–500 (2001)
11. Flamini, V., McDonald, C.A., Schumacher, L.B.: The Determinants of Commercial Bank Profitability in Sub-Saharan Africa. International Monetary Fund, Washington, DC (2009)
12. Majumdar, S.K.: The impact of size and age on firm-level performance: some evidence from India. Rev. Ind. Organ. **12**, 231–241 (1997)
13. Hvide, H.K., Møen, J.: Liquidity constraints and entrepreneurial performance. CEPR Discussion Paper No. DP6495
14. Robb, A.M., Robinson, D.T.: The capital structure decisions of new firms. Rev. Financ. Stud. **27**(1), 153–179 (2014)
15. Jensen, M.C., Meckling, W.H.: Theory of the firm: managerial behavior, agency costs and ownership structure. In: Corporate Governance, pp. 77–132. Gower (2019)
16. İltasÌ§, Y., DemirgünesÌ§, K.: Asset tangibility and financial performance: a time series evidence. Ahi Evran Univ. Sos. Bilim. Enst. Derg. **6**(2), 345–364 (2020)
17. Almeida, H., Campello, M.: Financial constraints, asset tangibility, and corporate investment. Rev. Financ. Stud. **20**(5), 1429–1460 (2007)
18. Kim, M., Ritter, J.R.: Valuing IPOs. J. Financ. Econ. **53**(3), 409–437 (1999)
19. Gompers, P.A., Lerner, J.: The Venture Capital Cycle. MIT Press (2004)
20. Ittner, C.D., Larcker, D.F.: Innovations in performance measurement: trends and research implications. J. Manag. Account. Res. **10**, 205 (1998)
21. Delen, D., Kuzey, C., Uyar, A.: Measuring firm performance using financial ratios: a decision tree approach. Expert Syst. Appl. **40**(10), 3970–3983 (2013)
22. Yousaf, M., Dey, S.K.: Best proxy to determine firm performance using financial ratios: a CHAID approach. Rev. Econ. Perspect. **22**(3), 219–239 (2022)
23. Ballings, M., Van den Poel, D., Hespeels, N., Gryp, R.: Evaluating multiple classifiers for stock price direction prediction. Expert Syst. Appl. **42**(20), 7046–7056 (2015)
24. Anand, V., Brunner, R., Ikegwu, K., Sougiannis, T.: Predicting profitability using machine learning (2019). Available at SSRN: https://ssrn.com/abstract=3466478

25. Toochaei, M.R., Moeini, F.: Evaluating the performance of ensemble classifiers in stock returns prediction using effective features. Expert Syst. Appl. **213**, 119186 (2023)
26. Ke, G., Meng, Q., Finley, T., Wang, T., Chen, W., Ma, W., Liu, T.Y.: LightGBM: a highly efficient gradient boosting decision tree. Adv. Neural Inf. Process. Syst. **30** (2017)
27. Yang, Y., Wu, Y., Wang, P., Jiali, X.: Stock price prediction based on XGBoost and LightGBM. In: E3S Web of Conferences, vol. 275, p. 01040 (2021). EDP Sciences
28. Huang, W., Nakamori, Y., Wang, S.Y.: Forecasting stock market movement direction with support vector machine. Comput. Oper. Res. **32**(10), 2513–2522 (2005)
29. Shin, K.S., Lee, T.S., Kim, H.J.: An application of support vector machines in bankruptcy prediction model. Expert Syst. Appl. **28**(1), 127–135 (2005)
30. Zhong, X., Enke, D.: Predicting the daily return direction of the stock market using hybrid machine learning algorithms. Financ. Innov. **5**(1), 1–20 (2019)
31. Kayakus, M., Tutcu, B., Terzioglu, M., TalasÌ§, H., Ünal Uyar, G.F.: ROA and ROE forecasting in iron and steel industry using machine learning techniques for sustainable profitability. Sustainability **15**(9), 7389 (2023)
32. Hosmer, Jr., D.W., Lemeshow, S., Sturdivant, R.X.: Applied Logistic Regression. Wiley (2013)
33. Breiman, L.: Classification and Regression Trees. Routledge (2017)
34. Breiman, L.: Random forests. Mach. Learn. **45**, 5–32 (2001)
35. Cortes, C., Vapnik, V.: Support-vector networks. Mach. Learn. **20**, 273–297 (1995)
36. Chiadamrong, N., Wattanawarangkoon, T.: Financial performance prediction model based on firms' internal capability determinants: evidence from listed firms in Thailand during the transition period of going public. Cogent Eng. **10**(1), 2216860 (2023)
37. Schölkopf, B.: Learning with Kernels: Support Vector Machines, Regularization, Optimization, and Beyond. MIT Press (2002)
38. Haykin, S.: Neural Networks: A Comprehensive Foundation. Prentice Hall PTR (1998)
39. LeCun, Y., Bengio, Y., Hinton, G.: Deep learning. Nature **521**(7553), 436–444 (2015)
40. Rumelhart, D.E., Hinton, G.E., Williams, R.J.: Learning representations by back-propagating errors. Nature **323**(6088), 533–536 (1986)
41. Bengio, Y., Grandvalet, Y.: No unbiased estimator of the variance of k-fold cross-validation. Adv. Neural Inf. Process. Syst. **16** (2003)
42. Kohavi, R.: A study of cross-validation and bootstrap for accuracy estimation and model selection. In: Proceedings of the 14th International Joint Conference on Artificial Intelligence (IJCAI), vol. 2, pp. 1137–1145 (1995)
43. Bergstra, J., Bengio, Y.: Random search for hyper-parameter optimization. J. Mach. Learn. Res. **13**(2), 281–305 (2012)
44. Powers, D.M.: Evaluation: From precision, recall and F-measure to ROC, informedness, markedness and correlation (2020). arXiv:2010.16061
45. Bradley, A.P.: The use of the area under the ROC curve in the evaluation of machine learning algorithms. Pattern Recognit. **30**(7), 1145–1159 (1997)
46. Mandrekar, J.N.: Receiver operating characteristic curve in diagnostic test assessment. J. Thorac. Oncol. **5**(9), 1315–1316 (2010)
47. Saito, T., Rehmsmeier, M.: The precision-recall plot is more informative than the ROC plot when evaluating binary classifiers on imbalanced datasets. PLoS ONE **10**(3), e0118432 (2015)
48. Davis, J., Goadrich, M.: The relationship between precision-recall and ROC curves. In: Proceedings of the 23rd International Conference on Machine Learning, pp. 233–240 (2006)
49. Chinchor, N., Sundheim, B.M.: MUC-5 evaluation metrics. In: Proceedings of the Fifth Message Understanding Conference (MUC-5), Baltimore, Maryland, 25–27 Aug 1993
50. McHugh, M.L.: Interrater reliability: the kappa statistic. Biochem. Med. **22**(3), 276–282 (2012)
51. Landis, J.R., Koch, G.G.: The measurement of observer agreement for categorical data. Biometrics **33**(1), 159–174 (1977)
52. Dormann, C.F., Elith, J., Bacher, S., Buchmann, C., Carl, G., Carré, G., Lautenbach, S.: Collinearity: a review of methods to deal with it and a simulation study evaluating their performance. Ecography **36**(1), 27–46 (2013)

53. Friedman, J.H.: Greedy function approximation: a gradient boosting machine. Ann. Stat. **29**(5), 1189–1232 (2001)
54. Caruana, R., Niculescu-Mizil, A., Crew, G., Ksikes, A.: Ensemble selection from libraries of models. In: Proceedings of the Twenty-First International Conference on Machine Learning, p. 18 (2004)
55. Lundberg, S.: A unified approach to interpreting model predictions (2017). arXiv:1705.07874
56. Fama, E.F., French, K.R.: Disappearing dividends: changing firm characteristics or lower propensity to pay? J. Financ. Econ. **60**(1), 3–43 (2001)
57. Brush, T.H., Bromiley, P., Hendrickx, M.: The free cash flow hypothesis for sales growth and firm performance. Strateg. Manag. J. **21**(4), 455–472 (2000)

Enhancing Corporate Bankruptcy Prediction with Machine Learning and Textual Analysis

Ba-Hung Nguyen, Hien Thu Thi Nguyen, and Van-Nam Huynh

Abstract Predicting corporate bankruptcy remains a core challenge in financial risk management. Traditional models, heavily reliant on static financial ratios, often fail to capture the nuanced and forward-looking signals embedded in unstructured corporate disclosures. This chapter proposes an enhanced approach that integrates machine learning techniques with textual analysis to improve both the accuracy and interpretability of bankruptcy prediction. We leverage firm-level financial reports—specifically the Management Discussion & Analysis (MD&A) sections—to extract meaningful textual features that complement numerical data. Using topic modelling techniques such as Latent Dirichlet Allocation and Dynamic Embedded Topic Models, we identify latent thematic structures that reflect managerial concerns over liquidity, innovation, regulation, and macroeconomic uncertainty. These topical features evolve in sync with major economic events, providing early warning signals of financial distress. When combined with traditional z-score indicators, these features significantly improve classification performance, especially in predicting rare default events. In parallel, we develop a sentiment-based classifier using a domain-specific financial dictionary to quantify tone at the word level. These sentiment scores are shown to enhance one-year-ahead predictive accuracy and offer more intuitive explanations of firm risk than deep learning approaches. Our experiments demonstrate that textual features—both topical and sentiment-based—provide predictive power comparable to or exceeding that of numerical indicators alone. By integrating

B.-H. Nguyen (✉) · H. T. T. Nguyen · V.-N. Huynh
School of Finance and Accounting, Industrial University of Ho Chi Minh City, Ho Chi Minh City, Vietnam
e-mail: nguyenbahung@iuh.edu.vn

H. T. T. Nguyen
e-mail: nguyenthuhien@iuh.edu.vn

V.-N. Huynh
School of Knowledge Science, Japan Advanced Institute of Science and Technology, Nomi, Ishikawa, Japan
e-mail: huynh@jaist.ac.jp

H. T. T. Nguyen et al. (eds.), *Data Science in Finance and Accounting*, Studies in Big Data 181, https://doi.org/10.1007/978-3-032-06179-9_7

textual analysis into machine learning workflows, this research not only improves bankruptcy prediction outcomes but also enriches our understanding of corporate behaviour under stress. The findings highlight the potential of unstructured data in credit modelling and suggest promising directions for future work, including multi-modal learning and fairness-aware financial AI.

Keywords Credit risk · Machine learning · Textual data · Topic modelling · Sentiment analysis

1 Introduction

Credit scores quantify default risk, yet most models rely on *static* financial ratios and ignore recent trends and real-time signals. Recent surveys [10, 42, 44] urge the inclusion of time-series and alternative (e.g. social-media) data. Text mining and social-media analytics now let us add forward-looking features, improving both cost and accuracy.

Following [23], credit-risk methods fall into two groups:

1. **Statistical models**. Discriminant analysis and logistic regression (LR) [2, 43] are simple and interpretable, but variable choice remains ad hoc and normality assumptions often fail [10].
2. **AI models**. Early work showed artificial neural networks (ANN) outperforming statistics [26]. Support-vector machines (SVM) [44], LR+ANN hybrids [41], and improved multilayer perceptrons (MLP) for imbalanced data [45] boost accuracy, but their opacity limits adoption [40].

Ensemble methods mitigate single-model limits. New hybrids surpass LR, ANN, and SVM [21]; early-warning systems thus benefit from NLP applied to conference calls, MD&A, and annual reports [40].

Despite progress, key gaps persist for financial *text* data. Deep learning has improved bankruptcy prediction [35], yet we still need to: (i) uncover latent topics, (ii) test word-level signals, and (iii) compare dictionary versus deep representations.

In this work, we summarize how we:

1. apply topic models to reveal hidden themes in filings and link them to default risk;
2. build a domain-specific sentiment classifier for stronger bankruptcy signals;
3. benchmark dictionary-based and deep representations to quantify their gap.

This work is primarily based on the findings and methodologies developed in earlier works [6, 38], which examined how financial narrative disclosures, particularly sentiment in MD&A and topic modelling from textual data, can enhance bankruptcy prediction. Specifically, this research focuses on the experiment with textual data and emphasizes how these new sources of data could improve both understanding of the nature of risk and predictability of the classifier. Specifically, Sect. 2 explores the topics in financial filings and discusses how their evolution is in line with crucial economic events and how we improve bankruptcy prediction by using topical textual features. Section 3 focuses on sentiment at the word level based on a domain-specific wordlist; In this chapter, we propose a financial dictionary-based sentiment classifier to construct new sentiment-based features for the bankruptcy prediction task. Section 4 concludes and discusses our current limitations and further research directions.

2 Textual Analysis in Accounting and Finance

Textual analysis has emerged as a powerful tool in accounting and finance research, facilitating the extraction of sentiment, tone, and latent topics from narrative disclosures. Recent studies have demonstrated that linguistic content in corporate documents can explain variation in financial outcomes such as risk, returns, and default probability.

Loughran and McDonald [34] reviewed recent developments in financial textual analysis, highlighting the importance of domain-specific dictionaries tailored to financial narratives. They showed that financial sentiment tools outperform general-purpose ones in capturing economically meaningful tone.

Huang et al. [20] applied topic modeling to analyst reports and found that optimistic language in these texts relates to stronger market responses and more accurate forecasts, underscoring analysts' role in information dissemination.

Lopez-Lira and Tang [32] examined risk disclosure tone in annual filings, revealing that negative tone correlates with increased stock volatility and lower returns. They quantified sentiment using sentence-level tagging techniques.

Jiang et al. [25] studied managerial narrative consistency, showing that firms repeating similar MD&A content across years were more likely to exhibit earnings surprises. This suggests that boilerplate language reflects greater underlying uncertainty.

Lawrence et al. [27] investigated forward-looking disclosures and demonstrated that positive sentiment in future-oriented language predicts improved firm performance. Their findings highlight the predictive value embedded in strategic narrative content.

These contributions establish the importance of linguistic signals in financial texts. Building on this literature, our study investigates how sentiment-based textual features from 10-K filings can improve the prediction of corporate default and complement traditional financial ratios. While these studies provide a broad foundation, our analysis extends this line of research through focused applications of financial sentiment and topic modeling for corporate default prediction, summarized from our prior works [6, 38].

3 Topic Modelling for Credit Risk

Public US firms submit 10-K and 10-Q filings to the SEC via EDGAR, including Management Discussion & Analysis (MDA) sections[1] where management outlines risks, performance, and future plans. These forward-looking texts offer signals that complement traditional ratios, which may suffer from window dressing [16, 18].

Most financial text analysis focuses on sentiment, analyst reports, or forecasting [20, 31, 39, 46]. We extend this by uncovering latent topics in MDAs using LDA and DETM [9, 14], and show that these features enhance bankruptcy prediction beyond z-scores when combined.

Using 20 years of 10-K filings, we extract 30 interpretable topics and analyze their evolution. Results show MDA topics track market changes and boost prediction performance over financial and NLP baselines.

3.1 Backgrounds

LDA [9] models latent topics via variational inference. DTM [8] adds temporal dynamics. ETM [13] embeds word semantics into topic discovery, handling noisy and rare terms. DETM [14] merges both, learning topic evolution through random walk priors and amortized RNN inference.

Bao and Datta [7] adapted LDA for sentence-level topics, identifying 30 risk types predictive of investor behavior outperforming supervised models [22]. Nguyen et al. [39] fused topics and sentiment to predict stock moves using social media.

Sentiment also links to firm outcomes: higher tone correlates with growth [25, 46]; negative sentiment predicts liquidation [16]; deep learning improves accuracy [35] but lacks interpretability. We address these gaps using DETM to model interpretable MDA topics for bankruptcy prediction.

3.2 Data and Methods

3.2.1 Accounting Data

We collect 1997–2017 US firm data from WRDS (via University of Edinburgh), excluding finance and utilities (SIC 6000–6999, 4900–4949). Bankruptcy is flagged by Chap. 7 or 11 filings.[2] SMEs are defined as firms with sales $\leq$ \$65 million [4].

[1] https://www.sec.gov/corpfin/cf-manual/topic-9.

[2] https://www.sec.gov/reportspubs/investor-publications/investorpubsbankrupt.html.

Table 1 MDAs extracted from the filings

Year	#MDAs	#NF	#unc	#omitted	#Sent.	Mean	Median	#tokens
1997	8711	422	7	561	740208	85	64	47328
1998	8931	454	4	563	862791	96.6	73	51372
1999	8934	391	2	611	1044325	116.9	89	54665
2000	9508	458	8	580	1065612	112.1	79	57295
2001	9447	424	1	510	1131682	119.8	83	59454
2002	10179	434	2	806	1488084	146.2	89	64979
2003	11878	419	6	1236	2107816	177.5	117	74079
2004	12124	390	4	1496	2274859	187.6	115	76837
2005	12475	635	2	2016	2501571	200.5	120	81210
2006	12251	678	5	1863	2472559	201.8	135	81163
2007	12087	485	5	1840	2526728	209	145	83147
2008	11432	443	6	1470	2519519	220.4	156	83350
2009	9919	366	4	769	2525077	254.6	189	79620
2010	9165	190	3	676	2405854	262.5	199	78967
2011	8840	162	2	659	2290261	259.1	193	78147
2012	8393	175	1	693	2214756	263.9	195	75322
2013	8105	186	1	677	2183919	269.5	203	74772
2014	8084	184	1	751	2193408	271.3	202	76434
2015	7985	182	1	912	2181273	273.2	204	76066
2016	7589	158	2	1081	2077774	273.8	201	74669
2017	7248	184	1	1113	1931412	266.5	192	71713
Total	203,285	7,420	68	20,833	40,739,488			

#MDAs: number of filings with MDA; #NF: without MDA; #unc: untraceable MDAs; #omitted: omitted MDAs; #Sent.: total sentences; mean/median: sentence count; #tokens: total unique words

Table 2 Number of liquidations in two corporate segments

	non-SME	SME	All
Total	181	134	315

3.2.2 10-K Filings

10-K and 10-KSB filings are retrieved from EDGAR.[3] We remove tables, figures, PDFs, and extract only MDA sections. 10-K405/10-KT filings are excluded due to timing or format shifts. Table 1 summarizes MDA data extracted across years.

In the final stage, we merge financial data with SEC filings by aligning CIK identifiers and fiscal year-ends. The resulting dataset includes 51,128 firm-year observations, with 213 liquidations (approx. 0.42%). The breakdown of liquidations by business segment is as follows (Table 2).

[3] https://www.sec.gov/edgar/aboutedgar.htm.

3.3 Topic Modellings

LDA [9] is a generative probabilistic model that represents documents as mixtures of latent topics, where each topic is a distribution over terms. Blei and Lafferty [8] later extended this to dynamic topic models (DTM), allowing topic distributions to evolve over time using state-space models on the natural parameters of multinomials. Topic inference is performed via variational methods using Kalman filters or non-parametric wavelet regression.

DTM improves on LDA by incorporating document order. While words remain exchangeable within documents, DTM respects the sequence of documents across time slices (e.g., quarterly, yearly), modeling topics in each slice as evolving from the previous one.

We apply online variational inference [19] on bigram tokens to estimate the optimal number of topics, then use dynamic embedding topic modelling to track topic evolution over time. The next section introduces key ideas behind dynamic topic modelling in embedding spaces.

3.4 Bankruptcy Prediction Feature Sets

We evaluate the predictive value of hidden MDA topics by comparing them with standard and baseline features. The feature sets include:

- S1: Altman z-score—5 factors for general corporates [1], and 5 for SMEs [5]
- S2: Dictionary-based sentiment count (normalized by MDA length) using Loughran and McDonald lexicon [33]
- S3: *doc2vec* representations of MDA [28]
- S4: Topic vectors using LDA
- S5: Topic vectors using DETM

We define the S1 and S2 features as follows (Table 4).

The construction of textual features, experimental setup, and comparative analysis with traditional financial ratios presented in this study follow the methodology outlined in our original research [6, 38].

3.5 Experimental Setup

We remove words appearing in over 70% of documents, standard stop words, and low-frequency words (fewer than 10 documents). To select the optimal number of topics, we run 10 epochs of online LDA [19], varying topic counts from 10 to 100 and measuring perplexity on a held-out set. We then apply LDA and DETM to the full dataset (1997–2017) to extract and track topic dynamics.

Table 3 Altman's 5-factor

Notation	Formula
Panel I	**z-score 5-factor** [2]
Z1	Working Capital/Total Assets
Z2	Retained Earnings/Total Assets
Z3	EBIT/Total Assets
Z4	Market Value of Equity/Total Liabilities
Z5	Sales/Total Assets
Panel II	**SME 5-factor** [5]
A1	Cash Flow from Operating Activities/Current Liabilities
A2	Short Term Debt/Equity Book Value
A3	Cash/Total Assets
A4	EBIT/Interest Expenses
A5	Account Receivable/Liabilities

Table 4 Loughran and Mcdonald [33] wordlists

Wordlist	Sample words
Positive	Enthusiastically, assures, improve, empower, complimented
Negative	Investigating, complaining, confusion, severe, exculpations
Uncertainty	Risking, anticipated, hidden, unobservable, imprecision
Litigious	Prosecute, referenda, presumptively, licensable, sequestrator
Modal Strong	Highest, strongly, will, unequivocally, lowest

Table 5 DETM settings

Component	Setting
Word2Vec [37]	Skip-gram, 300 dimensions
Batch size	200
Activation	ReLU
Number of hidden layers	2 (800 nodes each)
Drop-out	0.1
Epochs	100

For bankruptcy prediction, we use an out-of-sample, out-of-time strategy. A 10-year rolling window trains logistic regression with l_2 regularization; the next year serves as the test set. We repeat this process annually and report both yearly and average results.

We configure DETM following [14] with the parameters below (Table 5).

For final prediction, we use logistic regression with l_2 regularization. We set inverse strength λ between 0.001 and 0.01; smaller λ applies stronger penalties.

Logistic regression remains a widely used model in bankruptcy prediction due to its simplicity and interpretability, with performance comparable to advanced classifiers [3, 29].

3.6 Experimental Results

3.6.1 Number of Topics

Figure 1 shows perplexity bounds for topic numbers from 5 to 100. We train LDA models [19] using asymmetric α (learned from the corpus), 10 epochs, and a 100-iteration limit. Perplexity drops sharply at $k = 30$, indicating that 30 is a suitable topic count. We use this value for both full-topic exploration and moving-window-based prediction models.

3.6.2 Topics in MDA and Their Evolution

MDA sections reflect not only financial performance but also emphasize new products, partnerships, strategic plans, and M&A activities. We use this large-scale dataset to build a dynamic topic representation over time using unsupervised learning. Figure 2 presents word clouds for selected LDA topics (word size reflects word probability in the topic):

The topics include partnerships, R&D, energy prices, investment loss, retail activity, and taxes. DETM tracks how these topics change over time using embeddings and smooth transitions. Figure 3 shows topic-word dynamics via $\beta = \text{softmax}(\rho^\top \alpha)$.

Key observations:

- Topic *investment loss* peaks in crisis periods with words like 'recession', 'turmoil', and 'unemployment'.

Fig. 1 Bound of perplexity on the test set

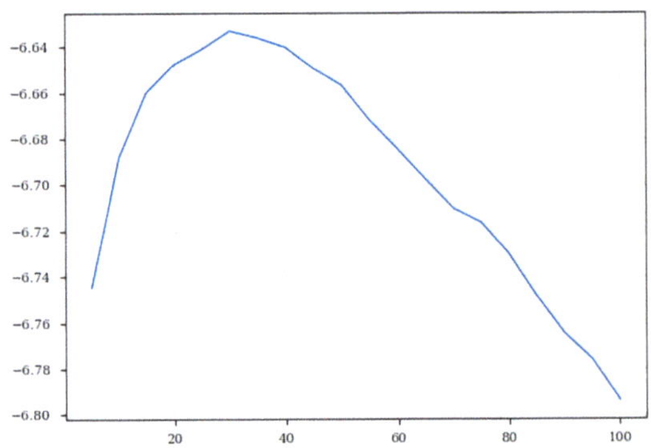

Fig. 2 Topics discovered by LDA

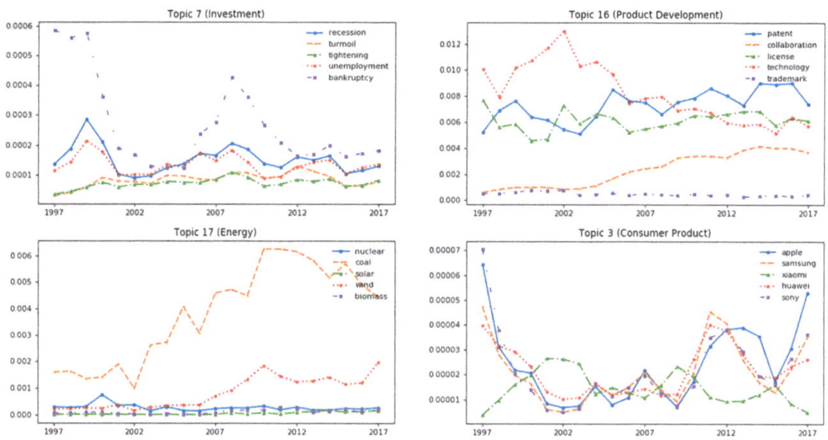

Fig. 3 Evolution of topics

- Topic *product development* increases with terms like 'collaboration', 'patent', and 'license'.
- Topic *energy* shifts from 'coal' to 'wind' and 'solar' following the 2015 coal production decline[4]
- Topic *consumer product* reveals shifts among major electronics manufacturers.

We next evaluate how these topic features improve bankruptcy prediction.

3.6.3 Predictive Performance

We assess model predictions using the following metrics:

- **Precision:** $P = Pr(Y = 1|\hat{Y} = 1)$
- **Recall:** $R = Pr(\hat{Y} = 1|Y = 1)$
- **AUC:** $\text{AUC} = p\big(score(x^+) > score(x^-)\big)$
- **AUPRC:** $\text{AUPRC} = \sum_n (R_n - R_{n-1})P_n$
- **Brier Score:** $BS = \frac{1}{N}\sum_{t=1}^{N}(predict_t - o_t)^2$

We compare LDA and DETM topic features with three baselines: Altman's z-score [3], sentiment word count [33], and doc2vec [28]. AUPRC is favored for imbalanced data [11, 12, 17]. AP equals the positive class rate under random prediction.

From 2007–2016, we observe:

- **Brier Score:** No clear winner due to class imbalance.
- **AUC & AUPRC:** LDA and DETM outperform dictionary-based and doc2vec features, which lack topic structure.
- **Topic versus Z-score:** Topic features reach AUC of 83% versus 84%, and AUPRC of 0.097 versus 0.103, without using numerical or accounting data.

These trends suggest that combining topic and financial features improves predictions. Table 6 shows results for combined models.

DETM improves AUC by 1.55%, while LDA achieves a 3.89% gain. However, BS and AUPRC decline in both models, especially in 2015–2016 where z-score alone achieves top AUC (99%, 98%). The 20-year window may miss full economic cycles, limiting topic signal differentiation. Expanding the time span could provide stronger topic-based predictors in future work.

3.7 *Discussions*

Our approach analyzes MDA from over 189k filings containing 38 million words and produces strong results. We identify 30 latent topics in MDA that reflect key business

[4] U.S. coal production dropped by over 10% in 2015 to 897 million short tons, the lowest since 1986. Source: US Energy Information Administration, https://www.eia.gov/todayinenergy/detail.php?id=28732.

Table 6 Comparison of z-score and combined feature set using LDA and DETM

z-score				LDA + z-score			DETM + z-score		
	AUC	AUPRC	BS	AUC	AUPRC	BS	AUC	AUPRC	BS
2007	0.8217	0.0110	0.0028	0.8231	0.0250	0.1501	0.8765	0.0274	0.0028
2008	0.7946	0.0525	0.0072	0.8744	0.0936	0.1644	0.8651	0.0785	0.0071
2009	0.8179	0.0023	0.0005	0.8412	0.0027	0.1666	0.7241	0.0016	0.0005
2010	0.9268	0.0459	0.0013	0.9710	0.1298	0.1363	0.9715	0.1485	0.0013
2011	0.7850	0.0756	0.0025	0.8165	0.0947	0.1315	0.7835	0.0650	0.0025
2012	0.8271	0.0036	0.0009	0.9242	0.0145	0.1544	0.9483	0.0459	0.0008
2013	0.7039	0.0203	0.0013	0.8075	0.0061	0.1510	0.7033	0.0047	0.0013
2014	0.8299	0.0098	0.0027	0.8609	0.0283	0.1544	0.8322	0.0102	0.0027
2015	0.9887	0.3044	0.0029	0.9599	0.1429	0.1630	0.9586	0.2456	0.0028
2016	0.9756	0.5098	0.0010	0.9717	0.1088	0.1464	0.9730	0.0382	0.0010
Mean	0.8471	0.1035	0.0023	0.8850	0.0646	0.1518	0.8636	0.0666	0.0023

dimensions such as ‘energy’, ‘partnership’, ‘research and development’, and ‘loan and interest rate’. Topic-word trends align with major economic events, including the 2008 financial crisis, the sharp decline in coal output in 2015, and market shifts among electronics producers.

The rise of window dressing practices across sectors [16, 18] increases the demand for alternative data to complement traditional credit scoring. Forward-looking textual data like MDA addresses this gap. We demonstrate that textual features from MDA support out-of-time and out-of-sample bankruptcy prediction. These features deliver performance comparable to z-score benchmarks. Furthermore, combining textual and z-score features enhances prediction accuracy.

4 Sentiment Analysis of Textual Data

This section proposes new word-level textual features for default prediction and evaluates their predictive power. We detect sentiment words in MDA sentences using a dictionary-based financial sentiment classifier and average the sentiment per firm-year. Results show that combining these features with traditional accounting models improves predictability.

Zhou [46], Jiang et al. [25] find that higher managerial sentiment correlates with lower earnings disruption and stronger investment growth. In default prediction, [15] link negative sentiment in bank reports with delisting risk, while [35] use CNN with word embeddings to improve forecast accuracy. However, their work lacks coverage on non-bank sectors, interpretability, and out-of-sample evaluation. We address this by combining simple dictionary-based sentiment with traditional models.

Table 7 The [33] wordlists

Wordlist	Sample words
Positive	Enthusiastically, assures, improve, empower, complimented
Negative	Investigating, complaining, confusion, severe, exculpations
Uncertainty	Risking, anticipated, hidden, unobservable, imprecision
Litigious	Prosecute, referenda, presumptively, licensable, sequestrator
Modal strong	Highest, strongly, will, unequivocally, lowest
Modal weak	Occasionally, perhaps, somewhat, suggest, nearly

We use 10-K filings from 1997–2017 from EDGAR. The goal is to evaluate how sentiment in MDAs predicts default. We segment firms, construct sentiment features, and assess models using stratified k-fold validation and one-year-ahead predictions.

4.1 Experimental Settings

The construction of textual features, experimental setup, and comparative analysis with traditional financial ratios presented in this study follow the methodology outlined in our original research [6, 38]. First, we use logistic regression for baseline, text-only, and hybrid models due to its interpretability and popularity in credit scoring [30]. Altman z-score features [1, 5] serve as the baseline. Textual features come from word count-based vectorization using financial sentiment dictionaries. We compare six performance metrics across all firms, SMEs (sales $\leq$ 65M), and non-SMEs.

4.1.1 Financial Ratios

We follow [1, 5] for calculating five financial ratios for general and SME models. See Table 3 for definitions.

4.1.2 Financial Wordlist

Loughran and Mcdonald [33] identify that traditional sentiment dictionaries mislabel financial terms (e.g., depreciation, liability). They introduce domain-specific lists for positive, negative, uncertainty, litigious, strong modal, and weak modal terms. See Table 7 for examples.

Table 8 Performance of dictionary-based financial classifier

Accuracy	0.7380			
	Negative Sent.	Positive Sent.	Neutral Sent.	Total
Precision	0.5779	0.7913	0.7289	
Recall	0.8161	0.7952	0.5546	
F1-score	0.6759	0.7930	0.6284	
Support	303	570	1391	2264

4.2 Dictionary-Based Sentiment Classifier

We define a rule-based sentiment classifier based on [24, 33]. It uses:

1. Majority voting on sentiment words per sentence.
2. Polarity shift with contrastive conjunctions (e.g., 'however', 'but').
3. Valence adjustment using n-gram degree adverbs.

The classifier applies a lexicon and weights based on word context. The full algorithm is shown in Algorithm 1.

Algorithm 1 Dictionary-based financial sentiment analysis

Input: $s = \{w_1, w_2, .., w_m\}$, a financial sentence of m words.
Parameter: n, n-gram preceding a lexical feature.
Output: $score$, a weighted composite sentiment score.

```
initiate sentiment list sentiment ← [ ]
for each word w_m ∈ s do
  set val(w_m) ← 0
  if w_m ∈ dict_lex then
    val(w_m) ← dict_lex[w_m]
  end if
  for i ∈ [1, n] do
    if w_{m-i} ∈ dict_da then
      modify val(w_m)
    end if
  end for
  add val(w_m) to sentiment
end for
if a word w_m ∈ list_cc then
  update sentiment
end if
compute final score from sentiment
return score
```

We evaluate the classifier on 2264 manually annotated financial sentences [36]. A threshold of ±0.05 determines positive, negative, or neutral polarity. Table 8 shows results.

Table 9 Error in negative sentence prediction

ID	Negative sentence	Predict
29	Profitability (EBIT%) was 13.6%, compared to 14.3% in Q2 2009	1
31	Shares in Royal and Sun Alliance continued to slide back from a 12-month high...	1
38	Return on investment ROI was 4.1% compared to 43.8% in the first half of 2008	0
45	Diluted earnings per share (EPS) stood at EUR 0.25 versus EUR 0.42	0
69	Oyj posted a net profit of 7.99 m EUR for 2006, compared to 8.4 m EUR for 2005	1
75	Operating profit was EUR 1.6 mn in 2005 compared to EUR 5.9 mn in 2004	0
141	Uponor OYJ cut its full-year sales growth forecast to 6 pct from 10 pct ...	1
277	In the building and home improvement trade, sales decreased by 22.5%	1

Table 10 Textual features

CIK	Filing ID	pneg	ppos	...	PCNEG	PCPOS	...
151629	6	0.011	0.025	...	16	23	...
708788	17	0.002	0.067	...	54	71	...

CIK is the Central Index Key assigned by the SEC

Table 9 illustrates common misclassifications. Most errors stem from numerical changes, missing context, or misclassified lexicon (Table 10).

4.2.1 Textual Features

We extract three sets of textual features:

1. **Entire filing:** Word percentages - *pneg*, *ppos*, *punc*, *plit*, *pmods*, *pmodw*; and composite score *senti* = *ppos* - *pneg*.
2. **MDA section:** Word counts - *NEG*, *POS*, *UNC*, *LIT*, *M1*, *M3*.
3. **MDA section:** Sentence percentages - *PCNEG*, *PCPOS*, *PCNEU*.

4.2.2 Predictive Measurement

To evaluate default prediction, we use AUC, AUPRC, precision, recall, and Brier score. AUPRC is recommended for imbalanced datasets [12]. Table 11 shows the distribution of defaults across firm segments.

Table 11 Number of defaults in two data segments

Default	Business type	Count	Percentage (%)
False	non-SME	28980	99.612
	SME	21972	99.714
True	non-SME	113	0.388
	SME	63	0.286

4.3 Results

We compare regularized logistic regression models using (i) financial features, (ii) textual features, and (iii) both. Tables 12 and 13 report model estimates. Table 14 shows predictive performance under stratified 10-fold cross-validation and one-year-ahead forecasting.

4.3.1 Regression

Tables 12 and 13 summarize logistic regression results for SMEs, non-SMEs, and the full sample. Each model includes controls: employee count (EMP), size (LNATSQ), profitability (ROA, ROE), and industry sector (SIC). Financial variables follow [1, 5]. Combined models include both sets.

Financial variables perform consistently with existing studies [5]. Textual variables improve model fit. In SMEs, uncertainty (punc) and sentiment difference (senti) reduce default risk. More positive MDA sentences further lower default odds.

In non-SMEs, PCNEG shows negative association with default, while senti remains significant. In the full sample, PCNEG positively correlates with default risk. PCPOS and senti reduce default probability. Results support findings in [15].

Textual data offer meaningful signals and enhance model performance when combined with financial indicators. Next, we compare predictive accuracy.

4.3.2 Predictive Power Comparison

We compare combined models (financial + textual features) against models using only Altman factors under three data balancing strategies: no sampling, undersampling, and oversampling. We apply 10-fold stratified cross-validation with grid-search to select logistic regression regularization parameters, separating evaluations for SME, non SME, and full samples.

To address classifier bias in imbalanced datasets (IDS) where accuracy can mask poor performance on minority classes [10], we include undersampling and

Table 12 Regression models

	SME			non-SME		
	Altman	Text	Combined	Altman	Text	Combined
Intercept	−3.703***	−3.122**	−2.411	−3.133**	−3.465***	−2.285
	(1.034)	(1.359)	(1.467)	(1.527)	(1.334)	(1.403)
A1	1.410*		1.280*			
	(0.729)		(0.754)			
A2	−0.927		−0.894			
	(0.616)		(0.605)			
A3	−1.187**		−0.695			
	(0.477)		(0.486)			
A4	−0.558		−0.605			
	(0.505)		(0.524)			
A5	−0.122		−0.175			
	(0.581)		(0.597)			
Z1				−3.810***		−3.407***
				(0.564)		(0.572)
Z2				−0.183		−0.003
				(0.561)		(0.563)
Z3				−1.353**		−1.226*
				(0.644)		(0.642)
Z4				−0.345		−0.567
				(0.362)		(0.369)
Z5				0.664		0.537
				(0.426)		(0.435)
PCNEG		0.658	0.623		0.935**	0.827*
		(0.552)	(0.554)		(0.435)	(0.440)
PCPOS		−1.209**	−1.158**		−0.408	−0.283
		(0.532)	(0.535)		(0.423)	(0.431)
plit		−0.637	−0.471		−0.024	0.179
		(0.603)	(0.610)		(0.544)	(0.554)
pmods		0.315	0.341		0.556	0.473
		(0.502)	(0.506)		(0.375)	(0.380)
ppos		−0.307	−0.308		−0.052	0.192
		(0.643)	(0.644)		(0.439)	(0.448)
punc		−2.412***	−2.271***		−2.502***	−2.059***
		(0.634)	(0.640)		(0.569)	(0.580)
senti		−2.152***	−2.006***		−0.584	−0.612
		(0.663)	(0.662)		(0.495)	(0.506)
SENTI_MDA		−0.408	−0.418		0.048	0.111
		(0.589)	(0.585)		(0.488)	(0.492)

(continued)

Table 12 (continued)

	SME			non-SME		
	Altman	Text	Combined	Altman	Text	Combined
EMP	−0.437	−0.605	−0.559	0.775	0.571	0.263
	(0.575)	(0.560)	(0.575)	(0.608)	(0.564)	(0.643)
LNATSQ	1.381**	1.818***	1.860***	1.785***	2.078***	2.187***
	(0.611)	(0.676)	(0.674)	(0.669)	(0.585)	(0.705)
ROA	−2.080**	−1.233**	−1.593*	−8.056***	−8.713***	−7.351***
	(0.845)	(0.596)	(0.845)	(1.024)	(0.845)	(1.053)
ROE	−0.484	0.376	−0.510	0.770***	1.220***	0.686***
	(0.590)	(0.434)	(0.589)	(0.264)	(0.248)	(0.266)
SIC_1	−0.998	−0.661	−0.628	−0.705	−0.697	−1.252
	(0.877)	(1.040)	(1.054)	(1.518)	(1.102)	(1.140)
SIC_2	0.195	0.281	0.448	−2.217	−2.376*	−2.922*
	(1.091)	(1.226)	(1.223)	(1.863)	(1.436)	(1.497)
SIC_3	−0.976	−0.442	−0.310	−0.461	−1.246	−1.265
	(0.823)	(1.002)	(1.020)	(1.480)	(1.056)	(1.087)
SIC_4	0.148	0.519	0.546	−0.670	−1.002	−1.532
	(0.861)	(1.030)	(1.044)	(1.492)	(1.072)	(1.103)
N	22035	22035	22035	29093	29093	29093
AIC	857.4562	824.7012	826.5098	1092.4631	1118.9101	1073.2403
BIC	969.4616	960.7078	1002.5183	1208.3587	1259.6404	1255.3619
LLR	34.3774	73.1323	81.3238	415.5929	395.1459	450.8157
R-squared	0.0398	0.0847	0.0941	0.2808	0.2670	0.3046

N is the number of observations and LLR is the log-rank test statistics. ***, **, and * indicate 99, 95 and 90% significant levels, respectively. EMP is the number of employees (thousand). LNATSQ is the natural logarithm of total assets squared. ROA and ROE are the return on assets and return on equity. SIC_i is the ith SIC category with SIC_0 as the reference level

oversampling during modelling. For practical relevance, Basel III validation compliance,[5] and alignment with scorecard design, we also report one-year-ahead prediction results using a five-year rolling window.

4.3.3 Out-of-Sample Cross-Validation Prediction Performance

Figures 4 to 6 present performance distributions using six metrics defined in Sect. 3.5. Each figure includes three rows: original data (top), undersampling (middle), and oversampling (bottom). Segment-specific results are shown for SMEs (Fig. 4), non-SMEs (Fig. 5), and full sample (Fig. 6). Table 14 lists average 10-fold scores for all models and metrics.

[5] The IRB Use Test: Background and Implementation, Basel Committee on Banking Supervision, https://www.bis.org/publ/bcbs_nl9.pdf, accessed September 25th, 2019.

Table 13 Regression models (cont.)

	SME			All sample		
	Altman	Text	Combined	Altman	Text	Combined
Intercept	−3.703***	−3.122**	−2.411	−3.004***	−3.577***	−2.125**
	(1.034)	(1.359)	(1.467)	(0.827)	(0.945)	(1.030)
A1	1.410*		1.280*	0.907		1.123
	(0.729)		(0.754)	(0.707)		(0.714)
A2	−0.927		−0.894	0.131		0.182
	(0.616)		(0.605)	(0.344)		(0.342)
A3	−1.187**		−0.695	−1.680***		−1.531***
	(0.477)		(0.486)	(0.342)		(0.348)
A4	−0.558		−0.605	−2.451***		−2.429***
	(0.505)		(0.524)	(0.661)		(0.681)
A5	−0.122		−0.175	−1.080**		−1.247**
	(0.581)		(0.597)	(0.486)		(0.504)
Z1				−1.003**		−0.486
				(0.439)		(0.438)
Z2				−0.261		0.025
				(0.535)		(0.540)
Z3				−0.932		−0.928
				(0.722)		(0.718)
Z4				−0.650**		−0.776***
				(0.293)		(0.300)
Z5				1.148***		0.880***
				(0.329)		(0.339)
PCNEG		0.658	0.623		1.061***	0.920***
		(0.552)	(0.554)		(0.332)	(0.336)
PCPOS		−1.209**	−1.158**		−0.752**	−0.765**
		(0.532)	(0.535)		(0.329)	(0.331)
plit		−0.637	−0.471		−0.263	−0.120
		(0.603)	(0.610)		(0.393)	(0.400)
pmods		0.315	0.341		0.452	0.525*
		(0.502)	(0.506)		(0.314)	(0.318)
ppos		−0.307	−0.308		−0.287	0.071
		(0.643)	(0.644)		(0.362)	(0.368)
punc		−2.412***	−2.271***		−2.689***	−2.440***
		(0.634)	(0.640)		(0.411)	(0.418)
senti		−2.152***	−2.006***		−1.238***	−1.327***
		(0.663)	(0.662)		(0.396)	(0.401)
SENTI_MDA		−0.408	−0.418		−0.186	−0.226
		(0.589)	(0.585)		(0.372)	(0.373)

(continued)

Table 13 (continued)

	SME			All sample		
	Altman	Text	Combined	Altman	Text	Combined
EMP	−0.437	−0.605	−0.559	0.081	−0.181	−0.302
	(0.575)	(0.560)	(0.575)	(0.492)	(0.499)	(0.520)
LNATSQ	1.381**	1.818***	1.860***	3.835***	3.983***	3.914***
	(0.611)	(0.676)	(0.674)	(0.555)	(0.545)	(0.599)
ROA	−2.080**	−1.233**	−1.593*	−5.101***	−5.562***	−4.461***
	(0.845)	(0.596)	(0.845)	(0.987)	(0.523)	(0.979)
ROE	−0.484	0.376	−0.510	0.296	0.682***	0.326
	(0.590)	(0.434)	(0.589)	(0.340)	(0.226)	(0.342)
SIC_1	−0.998	−0.661	−0.628	−0.982	−0.735	−0.818
	(0.877)	(1.040)	(1.054)	(0.742)	(0.767)	(0.777)
SIC_2	0.195	0.281	0.448	−0.963	−1.412	−1.216
	(1.091)	(1.226)	(1.223)	(0.923)	(0.946)	(0.946)
SIC_3	−0.976	−0.442	−0.310	−0.752	−0.923	−0.720
	(0.823)	(1.002)	(1.020)	(0.708)	(0.743)	(0.748)
SIC_4	0.148	0.519	0.546	−0.424	−0.448	−0.488
	(0.861)	(1.030)	(1.044)	(0.725)	(0.756)	(0.764)
N	22035	22035	22035	51128	51128	51128
AIC	857.4562	824.7012	826.5098	2008.0285	1990.2131	1930.4706
BIC	969.4616	960.7078	1002.5183	2176.0281	2140.5285	2169.2069
LLR	34.3774	73.1323	81.3238	377.7695	391.5849	471.3273
R-squared	0.0398	0.0847	0.0941	0.1609	0.1668	0.2008

N is the number of observations and LLR is the log-rank test statistics. ***, **, and * indicate 99, 95 and 90% significant levels, respectively. EMP is the number of employees (thousand). LNATSQ is the natural logarithm of total assets squared. ROA and ROE are return on assets and return on equity. SIC_i is the ith SIC category with SIC_0 as the reference level

Combined models consistently outperform Altman models in AUC across all segments and sampling strategies. Without sampling, both models show weak recall, precision, and F1 scores due to class imbalance.

For SMEs, no sampling leads to no clear gain in recall, precision, or F1, but AUC improves. With balancing, recall and AUC improve, especially under the combined model. Brier scores also show reductions.

For non-SMEs, improvements are similar. Combined models increase recall, F1, and AUC, particularly with sampling. Altman-only models remain less robust in all imbalanced settings.

Across the full sample, AUC, F1, AUPRC, and Brier scores improve when adding textual features. These gains suggest that textual features offer incremental predictive power beyond traditional financial ratios.

We summarize model performance and test for significant improvements when including textual features. Table 14 reports results for both models under three

Table 14 Predictive power comparison

	Altman			Combine		
	N	U	O	N	U	O
Panel I: SME						
AUC	0.6278	0.6075	0.6397	0.7295***	0.7227***	0.7238***
F1	_	0.3744	0.4085	_	0.4080***	0.4280***
R	0.0000	0.5792	0.5822	0.0000	0.6762***	0.6152*
P	0.0000	0.0044	0.0055	0.0000	0.0059***	0.0066***
AUPRC	0.0149	0.0173	0.0138	0.0172	0.0214***	0.0117***
BS	−0.0029	−0.2394	−0.2151	−0.0028***	−0.2159***	−0.1821***
Panel II: non-SME						
AUC	0.9108	0.9046	0.9086	0.9206***	0.9115***	0.9128***
F1	_	0.4598	0.4684	_	0.4668***	0.4780***
R	0.0000	0.8272	0.8356	0.0000	0.8314	0.8382***
P	0.0000	0.0193	0.0197	0.0000	0.0195***	0.0216***
AUPRC	0.1286	0.1153	0.1184	0.1405***	0.1231***	0.1180***
BS	−0.0037	−0.1466	−0.1212	−0.0037	−0.1383***	−0.1110***
Panel III: All sample						
AUC	0.8253	0.8123	0.8271	0.8545***	0.8457***	0.8542***
F1	_	0.4244	0.4277	_	0.4381***	0.4506***
R	0.0000	0.7750	0.7475	0.0000	0.7761	0.7887***
P	0.0000	0.0105	0.0106	0.0000	0.0113***	0.0130***
AUPRC	0.0640	0.0397	0.0375	0.0754***	0.0529***	0.0550***
BS	−0.0034	−0.1885	−0.1700	−0.0034	−0.1721***	−0.1467***

N is no sampling, U is undersampling, and O is oversampling strategy. If both recall and precision are zero, the F1 score is ill-defined and reported as '_'. ***, **, and * indicate significance under repeated-measures t-test at 99%, 95%, and 90% levels

sampling strategies. Textual features consistently increase AUC across all data segments and balancing settings. For most metrics, especially under undersampling and oversampling, the combined model improves predictive power. One exception is the AUPRC in the SME oversampling case, where performance slightly drops.

The combined model achieves an AUC of 0.854 on the original full dataset—close to the 0.856 reported by [35], who used a deep learning model with text, financial, and market data. This result highlights the practical value of interpretable models using textual features, without the complexity and resource demands of deep learning.

We quantify performance gains between the Altman and combined models in Table 15. The SME segment shows the largest improvement—AUC increases by 8.4–11.5%, recall by up to 9.7%, and F1 score by 2–3.4%. Improvements in Brier scores range from 0 to 3.3% across all samples. Although oversampling slightly reduces AUPRC in the SME segment, textual features still contribute positively in most cases.

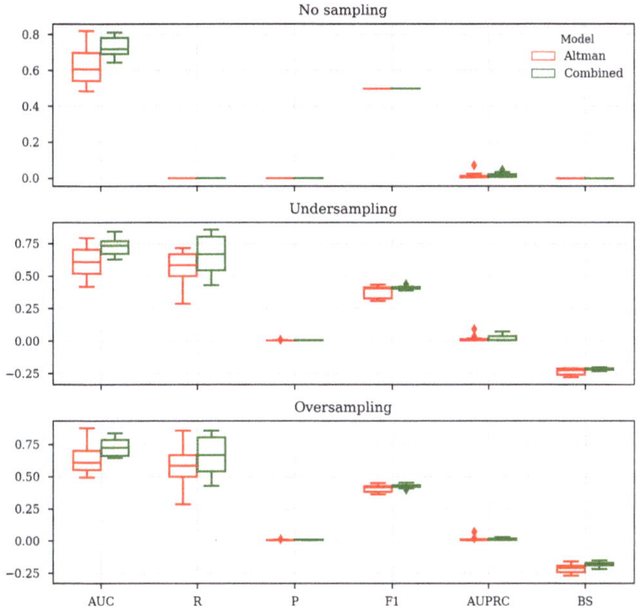

Fig. 4 Box plots of performance metrics, SME segment

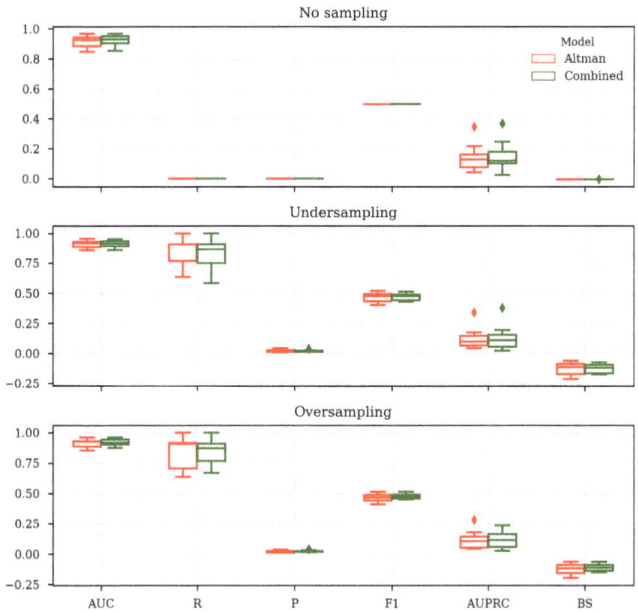

Fig. 5 Box plots of performance metrics, non-SME segment

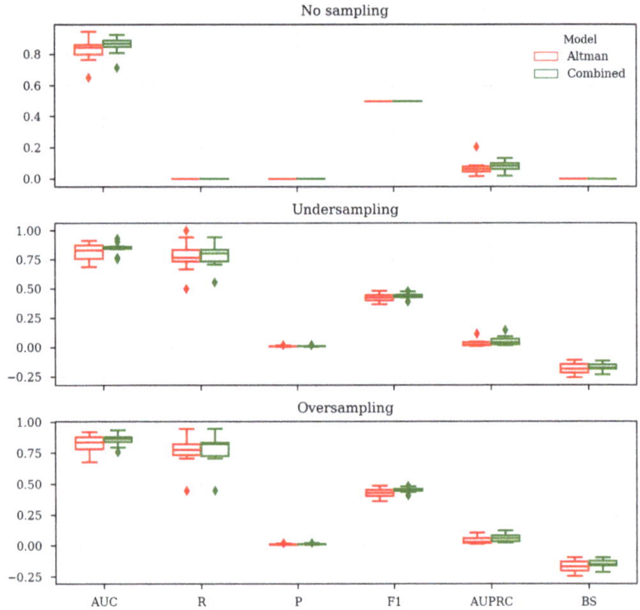

Fig. 6 Box plots of performance metrics, all samples

4.3.4 One-Year Ahead Prediction

We present one-year-ahead rolling window predictions using four consecutive years of data to train and predict defaults in the fifth year. For instance, models trained on 1997–2000 data predict 2001 defaults; then the window shifts forward one year (e.g., 1998–2001 data for 2002 prediction), continuing until 2017. The predictive power is assessed using the same six metrics as in previous sections, comparing Altman-only versus combined models, and evaluating statistical significance and performance gains.

Figure 7 shows the AUC trends over time for SME, non-SME, and full samples. The solid lines represent average AUCs, and the shaded bands indicate ± 1 standard deviation.

In Fig. 7a the SME segment shows performance improvements from 2001–2012 when using combined features. However, from 2013 onward, AUC gains diminish due to limited default events: no SME defaults were recorded in 2010 and 2011, and only 1–2 defaults per year occurred from 2012–2017. For the non-SME group (Fig. 7b), AUC ranges from 0.6 to 0.9, with the combined model outperforming Altman consistently from 2007 onwards. In the full sample (Fig. 7c), substantial AUC gains occur between 2003–2012.

Aggregate one-year-ahead results are summarized in Table 16. The combined model significantly outperforms the Altman model in most settings, particularly for AUC, F1, AUPRC, and Brier Score metrics.

Table 15 Performance gain (%)

	No sampling	Undersampling	Oversampling
Panel I: SME sample			
AUC	10.2	11.5	8.4
F1	–	3.4	2.0
R	–	9.7	3.3
P	–	0.1	0.1
AUPRC	0.2	0.4	−0.2
BS	–	2.3	3.3
Panel II: non-SME sample			
AUC	1.0	0.7	0.4
F1	–	0.7	1.0
R	–	0.4	0.3
P	–	–	0.2
AUPRC	1.2	0.8	–
BS	–	0.8	1.0
Panel III: All sample			
AUC	2.9	3.3	2.7
F1	–	1.4	2.3
R	–	0.1	4.1
P	–	0.1	0.2
AUPRC	1.1	1.3	1.8
BS	–	1.6	2.3

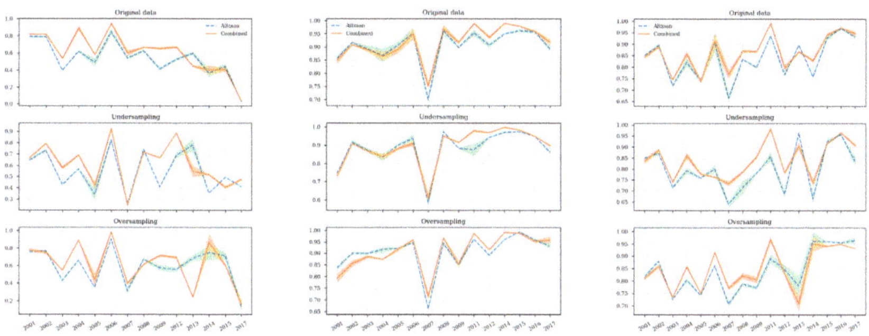

(a) SME (b) non-SME (c) All sample

Fig. 7 One-year-ahead prediction performance (AUC)

Table 16 Predictive power comparison, one-year-ahead prediction

	Altman			Combine		
	N	U	O	N	U	O
Panel I: SME						
AUC	0.5458	0.5474	0.5924***	0.6216***	0.6102***	0.6281***
F1	_	0.3423	0.4207	_	0.3589***	0.4398***
R	0.0000	0.5049	0.3717	0.0000	0.5990***	0.3599
P	0.0000	0.0031	0.0037	0.0000	0.0038***	0.0043**
AUPRC	0.0101	0.0081	0.0071	0.0187***	0.0123***	0.0136***
BS	−0.0028	−0.2572	−0.1810	−0.0028	−0.2527***	−0.1461***
Panel II: non-SME						
AUC	0.9045	0.8819	0.9047	0.9176***	0.8933**	0.9086***
F1	_	0.4556	0.4841	_	0.4583***	0.4986***
R	0.0000	0.7858	0.7450	0.0000	0.8310***	0.7857**
P	0.0000	0.0157	0.0229	0.0000	0.0168***	0.0312***
AUPRC	0.0965	0.0832	0.1183	0.1517***	0.1359***	0.1387***
BS	−0.0042	−0.1676	−0.0954	−0.0042	−0.1606***	−0.0809***
Panel III: All sample						
AUC	0.8343	0.7960	0.8357	0.8611***	0.8352***	0.8468***
F1	_	0.4261	0.4481	_	0.4373***	0.4608***
R	0.0000	0.7323	0.7252	0.0000	0.7868***	0.7063**
P	0.0000	0.0086	0.0107	0.0000	0.0099***	0.0121***
AUPRC	0.0552	0.0302	0.0529	0.0685***	0.0390***	0.0559**
BS	−0.0033	−0.1973	−0.1442	−0.0033	−0.1830***	−0.1242***

N is no sampling, U is undersampling, and O is oversampling strategy. If both recall and precision are zero, the F1 score is ill-defined and hence is reported as '_'. ***, **, and * indicate the corresponding repeated-measures t-test is significant at the 99, 95, and 90% levels, respectively

Our average AUC result is comparable to [35], which reports 0.842 using a CNN model on textual, accounting, and market features. We slightly outperform with 0.847, highlighting that dictionary-based textual features can yield competitive predictive power. This supports the case for interpretable, cost-efficient models as a viable alternative to deep learning in credit scoring.

5 Conclusions

This work synthesizes insights from the previous peer-reviewed contributions [6, 38] to provide a cohesive and accessible account of how textual analysis and machine learning improve bankruptcy prediction. Textual data play an increasingly important role in financial risk analysis, complementing traditional data to explain manager sentiment, analyst behavior, and asset pricing [20, 31, 46]. This study shows that textual features significantly enhance accounting-based default prediction models. Using over 50,000 US-listed firm observations from 1997 to 2017, we demonstrate the predictive power of simple word-count sentiment features from full filings and MDA sections.

Despite challenges from imbalanced default samples across SME, non-SME, and all firms, textual features consistently improve performance. Our results, comparable to those of deep learning approaches such as [35], highlight the value of interpretable, low-cost alternatives. Specifically, positive sentence ratios and uncertainty terms in MDAs associate positively with default risk, while higher overall **senti** scores (positive minus negative terms) reduce it. Conversely, more negative MDA content raises default likelihood.

Performance gains are robust across six metrics and sampling strategies. AUC increases by 8.4–11.5% for SMEs, with recall gains of 3.3–9.7%. One-year-ahead prediction also shows AUC improvements of 1.1–7.6% across segments, confirming the out-of-time predictability of textual features.

Future work can expand on several fronts. First, incorporating more textual sources to capture macroeconomic cycles, and refining resampling strategies to better address imbalance. Further research may also explore advanced feature engineering and optimal training horizons.

Finally, while our sentiment method uses basic ± 1 weights for word polarity, this simplification ignores intensity and context. Developing graded or contextual sentiment scoring—especially at sentence or section levels (e.g., MDAs)—is a key direction. Though costly, annotating financial text to validate dictionary-based classifiers would improve both accuracy and interpretability. This study provides a foundation for integrating textual cues into practical, explainable credit risk models.

References

1. Altman, E.I.: Financial ratios, discriminant analysis and the prediction of corporate bankruptcy. J. Financ. **23**(4), 589–609 (1968)
2. Altman, E.I., Haldeman, R.G., Narayanan, P.: ZETATM analysis a new model to identify bankruptcy risk of corporations. J. Bank. Financ. **1**(1), 29–54 (1977)
3. Altman, E.I., Iwanicz-Drozdowska, M., Laitinen, E.K., Suvas, A.: Financial distress prediction in an international context: a review and empirical analysis of Altman's z-score model. J. Int. Financ. Manag. Account. **28**(2), 131–171 (2017)
4. Altman, E.I., Sabato, G.: Modelling credit risk for SMEs: evidence from the U.S. market. Abacus **43**(3), 332–357 (2007)
5. Altman, E.I., Sabato, G., Wilson, N.: The value of non-financial information in SME risk management. J. Credit. Risk **6**(2), 95–127 (2010)
6. Ba-Hung, N., Van-Nam, H.: Textual analysis and corporate bankruptcy: a financial dictionary-based sentiment approach. J. Oper. Res. Soc. **73**(1), 102–121 (2022)
7. Bao, Y., Datta, A.: Simultaneously discovering and quantifying risk types from textual risk disclosures. Manage. Sci. **60**(6), 1371–1391 (2014)

8. Blei, D.M., Lafferty, J.D.: Dynamic topic models. In: Proceedings of the 23rd international conference on Machine learning - ICML '06, pp. 113–120, Pittsburgh, Pennsylvania. ACM Press (2006)
9. Blei, D.M., Ng, A.Y., Jordan, M.I.: Latent Dirichlet allocation. J. Mach. Learn. Res. **3**, 993–1022 (2003)
10. Chen, N., Ribeiro, B., Chen, A.: Financial credit risk assessment: a recent review. Artif. Intell. Rev. **45**(1), 1–23 (2016)
11. Crone, S.F., Finlay, S.: Instance sampling in credit scoring: an empirical study of sample size and balancing. Int. J. Forecast. **28**(1), 224–238 (2012)
12. Davis, J., Goadrich, M.: The relationship between Precision-Recall and ROC curves. In: Proceedings of the 23rd International Conference on Machine Learning, pp. 233–240. ACM (2006)
13. Dieng, A.B., Ruiz, F.J., Blei, D.M.: Topic modeling in embedding spaces (2019). arXiv preprint arXiv:1907.04907
14. Dieng, A.B., Ruiz, F.J.R., Blei, D.M.: The Dynamic Embedded Topic Model (2019). arXiv:1907.05545 [cs, stat]
15. Gandhi, P., Loughran, T., McDonald, B.: Using Annual Report Sentiment as a Proxy for Financial Distress in U.S. Banks. SSRN Scholarly Paper ID 2905225, Social Science Research Network, Rochester, NY (2018)
16. Gandhi, P., Loughran, T., McDonald, B.: Using annual report sentiment as a proxy for financial distress in U.S. banks. J. Behav. Financ. **20**(4), 424–436 (2019)
17. García, V., Marqués, A.I., Sánchez, J.S.: Exploring the synergetic effects of sample types on the performance of ensembles for credit risk and corporate bankruptcy prediction. Inf. Fusion **47**, 88–101 (2019)
18. Guan, L., He, S.D., McEldowney, J.: Window dressing in reported earnings. Com. Lending Rev. **23**, 26 (2008)
19. Hoffman, M., Bach, F.R., Blei, D.M.: Online learning for latent Dirichlet allocation. In: Lafferty, J.D., Williams, C.K.I., Shawe-Taylor, J., Zemel, R.S., Culotta, A. (eds.) Advances in Neural Information Processing Systems, vol. 23, pp. 856–864. Curran Associates Inc. (2010)
20. Huang, A.H., Lehavy, R., Zang, A.Y., Zheng, R.: Analyst information discovery and interpretation roles: a topic modeling approach. Manage. Sci. **64**(6), 2833–2855 (2018)
21. Huang, G., Huang, G.-B., Song, S., You, K.: Trends in extreme learning machines: a review. Neural Netw. **61**, 32–48 (2015)
22. Huang, K.-W., Li, Z.: A multilabel text classification algorithm for labeling risk factors in SEC form 10-K. ACM Trans. Manage. Inf. Syst. **2**(3), 1–19 (2011)
23. Huang, Z., Chen, H., Hsu, C.-J., Chen, W.-H., Wu, S.: Credit rating analysis with support vector machines and neural networks: a market comparative study. Decis. Support Syst. **37**(4), 543–558 (2004)
24. Hutto, C.J., Gilbert, E.: VADER: a parsimonious rule-based model for sentiment analysis of social media text. In: Eighth International AAAI Conference on Weblogs and Social Media (2014)
25. Jiang, F., Lee, J., Martin, X., Zhou, G.: Manager sentiment and stock returns. J. Financ. Econ. **132**(1), 126–149 (2019)
26. Jo, H., Han, I., Lee, H.: Bankruptcy prediction using case-based reasoning, neural networks, and discriminant analysis. Expert Syst. Appl. **13**(2), 97–108 (1997)
27. Lawrence, A., Rogo, R., Zhang, P.: Forward-looking language in disclosures. Contemp. Account. Res. **37**(4), 2504–2541 (2020)
28. Le, Q., Mikolov, T.: Distributed representations of sentences and documents. Presented at the (2014)
29. Lessmann, S., Baesens, B., Seow, H.-V., Thomas, L.C.: Benchmarking state-of-the-art classification algorithms for credit scoring: an update of research. Eur. J. Oper. Res. **247**(1), 124–136 (2015)
30. Lin, S.-M., Ansell, J., Andreeva, G.: Predicting default of a small business using different definitions of financial distress. J. Oper. Res. Soc. **63**(4), 539–548 (2012)

31. Lopez Lira, A.: Risk factors that matter: textual analysis of risk disclosures for the cross-section of returns. SSRN J. (2019)
32. Lopez-Lira, A., Tang, Y.: Risk disclosure and asset prices. Account. Rev. **95**(6), 257–287 (2020)
33. Loughran, T., Mcdonald, B.: When is a liability not a liability? textual analysis, dictionaries, and 10-Ks. J. Financ. **66**(1), 35–65 (2011)
34. Loughran, T., McDonald, B.: Textual analysis in accounting and finance: a survey. J. Account. Res. **56**(2), 341–388 (2018)
35. Mai, F., Tian, S., Lee, C., Ma, L.: Deep learning models for bankruptcy prediction using textual disclosures. Eur. J. Oper. Res. **274**(2), 743–758 (2019)
36. Malo, P., Sinha, A., Korhonen, P., Wallenius, J., Takala, P.: Good debt or bad debt: detecting semantic orientations in economic texts. J. Am. Soc. Inf. Sci. **65**(4), 782–796 (2014)
37. Mikolov, T., Sutskever, I., Chen, K., Corrado, G.S., Dean, J.: Distributed representations of words and phrases and their compositionality. In: Neural Information Processing Systems (2013)
38. Nguyen, B.-H., Kiyoaki, S., Huynh, V.-N.: Topics in financial filings and bankruptcy prediction with distributed representations of textual data. In: Machine Learning and Knowledge Discovery in Databases. Applied Data Science and Demo Track: European Conference, ECML PKDD 2020, Ghent, Belgium, Sept 14–18, 2020, Proceedings, Part V, pp. 306–322. Springer (2021)
39. Nguyen, T.H., Shirai, K., Velcin, J.: Sentiment analysis on social media for stock movement prediction. Expert Syst. Appl. **42**(24), 9603–9611 (2015)
40. Sun, J., Li, H., Huang, Q.-H., He, K.-Y.: Predicting financial distress and corporate failure: a review from the state-of-the-art definitions, modeling, sampling, and featuring approaches. Knowl.-Based Syst. **57**, 41–56 (2014)
41. Tsai, C.-F., Chen, M.-L.: Credit rating by hybrid machine learning techniques. Appl. Soft Comput. **10**(2), 374–380 (2010)
42. Tsai, C.-F., Hung, C.: Modeling credit scoring using neural network ensembles. Kybernetes **43**(7), 1114–1123 (2014)
43. Wiginton, J.C.: A note on the comparison of logit and discriminant models of consumer credit behavior. J. Financ. Quant. Anal. **15**(3), 757–770 (1980)
44. Wu, H.-C., Hu, Y.-H., Huang, Y.-H.: Two-stage credit rating prediction using machine learning techniques. Kybernetes **43**(7), 1098–1113 (2014)
45. Zhao, Z., Xu, S., Kang, B.H., Kabir, M.M.J., Liu, Y., Wasinger, R.: Investigation and improvement of multi-layer perceptron neural networks for credit scoring. Expert Syst. Appl. **42**(7), 3508–3516 (2015)
46. Zhou, G.: Measuring investor sentiment. Annu. Rev. Financ. Econ. **10**, 239–259 (2018)

AI-Enhanced Investing: Sentiment Analysis, Strategy Design, and Automation

Mikulas Antonik, Josef Novotny, and Petr Hajek

Abstract This chapter explores the transformative role of artificial intelligence (AI) in investment decision-making, focusing on market sentiment analysis, strategy development and trading automation. Through a series of experiments using both proprietary and open source platforms, the study demonstrates how AI tools, ranging from large language models to custom-built sentiment analyzers, can improve portfolio construction and short-term trading performance. Key findings reveal that, while AI-generated strategies offer improved risk-return profiles, their effectiveness is closely tied to data quality, model interpretability, and the integration of human oversight. Notably, short-selling strategies and DeepSeek-based trading models produced promising results; however, technical limitations such as incomplete API integration and constrained cloud infrastructure posed challenges to achieving full automation. The chapter highlights the practical trade-offs between custom and commercial AI solutions, and suggests ways to improve model robustness, regulatory compliance, and strategic alignment with investor objectives in future research. To sum up, the findings emphasize the increasing potential of AI as a decision support and execution tool in modern investment practice.

Keywords Artificial intelligence · Algorithmic trading · Portfolio optimization · Sentiment analysis · Trading automation

1 Introduction

Artificial Intelligence (AI) has evolved rapidly into a foundational technology across industries, with the finance and investment sectors emerging as the primary beneficiaries of its transformative capabilities [1, 2]. Integrating AI into investment decision-making represents a paradigm shift from heuristics and manual analysis

M. Antonik · J. Novotny · P. Hajek (✉)
Faculty of Economics and Administration, University of Pardubice, Pardubice, Czech Republic
e-mail: petr.hajek@upce.cz

H. T. T. Nguyen et al. (eds.), *Data Science in Finance and Accounting*, Studies in Big Data 181, https://doi.org/10.1007/978-3-032-06179-9_8

to data-driven, algorithmically optimized strategies. This transformation is driven by advances in computing power, the increasing availability of high-frequency financial data, and breakthroughs in machine learning (ML), deep learning (DL) and natural language processing (NLP) [3, 4].

One particularly impactful application of AI in finance is algorithmic trading [5, 6], where intelligent agents autonomously execute trades based on statistical signals, historical patterns and real-time market indicators. These systems can identify market inefficiencies and exploit short-term opportunities with a speed and precision that human traders cannot match. New methods such as reinforcement learning and adaptive algorithms allow trading models to adapt to changing market conditions, enhancing resilience and reducing exposure to systemic shocks [7, 8].

The increasing use of AI in trading has had a significant impact on market microstructure, improving liquidity and narrowing spreads [6]. However, it also raises important challenges regarding market stability, regulatory oversight and the transparency of autonomous financial agents [9]. These issues are particularly pertinent in periods of heightened volatility, prompting an ongoing discourse on the ethical and legal implications of AI in capital markets.

Recent literature has expanded on various facets of AI in finance, including predictive modelling using long short-term memory (LSTM) networks and hybrid neural networks [10, 11], AI-guided portfolio optimization [12], and conversational AI for financial services [13]. Despite these advances, however, several gaps persist. Notably, there is limited research addressing the practical deployment and comparative evaluation of accessible AI tools for institutional and retail investors alike. Furthermore, the real-world integration of AI-based sentiment analysis, algorithmic strategy generation and automated trading workflows is under-explored, particularly from the perspectives of systems implementation and cost efficiency.

This chapter aims to address these gaps by providing a comprehensive, evidence-based analysis of AI-driven tools and platforms used for investment decision-making. The chapter makes three key contributions:

- Conceptual framework: It outlines the key AI methodologies relevant to investment contexts, such as algorithmic trading, market sentiment analysis and predictive modelling.
- Empirical assessment: It evaluates the practical effectiveness of selected AI tools, such as large language models (LLMs), sentiment analytics and trading strategy generators, through real-world case studies and performance metrics.
- Strategic recommendations: It provides investors, analysts and financial managers with actionable guidance on how to deploy AI systems in investment workflows, ensuring transparency, accountability and ethical compliance.

The chapter begins with an overview of AI applications in finance, followed by a discussion of technological enablers and regulatory considerations based on the literature. The empirical section presents multi-stage experiments that evaluate sentiment-based predictions and AI-generated trading strategies using platforms such as TradingView. It also presents an autonomous trading bot prototype that was implemented on Google Cloud. Comparative analyses of commercial tools (e.g., AltIndex) versus custom-built ones further highlight trade-offs in cost, accessibility, and control.

The chapter concludes with synthesized insights and strategic implications for the future of AI-assisted investment management. By bridging the gap between theory and practice, this chapter makes a valuable contribution to both scholarly understanding and professional practice in the era of AI-guided financial intelligence.

2 Literature Review: The Transformative Impact of Artificial Intelligence on Investment Practices

2.1 The Evolving Role of Artificial Intelligence in Financial Services

In recent years, AI has emerged as a transformative technology within the financial sector, fundamentally altering investment decision-making, asset management, and client interaction. AI is being used more and more to streamline operational processes, reduce costs, and improve the accuracy and responsiveness of financial services. According to a Deloitte global survey (2023),[1] approximately 79% of financial institutions have implemented AI technology, underscoring its strategic importance. Market forecasts estimate that the global AI market in financial services will reach USD 45.6 billion by 2027, signaling an unprecedented level of integration between advanced analytics and investment operations.

AI's integration into finance goes beyond automation, enabling the dynamic analysis of complex datasets and providing actionable insights that improve strategic and tactical decision-making. Its applications span a broad spectrum, including credit scoring, payment systems, fraud detection, and regulatory compliance.[2] However, these advancements are accompanied by growing concerns about data security, model bias, and algorithmic opacity. These challenges require careful governance and oversight.

2.2 AI-Guided Algorithmic Trading

Algorithmic trading, also referred to as automated or black-box trading, is one of the most prominent applications of AI in finance. It involves using pre-defined sets of instructions based on time, price, volume, or mathematical models to execute trades. These algorithms operate at speeds and frequencies that are unattainable for

[1] https://www.deloitte.com/global/en/about/press-room/gen-ai-survey.html.

[2] https://interfaz.io/2024/08/the-revolution-of-ai-in-financial-services/.

human traders, and they have become central to modern market infrastructure [5]. Algorithmic trading is generally categorized as execution algorithms, proprietary black-box strategies, or high-frequency trading (HFT) systems. However, in practice, these categories often overlap [14].

Since the 1970s, the evolution of algorithmic trading has been propelled by increasing digitalization and computing power. Early models relied on basic, rule-based strategies, whereas contemporary, AI-enhanced algorithms use deep learning architectures to detect nonlinear market patterns and adapt dynamically to real-time market changes. Integrating AI enables algorithms to process and learn from vast, heterogeneous data streams, including social media sentiment, macroeconomic indicators, and geopolitical news [15, 16]. This expands the scope of algorithmic trading to alternative assets, such as cryptocurrencies and commodities [5, 17].

Recent advancements in predictive modeling have significantly improved the accuracy of financial forecasting. Machine learning techniques, particularly recurrent neural networks (RNNs) and their LSTM variants, have demonstrated an impressive capacity to identify temporal dependencies and hidden patterns in financial time series data. Prabakar et al. [10] introduced the SMS-ILS system, which uses LSTM-RNN networks, deep belief networks (DBNs), and grey wolf optimization (GWO), achieving over 97% accuracy in predicting stock price movements. Despite their computational intensity, such hybrid models offer a promising approach to precise, asset-specific forecasting.

Concurrent research by [18] expanded upon this by combining LSTMs with support vector machines (SVMs) to address volatility in intraday and daily trading environments. While these models demonstrate strong performance in trend detection, precisely predicting price levels remains challenging.

2.3 Portfolio Optimization and Predictive Analytics

In recent years, AI has transformed the way portfolios are optimized, enabling more flexible, adaptive and data-intensive approaches [12]. Traditional models often rely on assumptions of linearity, normal distributions and static covariances, which can limit their effectiveness in volatile and complex markets. AI models, particularly those based on deep learning and probabilistic reasoning, offer powerful alternatives that can capture nonlinear dependencies and temporal dynamics in financial data. For example, [19] showed that attention-based deep learning models can dynamically adjust asset allocations by identifying hidden relationships between macroeconomic indicators, asset returns and risk factors. These models outperformed conventional mean-variance frameworks across multiple time horizons.

AI-based portfolio optimization also benefits from advances in reinforcement learning (RL), which treats the allocation task as a sequential decision-making process under uncertainty. Huang et al. [20], for example, proposed a deep Q-network-based approach that continuously learns optimal portfolio weights through market interaction, effectively navigating dynamic risk environments and transaction costs.

Other studies, such as those by [21], have explored policy gradient and actor-critic methods that optimize the trade-off between return maximization and drawdown control. These models can adapt more effectively to market regime shifts and investor-specific constraints than static or rule-based strategies can, marking a shift towards intelligent, autonomous portfolio management systems.

In addition to optimizing returns and risks, AI now enables the construction of multi-objective portfolios that incorporate environmental, social and governance (ESG) preferences, liquidity constraints and behavioral risk metrics. Explainable AI (XAI) techniques are being integrated more and more to provide transparency in model decisions, which is crucial for compliance, auditing and investor trust. For instance, [22] used SHAP (Shapley Additive Explanations) to analyze the impact of individual assets and macroeconomic factors on portfolio performance, enabling portfolio managers to validate their decisions and recognize potential biases. Furthermore, federated learning is attracting attention as it enables collaborative portfolio modelling without compromising data privacy, which is a critical concern for financial institutions [23]. These innovations position AI not only as a performance enhancer, but also as a cornerstone of transparent and sustainable asset management. This marks a shift towards intelligent, autonomous portfolio management systems.

2.4 *Conversational AI and Chatbot Applications*

Another rapidly expanding frontier in AI-enabled finance is the deployment of conversational agents, also known as financial chatbots. These chatbots leverage NLP and machine learning to provide real-time market intelligence, customer support, and investment guidance. These systems can analyze large volumes of financial data and deliver personalized recommendations to investors on various communication platforms [13].

Notable examples include Bank of America's Erica, an AI-powered assistant that offers banking insights and financial planning tools, and JPMorgan Chase's COIN. COIN uses AI to automate legal document analysis, a task that used to require over 360,000 h of manual labor annually.[3] Goldman Sachs has developed AI-powered investment advisors that can tailor portfolio suggestions to client profiles, preferences, and risk tolerance.[4] These developments indicate a transition from static advisory services to dynamic, AI-driven financial ecosystems.

As these systems mature, future generations of chatbots are expected to incorporate emotional intelligence, blockchain-enabled transaction verification, and multimodal user interfaces, including voice and video communication. These advance-

[3] https://medium.com/@the_AI_ZONE/how-jpmorgan-chases-coin-is-revolutionizing-financial-operations-with-ai-120a2938dab7.

[4] https://www.stockgeist.ai/financial-chatbots-a-personalized-and-intelligent-stock-market-guide/.

ments promise to elevate client engagement and democratize access to sophisticated financial tools.

2.5 *Ethical and Regulatory Considerations*

Although AI offers substantial benefits to financial services, its implementation raises complex ethical and regulatory questions. Transparency, explainability, and accountability are increasingly viewed as prerequisites for trustworthy AI systems. The CFA Institute[5] emphasizes the importance of investment professionals adhering to ethical standards and ensuring that AI-driven recommendations can be clearly explained to clients. Collaboration between financial experts and technologists is essential to ensure that AI systems align with fiduciary responsibilities and investor protection.

In terms of regulation, the European Union introduced the AI Act, the world's first comprehensive legal framework for AI governance. The Act classifies AI systems based on risk and mandates requirements such as human oversight, algorithm transparency, data protection, and impact assessment. The Act introduces obligations for financial institutions to monitor and document AI systems used in high-stakes decision-making contexts, such as lending or investment management. Noncompliance may result in significant financial penalties of up to 6% of global annual turnover [24]. This evolving regulatory landscape underscores the need for interdisciplinary approaches that integrate technical innovation with ethical foresight and legal compliance.

3 Methodology

This chapter outlines the methodological framework adopted for the empirical investigation of AI tools in the context of investment decision-making. The study primarily aimed to evaluate the effectiveness and practical utility of selected AI-driven systems in supporting professional and non-professional investors in forecasting stock price movements and generating actionable investment recommendations (buy, sell, or hold). The methodology combines qualitative and quantitative approaches to evaluate prediction accuracy, trading performance, and technological feasibility.

[5] https://rpc.cfainstitute.org/research/reports/2022/ethics-and-artificial-intelligence-in-investment-management-a-framework-for-professionals.

3.1 Selection of Analyzed Companies

The first stage of the research involved selecting representative companies whose stock data would be used to test the AI tools. The following criteria guided the selection process:

- Market relevance: Only publicly traded companies with high trading volumes and significant market capitalization on U.S. exchanges were included to ensure data richness and generalizability.
- Media and data availability: Companies frequently covered in financial media were prioritized to enable effective sentiment analysis using text-based AI models.
- Sectoral diversification: To test the robustness of AI models across different economic sectors, firms were selected from distinct industries, including technology (Apple and NVIDIA), automotive (Tesla), and energy/innovation (OKLO).

Thus, the selected firms–NVIDIA (NVDA), Tesla (TSLA), Apple (AAPL), and OKLO–provided a balanced and diverse testbed for cross-sectoral analysis.

3.2 Investment Horizon Design

To capture the temporal dynamics of AI prediction capabilities, the analysis was structured across three investment horizons:

- Short-term (1 month): Focused on testing the responsiveness of AI models to market volatility and news-driven price fluctuations.
- Medium-term (3 months): This horizon assessed the models' ability to maintain consistency in signal generation across a quarterly cycle.
- Long-term (6 Months): Evaluated the stability and durability of AI-generated predictions over an extended period.

Additionally, two complementary use cases were defined.

- Sentiment analysis horizon: A one-month window was used for text sentiment analysis to identify correlations between media tone and short-term price behavior.
- Automated trading case study: A two-week trading experiment was conducted using the Zapier automation platform to test the operational feasibility of AI-generated signals in a real-time, semi-automated trading environment. The duration was limited by the Zapier platform's trial constraints.

3.3 Data Collection and Preprocessing

Data were sourced from publicly available financial APIs and media platforms to ensure reproducibility and cost-efficiency. The key data types and sources are listed below:

- Historical price data: Daily stock prices were retrieved from TradingView, Yahoo Finance, and Google Finance. These datasets supported technical analysis and time-series forecasting.
- Media and sentiment data: News articles and social media posts were collected using Google Cloud Platform APIs. LLMs, including Claude and OpenAI's GPT models, processed the raw text data to extract sentiment scores and detect shifts in market tone.

Data preprocessing involved standardizing time formats, removing noise from text corpora, and normalizing price series to make them comparable across companies and time periods.

3.4 Analytical Procedure

The analysis followed a structured, multi-step approach:

1. Automated sentiment analysis
   AI language models evaluated the polarity of the collected media content (positive, negative, or neutral). Then, these sentiment scores were cross-referenced with actual stock price movements to assess the predictive relevance of news sentiment.
2. Development and backtesting of trading strategies
   AI tools, including ChatGPT, Claude, and Deep Seek, were used to generate trading strategies based on technical indicators (e.g., MACD, RSI, and EMA). These strategies were then implemented and backtested in TradingView using Pine Script over the defined investment horizons. Performance metrics included cumulative returns, the Sharpe ratio, and maximum drawdown.
3. An automated trading bot was implemented
   A proof-of-concept AI trading bot was built using the Google Cloud Platform. It leverages the Pub/Sub messaging infrastructure and LSTM-based neural networks for signal generation. Trading execution was simulated via Alpaca's brokerage API. We evaluated the system's viability in terms of latency, scalability, and signal-to-trade conversion accuracy.
4. Real-time semi-automated trading experiment
   A short-term trading trial was conducted using Zapier automation to execute 40 trades manually based on real-time sentiment signals. Though limited by platform constraints, this case study offered insight into the practical application of AI signals in live decision-making environments.

3.5 Overview of Large Language Models for Programming Support

Modern LLMs play a pivotal role in enabling the development of AI-powered trading systems, particularly for users who seek to minimize costs and rely on open-access or low-code tools. These models assist with coding, debugging, data analysis, and sentiment extraction, making them highly valuable in the design and automation of investment strategies.

3.5.1 ChatGPT (OpenAI)

ChatGPT, developed by OpenAI, is a versatile transformer-based language model capable of generating human-like text, analyzing structured and unstructured data, and supporting programming tasks across various domains. Within the context of this study, ChatGPT was employed to automate the generation of trading signals and refine trading strategies based on technical and sentiment inputs. Using the OpenAI API, the model was integrated into platforms such as TradingView and Zapier, where it assisted in code optimization, Pine Script generation, and the real-time refinement of algorithmic logic. The model's contextual reasoning capabilities made it suitable for adjusting trading rules dynamically in response to market data.

3.5.2 Gemini (Google)

Gemini, Google's multimodal AI model, extends beyond textual capabilities to interpret and generate outputs based on images and code. This made it particularly useful for tasks requiring the combination of structured price data, chart analysis, and natural language interpretation. Within the scope of this study, Gemini was utilized to assist in the development and debugging of automated trading strategies. Its integration via Google's API allowed for rapid prototyping and testing of algorithmic systems, and it provided technical support in resolving coding-related issues related to market signal processing.

3.5.3 DeepSeek

DeepSeek is an emerging open-source LLM that provides a cost-efficient and customizable alternative to proprietary AI tools. Due to its open-source nature, DeepSeek allowed for enhanced flexibility in scripting and fine-tuning model behavior, especially in the context of rule-based strategy generation in TradingView. Its low operational overhead and adaptable architecture made it suitable for investors and developers seeking affordable and scalable AI-powered programming assistance.

3.5.4 Claude (Anthropic)

Claude, developed by Anthropic, is a robust and safety-oriented LLM tailored for applications requiring high accuracy and resilience. With strong capabilities in text classification, document summarization, and code generation, Claude was applied to the automated sentiment analysis of financial news articles. Integrated into a Google Apps Script environment via API, Claude processed daily news from RSS feeds and classified them into sentiment categories. This model's interpretability and contextual precision made it suitable for detecting subtle sentiment cues in financial discourse and evaluating their correlation with stock price fluctuations.

These models were assessed based on their ability to:

(i) assist in developing algorithmic trading logic,
(ii) perform sentiment-based market analysis,
(iii) optimize trading decisions, and
(iv) integrate within semi-automated financial workflows.

The comparative evaluation focused on identifying practical strengths, limitations, and implementation pathways for different types of investors.

This methodology provides a framework for evaluating the applicability of AI in investment contexts. By combining sentiment analysis, technical forecasting, and real-time trading implementation, the research provides a multidimensional view of how AI can improve investor decision-making processes. Despite technical and temporal constraints, the experimental design allows for quantitative benchmarking and qualitative insight into the opportunities and challenges of AI-based financial intelligence systems.

4 Empirical Analysis and Comparison of Selected AI Tools and Their Impact on Investment Decision-Making

This section provides a detailed empirical analysis of selected AI tools that are both widely accessible and cost-effective, assessing their effectiveness in supporting investment decision-making under realistic market conditions. The main objective is to evaluate AI-driven solutions that are applicable not only to institutional investors but also to retail investors, with a focus on their ability to contribute to strategic decision-making, sentiment analysis, and algorithmic trading. This comparative analysis aims to (i) distinguish between various tools and platforms, (ii) assess their functionalities and applications in financial modeling and prediction, and (iii) determine their relative performance based on a defined set of evaluation criteria. These criteria include predictive accuracy, technical feasibility, user-friendliness, and integration potential with trading platforms.

4.1 Sentiment Analysis Using AI: System Design and Implementation

One of the most promising and rapidly evolving applications of AI in finance is sentiment analysis, which aims to extract insights from unstructured text sources (e.g., news headlines, articles, social media) to assess market sentiment. In this study, a dedicated sentiment analysis system was developed to automatically classify media sentiment related to selected stocks–NVIDIA (NVDA), Tesla (TSLA), and OKLO–and evaluate its correlation with short-term price movements.

4.1.1 System Overview

The sentiment analysis pipeline was designed as a fully automated system operating within the Google Apps Script ecosystem, in combination with Google Sheets, Google News RSS feeds, and OpenAI API endpoints. The language model Claude was used to assess the sentiment of financial news headlines and assign numerical scores on a 1–100 scale, where lower values indicate negative sentiment and higher values denote optimism in media tone. The system operates daily at 14:45 CET, shortly before the U.S. stock market opens, to ensure timely sentiment assessment.

4.1.2 Workflow and Data Processing

The end-to-end workflow consists of the following components (see Fig. 1):

- Data acquisition: News headlines and summaries are fetched daily using Google News RSS feeds, filtered by company name.

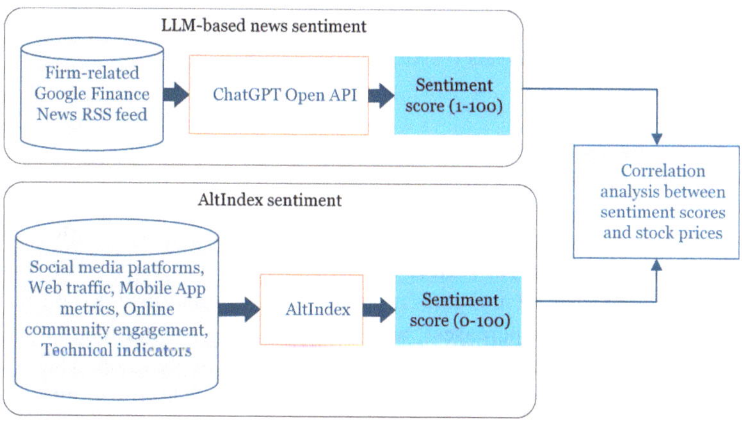

Fig. 1 Workflow of stock-related financial sentiment analysis

NVDA	Morning Sentiment Score	Closing Price	Sentiment Change	Price Change	Sentiment Correlations to Price
12/30/2024	65	137.49		0.35%	
12/31/2024	85	134.29	20	-2.33%	-0.7532324479
1/2/2025	75	138.31	-10	2.99%	0.2151386285
1/3/2025	85	144.47	10	4.45%	0.5525262212
1/6/2025	90	149.43	5	3.43%	0.4815165279
1/7/2025	90	140.14	0	-6.22%	0.4511550012
1/8/2025	70	140.11	-20	-0.02%	0.5175467323
1/10/2025	70	135.91	0	-3.00%	0.6175670889
1/13/2025	64	133.23	-6	-1.97%	0.6817865762
1/14/2025	64	131.76	0	-1.10%	0.6842257957
1/15/2025	65	136.24	1	3.40%	0.7069906246
1/16/2025	65	133.57	0	-1.96%	0.705513139
1/17/2025	71	137.71	6	3.10%	0.6751600749
1/21/2025	70	140.83	1	2.27%	0.2628915514
1/22/2025	50	147.07	-20	4.43%	-0.03962585947
1/23/2025	40	147.22	-10	0.10%	-0.06672930177
1/24/2025	60	142.62	20	-3.12%	0.2393260799
1/27/2025	45	118.42	-15	-16.97%	0.3423460891
1/28/2025	35	128.99	-35	8.93%	0.348620428
1/29/2025	60	123.70	25	-4.10%	0.4129391297
1/30/2025	45	124.65	-15	0.77%	0.490684313
1/31/2025	40	120.07	-5	-3.67%	0.5338108501
2/3/2025	45	116.66	5	-2.84%	0.1422502464
2/4/2025	65	118.65	20	1.71%	0.5060744762

Fig. 2 Illustration of sentiment analysis for NVDA in Google Sheets

- Data cleaning: Extracted text is pre-processed to remove extraneous characters and prepare it for model ingestion.
- Sentiment classification: Each headline is processed through Claude, which generates a sentiment score based on contextual interpretation.
- Data logging: Sentiment scores are automatically recorded in Google Sheets alongside the daily closing prices of the corresponding stock (sourced via the Google Finance function).
- Statistical evaluation: Pearson correlation coefficients are computed between sentiment scores and daily price changes. The results are visualized in tabular form and stored for further analysis (Fig. 2).

4.1.3 Statistical Analysis

The system performs correlation analysis to evaluate whether a statistically significant relationship exists between the sentiment score and daily stock returns. A Pearson coefficient near +1 indicates a strong positive relationship (i.e., favorable news is followed by price increases), while values near -1 suggest an inverse relationship. A coefficient near zero implies no meaningful association. The system is configured to update only on weekdays and during market hours, minimizing API usage and operational costs.

4.1.4 Advantages and Accessibility

By leveraging Google-native tools and avoiding reliance on cloud servers (e.g., AWS), the system maintains low operating costs and is accessible even to users without advanced programming expertise. The integration with Google Sheets allows for real-time data visualization, manual overrides, and transparent audit trails. This architecture supports a scalable and replicable methodology for investors and researchers aiming to explore sentiment-based market signals.

4.1.5 Empirical Results of AI-Based Sentiment Analysis and Stock Price Correlation

The automated sentiment analysis system designed in this study was used to evaluate the temporal relationship between media sentiment and the daily closing prices of selected equities–namely NVIDIA (NVDA), OKLO, and TESLA (TSLA). The results, summarized in Fig. 3, demonstrate that the correlation between sentiment scores and stock price movements is highly variable and lacks long-term consistency.

NVIDIA (NVDA)

The data reveal that the correlation between sentiment and stock price in the case of NVDA is inconsistent and subject to substantial temporal fluctuations. Notably, during the period from January 10 to January 17, 2025, both sentiment and stock price increased in parallel, with the Pearson correlation reaching +0.70, indicating a strong positive relationship during that window.

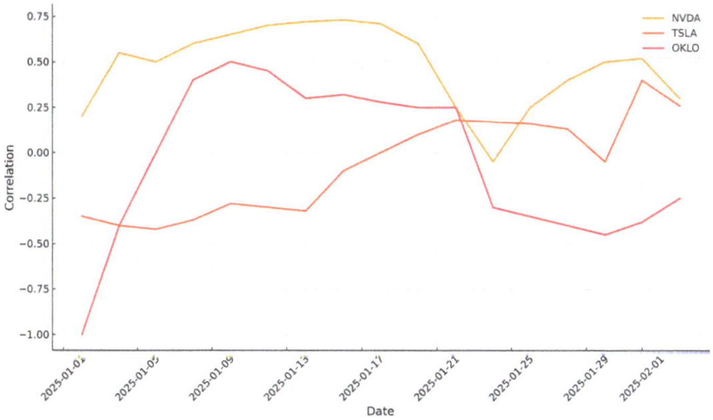

Fig. 3 Pearson's correlations between sentiment scores and stock prices

However, this relationship was not persistent. For instance, on January 22, sentiment dropped sharply by 20 points, yet NVDA's stock price increased by 4.43%, resulting in a near-zero correlation of −0.03. A similar pattern occurred the following day. These examples illustrate that although sentiment may occasionally align with price direction, it cannot be relied upon as a standalone predictor.

OKLO

Similar dynamics were observed in the case of OKLO. While the sentiment scores for OKLO generally ranged between 65 and 95, they did not consistently align with price changes. For example, on January 3, sentiment declined by 3 points, yet the stock surged by +24.71%. Conversely, on January 27, sentiment dropped by 5 points, accompanied by a substantial −25.61% drop in share price.

These observations indicate that other market factors may exert a stronger influence on price movements than sentiment alone. Notably, price changes tended to be more volatile than sentiment shifts. On January 21, sentiment increased by 14 points, and the stock rose by 19.96%, suggesting a degree of alignment. However, just days later on January 30, sentiment declined by 7 points while the stock paradoxically gained 18.73%, again highlighting the non-deterministic nature of sentiment-price relationships.

TESLA (TSLA)

As with the previous cases, the relationship between sentiment and price for TSLA was similarly unstable. Early in the observation period, sentiment-price correlation appeared negative. For instance, on January 3, sentiment fell by 3 points while the stock price rose by 8.22%. A comparable pattern emerged on January 15, with a 5-point drop in sentiment and a concurrent 8.04% increase in stock price.

Over time, a moderate positive correlation began to emerge. On January 31, a 15-point increase in sentiment was accompanied by a modest 1.08% increase in stock price. The strongest observed correlation occurred on February 4, when sentiment fell sharply by 25 points and the Pearson correlation reached 0.7567. This suggests that in certain market conditions, sentiment may temporarily dominate investor behavior; however, such periods were episodic rather than persistent.

4.1.6 System Assessment and Technical Constraints

The overall architecture, leveraging Claude (Anthropic), OpenAI API, Google Apps Script, and Google Sheets, proved to be functionally robust and cost-effective. The entire monthly sentiment analysis process incurred operational costs of only $0.88, making it accessible to retail users and researchers.

Nevertheless, several technical and methodological limitations were identified:

- Computational limitations: While Google Sheets provides a user-friendly and low-cost environment, it lacks the processing capacity for more sophisticated statistical modeling or large-scale data ingestion.
- Sentiment model dependencies: The accuracy of sentiment classification by AI models such as Claude depends heavily on the contextual nuance of input data and the quality of pre-training corpora. Ambiguity or sarcasm in financial headlines may reduce interpretability.
- Data access and frequency limits: Free-tier APIs for market data impose limits on query frequency and volume, which may introduce sampling bias or force the use of downsampled datasets.
- Temporal scope: The experiment was limited to a one-month period, which restricts the generalizability of findings, especially given the inherently non-stationary nature of financial time series.

Despite these constraints, the sentiment analysis system provided valuable insights into short-term sentiment-price dynamics. While no consistent predictive correlation was observed across the full time span, certain timeframes showed statistically meaningful associations—particularly during periods of heightened market sensitivity or media coverage intensity.

4.1.7 Implications and Recommendations

The empirical evidence indicates that sentiment, when used in isolation, is not a reliable predictor of stock price direction. Although correlations were occasionally strong on specific dates, the overall variability of these relationships limits their utility in real-time trading without additional context.

However, sentiment analysis retains potential value as a complementary indicator. In particular, it may enhance strategy robustness when combined with:

- Volume data to confirm market conviction,
- Technical indicators such as RSI, MACD, or Supertrend to contextualize sentiment signals,
- Fundamental indicators including earnings reports, P/E ratios, or macroeconomic metrics.

To improve accuracy and actionability, further iterations of the system should consider:

- Extending the time horizon to multiple months or quarters to capture seasonal trends and broader market cycles,
- Employing hybrid models that combine LSTM or transformer-based time series predictors with sentiment overlays,
- Leveraging cloud-based computing platforms for scalable analysis beyond spreadsheet environments,

- Incorporating multiple sentiment sources, including social media, financial blogs, and institutional research.

In conclusion, sentiment analysis, when properly implemented and integrated with broader analytical frameworks, can serve as a valuable tool for gauging market mood and detecting shifts in investor behavior. While it should not be used as a stand-alone decision criterion, it may significantly enhance the timing and risk management of trading strategies in volatile or momentum-driven markets.

4.1.8 Comparative Evaluation: Custom-Built Sentiment Analysis Versus AltIndex Platform

To further contextualize the results of the AI-driven sentiment analysis system developed in Google Apps Script, a comparative evaluation was conducted against a commercial solution AltIndex, a proprietary platform specializing in AI-based stock sentiment analytics using alternative data sources. This comparison aims to assess the trade-offs between building a customized low-cost solution versus using a ready-to-deploy professional platform optimized for financial analytics.

AltIndex represents a cutting-edge approach to alternative data analysis, offering sentiment-based predictions derived from non-traditional indicators such as job postings, website traffic, mobile app usage trends, social media reviews, and public discussion activity. These data streams are processed using machine learning models specifically trained for financial market applications.

Compared to the Google Script-based implementation, which relied exclusively on headlines from Google News RSS feeds, AltIndex processes a broader and more diverse set of inputs, increasing the potential reliability of its predictions. Additionally, AltIndex generates AI-driven buy/sell signals, tracks insider trading, and includes modules for price forecasting, technical analysis, and AI-curated stock picks, making it a more comprehensive solution for individual investors and institutional analysts alike.

To ensure objectivity, the premium version of AltIndex was evaluated during a 7-day trial period, enabling direct access to its historical sentiment datasets and analytics suite. The comparison included the same stocks analyzed previously, allowing for direct benchmarking of sentiment-price correlations, usability, and data granularity.

The objective was not only to compare prediction accuracy, but also to assess differences in technical complexity, cost-efficiency, implementation effort, and user control. While AltIndex offers immediate insights and minimal setup requirements, the custom-built solution provides deeper transparency and flexibility, albeit with increased demand for programming skills and ongoing system maintenance.

Across all three equities analyzed, both approaches revealed inconsistent correlations between sentiment and short-term stock price fluctuations, reinforcing earlier conclusions that sentiment, in isolation, is not a reliable standalone predictor.

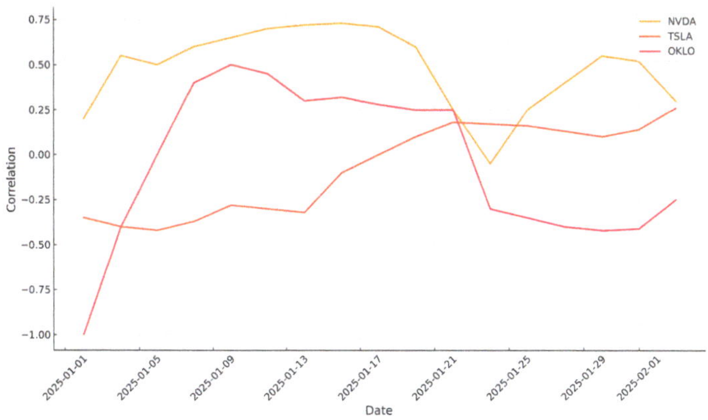

Fig. 4 Pearson's correlations between AltIndex sentiment and stock prices

NVIDIA (NVDA): As in the earlier experiment, the most notable divergence occurred on January 22, where sentiment dropped by 20 points, but the stock rose by +4.43%, leading to a correlation near zero. This anomaly was captured similarly by both AltIndex and the Google Script implementation, suggesting that additional explanatory variables are needed to model price dynamics effectively.

OKLO: On January 21, sentiment increased by 14 points and the stock gained +19.96%, reflecting a strong alignment. However, on January 27, a minor sentiment decline of 5 points coincided with a −25.61% crash in share price, highlighting once again that other factors, such as liquidity, trading volume, or external events, likely played a dominant role (see Fig. 4).

TESLA (TSLA): This stock exhibited the least stable relationship between sentiment and price changes across both platforms (see Fig. 4). As with the Google-based system, AltIndex detected fluctuating sentiment values without consistent alignment to market returns, reaffirming that sentiment on its own lacks sufficient predictive power over short-term intervals.

The key advantage of AltIndex is its turnkey functionality, ease of access to historical data, and availability of pre-calculated sentiment metrics (Table 1). These features are particularly useful for investors who prioritize speed, breadth of insight, and ease of interpretation. In contrast, the custom system built using Google Scripts, Claude, and OpenAI API offers greater flexibility and full process control, albeit at the cost of higher technical complexity and time investment.

From a practical standpoint, AltIndex is well-suited for users seeking a plug-and-play sentiment analysis solution without the need for coding or infrastructure management. It caters to analysts and traders who prefer high-level insights and decision support tools out-of-the-box. However, its subscription model and proprietary architecture may pose barriers for researchers or developers requiring granular control or integration with custom workflows.

Table 1 Comparative analysis between custom Google Script system and AltIndex platform

Feature	Custom Google Script system	AltIndex platform
Setup effort	High (manual coding, API setup)	Minimal
Cost	Very low (~$0.88/month)	High (subscription-based)
Control and transparency	Full	Limited (black-box architecture)
Sentiment sources	Financial news (RSS)	Alternative data (social, app trends, hiring)
Historical data access	Manual scraping or limited APIs	Built-in, extensive
Usability	Requires technical skills	User-friendly dashboard
Customizability	High (fully programmable)	Low

Conversely, the manual approach is more appropriate for technically skilled users, such as data scientists or algorithmic traders, who benefit from full visibility into the sentiment extraction pipeline and can extend the system with additional data sources or analytical modules.

While both systems ultimately reached similar conclusions–namely, that sentiment alone is not a robust predictor of short-term stock movements–they also highlight the value of AI in simplifying data acquisition and structuring decision-relevant information. Each method contributes uniquely to the broader goal of enhancing investor situational awareness through intelligent data processing. AltIndex offers convenience, data breadth, and faster insights, whereas the custom-built AI system offers cost control, technical transparency, and the opportunity for methodological experimentation. Ultimately, the choice between these approaches depends on the user's goals, technical capabilities, and resource constraints. For long-term strategy development, either method should be augmented with technical indicators, fundamental data, and macro-level context to improve reliability and predictive accuracy.

4.2 Development and Comparison of AI-Enhanced Investment Strategies in the TradingView Environment

This section focuses on the development, implementation, and evaluation of investment strategies using the TradingView platform in conjunction with AI-assisted programming, specifically utilizing two LLMs: OpenAI's ChatGPT and DeepSeek. TradingView is a widely used online platform for technical analysis that provides interactive charts, custom indicators, and automated strategy backtesting through its scripting language, Pine Script. To minimize cost, the study employed the platform's free-tier features, which still allow for advanced scripting and visualization capabilities.

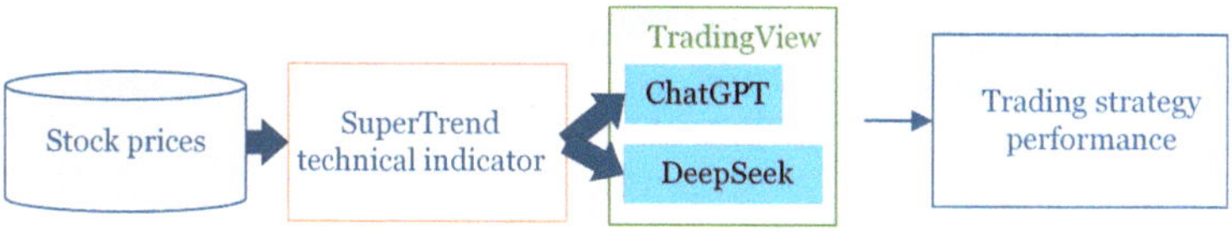

Fig. 5 Workflow of AI-enhanced investment strategies in TradingView

The primary goal was to design a trading strategy based on a commonly used technical indicator SuperTrend and to compare the effectiveness of strategy variations generated by ChatGPT and DeepSeek (Fig. 5). The SuperTrend indicator, known for its ability to track price trends and generate buy/sell signals, was selected due to its interpretability and suitability for trend-following approaches. Strategies were backtested on four selected equities: NVIDIA (NVDA), Tesla (TSLA), Oklo (OKLO), and Apple (AAPL). Testing spanned three time horizons: 1 month, 3 months, and 6 months, ending on February 5, 2025. These periods allowed the analysis of both short-term responsiveness and long-term stability.

4.2.1 Strategy Design Using ChatGPT

The SuperTrend indicator originally existed as a Pine Script v4 overlay that only visualized trends and highlighted potential entry points. The first step was to convert this indicator into a fully functional strategy script using Pine Script v5. ChatGPT was employed to rewrite the script, integrate `strategy.entry()` and `strategy.exit()` commands, and implement stop-loss conditions for capital protection.

During the coding phase, the AI model identified and resolved multiple issues, including deprecated function calls (e.g., replacing `max()` with `math.max()`) and type mismatch errors (e.g., assigning `bool` values to `float` variables). After successful debugging, the strategy executed long and short trades based on trend direction and was adapted to trade a fixed quantity of 10 shares per transaction, independent of capital constraints, for experimental consistency.

As can be seen in Table 2, TSLA yielded the highest net profit and profit factor, indicating relatively strong strategy performance. OKLO underperformed across all metrics, suggesting poor alignment between the strategy and the stock's price behavior.

Table 3 shows the results for the three-month backtest. TSLA remained the most profitable, but NVDA showed improvements. OKLO generated a net loss, and AAPL remained marginal. The low win rates and high drawdowns suggest the need for better risk management. Over six months (Table 4), TSLA again led in net profit and profitability metrics, while NVDA and AAPL produced modest gains. OKLO hovered near break-even. The declining win rates underscore that while profitable trades had strong average returns, the majority of trades were unsuccessful, highlighting the need for enhanced signal filtering.

Table 2 One-month backtest trading results

Metric	NVDA	TSLA	OKLO	AAPL
Net Profit (USD)	236	2262	24	565
Total Closed Trades	203	213	228	194
Win Rate (%)	16.75	15.96	10.53	15.89
Profit Factor	1.272	1.786	1.118	1.530
Max Drawdown (USD)	178.8	464	65	294
Avg Trade Profit (USD)	1.16	10.62	0.11	2.91
Avg Winning Trade (USD)	32.52	151.21	9.50	52.04
Avg Losing Trade (USD)	5.30	16.45	1.38	6.72

Table 3 Three-month backtest trading results

Metric	NVDA	TSLA	OKLO	AAPL
Net Profit (USD)	398	1454	-90	69
Total Closed Trades	192	168	136	177
Win Rate (%)	17.71	18.45	19.12	12.43
Profit Factor	1.535	1.339	0.612	1.060
Max Drawdown (USD)	182.1	1110	164.3	350
Avg Trade Profit (USD)	2.07	8.66	−0.67	0.39
Avg Winning Trade (USD)	33.60	185.45	5.51	54.74
Avg Losing Trade (USD)	4.93	32.29	2.92	7.57

Overall, the results suggest that the AI-assisted SuperTrend strategy is sensitive to the underlying asset's volatility and market behavior. While the integration of ChatGPT streamlined the development and debugging process, further refinement, such as position sizing algorithms, volatility filters, and machine learning-based signal confirmation, may be required for robust real-world deployment.

Table 4 Six-month backtest trading results

Metric	NVDA	TSLA	OKLO	AAPL
Net Profit (USD)	156.1	2901	−3.8	381
Total Closed Trades	180	169	85	175
Win Rate (%)	13.33	15.38	17.65	18.29
Profit Factor	1.183	1.730	0.978	1.270
Max Drawdown (USD)	365.9	1008	156.8	274.9
Avg Trade Profit (USD)	0.87	17.17	−0.04	2.18
Avg Winning Trade (USD)	42.12	264.55	11.05	56.18
Avg Losing Trade (USD)	5.90	28.61	3.08	10.17

Table 5 One-month trading performance—ADX filter strategy

Metric	NVDA	TSLA	OKLO	AAPL
Net Profit (USD)	282.4	2406	−90	−115.4
Total Closed Trades	66	56	40	57
Win Rate (%)	53.03	55.36	47.50	50.88
Profit Factor	1.193	1.591	0.753	0.931
Max Drawdown (USD)	584.08	1294.9	322.80	517
Avg Trade Profit (USD)	4.28	42.98	−2.25	−2.02
Avg Winning Trade (USD)	49.78	209.1	14.47	53.92
Avg Losing Trade (USD)	47.10	163.01	19.21	59.97

4.2.2 Enhanced SuperTrend Strategy with ADX Filter

Following the initial evaluation, it became evident that the baseline SuperTrend strategy lacked consistent profitability, particularly for assets with lower trend stability. To address this, an enhanced strategy variant incorporating the Average Directional Index (ADX) was proposed by ChatGPT. The ADX filter was introduced to identify strong trends and suppress weak market signals. The goal was to increase entry precision, reduce false positives, and improve long-term strategy robustness.

As presented in Table 5, the performance improved notably for TSLA, yielding a net profit of $2,406 and a win rate of 55.36%, with a profit factor of 1.591. NVDA

Table 6 Three-month trading performance—ADX filter strategy

Metric	NVDA	TSLA	OKLO	AAPL
Net Profit (USD)	−1.50	2960	145.80	1660
Total Closed Trades	48	59	28	45
Win Rate (%)	52.08	52.54	60.71	57.78
Profit Factor	0.999	1.444	1.699	2.301
Max Drawdown (USD)	583.4	2340	294.30	437.50
Avg Trade Profit (USD)	−0.03	50.17	5.21	36.89
Avg Winning Trade (USD)	56.5	310.3	20.84	112.93
Avg Losing Trade (USD)	61.48	237.9	20.85	67.16

Table 7 Six-month trading performance—ADX filter strategy

Metric	NVDA	TSLA	OKLO	AAPL
Net Profit (USD)	−468.60	4834	278	707
Total Closed Trades	42	57	18	60
Win Rate (%)	50.00	59.65	61.11	60.00
Profit Factor	0.684	1.797	4.290	1.416
Max Drawdown (USD)	960.4	2421.0	192.0	538.4
Avg Trade Profit (USD)	−11.16	84.80	15.44	11.80
Avg Winning Trade (USD)	48.34	320.6	32.95	66.90
Avg Losing Trade (USD)	70.65	263.75	12.07	70.87

recorded a modest improvement, while OKLO and AAPL generated losses, suggesting poor alignment between the strategy's mechanics and the underlying price behavior of those assets. Notably, drawdowns remained substantial (e.g., $1,294.9 for TSLA), indicating volatility exposure.

In a longer timeframe (Table 6), TSLA retained strong results ($2,960 net profit), while AAPL showed a significant turnaround, improving from prior underperformance to a $1,660 profit and a profit factor of 2.301. OKLO also turned profitable ($145.8), and NVDA stagnated near break-even (-$1.50), highlighting its lack of trend-following compatibility during the observed period.

Over six months (see Table 7), TSLA remained the top performer ($4,834 net profit), although accompanied by the highest capital drawdown ($2,421), emphasizing increased risk. OKLO outperformed expectations with a profit factor of 4.29, suggesting stable performance under this strategy. AAPL yielded $707, while NVDA suffered a -$468.6 loss, indicating inefficacy of the strategy for that stock in a long-term context.

4.2.3 Strategy Design Using DeepSeek

Using the same prompt methodology, the strategy was recreated and enhanced using the DeepSeek LLM. While the tool is open-source and offered higher customizability, the development experience was less seamless due to occasional server instability and slower response times. Nonetheless, the generated strategy introduced several technical refinements:

- Combination of SuperTrend with SMA/EMA filters
- Integration of ATR (Average True Range) for volatility-aware trade sizing
- Use of time-of-day constraints to filter market noise
- Implementation of dynamic stop-loss and take-profit levels

These enhancements were designed to better adapt to market volatility and optimize both risk control and profit realization. As a result, TSLA achieved a net profit of $569.8 with a moderate profit factor of 1.148, while AAPL had the highest profitability (40.16%) but a lower net return ($365) (Table 8). NVDA showed a decent gross profit ($404.7) but a very low profit factor (0.61), reflecting poor trade efficiency. OKLO again performed weakly, though the net loss (-$7.1) was lower than in prior strategies. The three-month timeframe in Table 9 revealed DeepSeek's strength. TSLA recorded a $3,160 profit, OKLO showed marked improvement with $240, and AAPL maintained consistency at $928. Notably, OKLO's profit factor reached 2.968, demonstrating this strategy's effectiveness when given a longer horizon. NVDA remained weak with just $114 net gain. TSLA remained the best performer ($2,562.2) also for the six-month period (Table 10), while NVDA fell into loss (-$157). OKLO and AAPL posted modest gains, but the effectiveness varied across assets. The profit factor for OKLO (1.919) and TSLA (1.641) was strong, though capital drawdowns were substantial (e.g., TSLA: $940), again suggesting the need for improved risk controls.

Table 8 One-month trading performance—DeepSeek strategy

Metric	NVDA	TSLA	OKLO	AAPL
Net Profit (USD)	404.7	569.8	−7.1	365
Total Closed Trades	115	130	128	122
Win Rate (%)	29.57	24.62	28.13	40.16
Profit Factor	0.61	1.148	0.966	1.243
Max Drawdown (USD)	459	1494	92.6	418
Avg Trade Profit (USD)	−3.52	4.38	0.06	2.99
Avg Winning Trade (USD)	23.18	137.93	5.57	38.17
Avg Losing Trade (USD)	24.85	75.37	2.93	25.51

Table 9 Three-month trading performance—DeepSeek strategy

Metric	NVDA	TSLA	OKLO	AAPL
Net Profit (USD)	114	3160	240	928
Total Closed Trades	107	85	70	98
Win Rate (%)	29.91	31.76	35.71	43.88
Profit Factor	1.135	1.835	2.968	1.747
Max Drawdown (USD)	391	892.9	81.1	507.2
Avg Trade Profit (USD)	1.07	37.18	3.44	9.48
Avg Winning Trade (USD)	30.07	257.0	14.53	50.51
Avg Losing Trade (USD)	20.20	135.1	4.08	36.56

Table 10 Six-month trading performance—DeepSeek strategy

Metric	NVDA	TSLA	OKLO	AAPL
Net Profit (USD)	−157.0	2562.2	106.0	70.6
Total Closed Trades	96	89	45	107
Win Rate (%)	29.17	23.60	33.33	28.04
Profit Factor	0.837	1.641	1.919	1.035
Max Drawdown (USD)	387.0	940.0	109.2	488.9
Avg Trade Profit (USD)	1.64	28.79	2.36	0.66
Avg Winning Trade (USD)	28.6	312.4	14.57	68.67
Avg Losing Trade (USD)	26.01	114.2	6.41	42.33

4.2.4 Comparative Evaluation of All Strategies

An aggregate evaluation of the three tested strategies reveals distinct performance tiers:

- DeepSeek strategy was most balanced in terms of profitability, stability, and adaptive trade logic. Its use of dynamic volatility filters and multi-indicator blending yielded robust performance, particularly for TSLA and AAPL. It also demonstrated lower volatility-adjusted losses compared to other models.
- SuperTrend + ADX (ChatGPT). This strategy showed strong raw performance, especially for TSLA, but suffered from greater drawdowns and capital risk. Therefore, it is more suitable for high-risk traders.
- Baseline SuperTrend (ChatGPT). While simpler and easier to implement, the original version lacked adequate predictive power and consistency, especially over

longer periods. Despite a lower drawdown, its profitability and trade efficiency were suboptimal, making it the least viable for sustained use.

These findings reinforce the critical importance of risk-adjusted strategy evaluation. While advanced AI models can assist in strategy generation, their real-world applicability depends on tuning, validation, and adaptation to market conditions. Although all strategies were backtested on historical data, which does not guarantee future performance, the methodologies used here have practical applications in automated trading bots, predictive model training, custom strategy development using any technical indicator, and low-code AI-assisted trading platform integration. This framework enables retail and professional traders alike to leverage AI not only for coding, but for end-to-end strategy lifecycle management, from indicator selection to signal generation, execution, and performance monitoring.

4.3 *Utilization of ChatGPT and Alpaca for Automated Trading*

The solution involved the integration of ChatGPT (AI-based sentiment analysis), the Zapier automation platform, and the Alpaca trading infrastructure. The primary goal was to build a semi-autonomous trading bot capable of making investment decisions based on the sentiment analysis of financial news. Due to the technical complexity of implementing full-scale technical indicators, the strategy pivoted toward news-based sentiment analysis as the core decision-making mechanism (Fig. 6).

4.3.1 Implementation Process

The first step was to set up an account on Alpaca, which provides API access for trading stocks and cryptocurrencies. Using a paper trading account, it was possible to simulate trades under real market conditions with virtual capital. The next step involved connecting Alpaca to Zapier, which would automate trading processes between various applications.

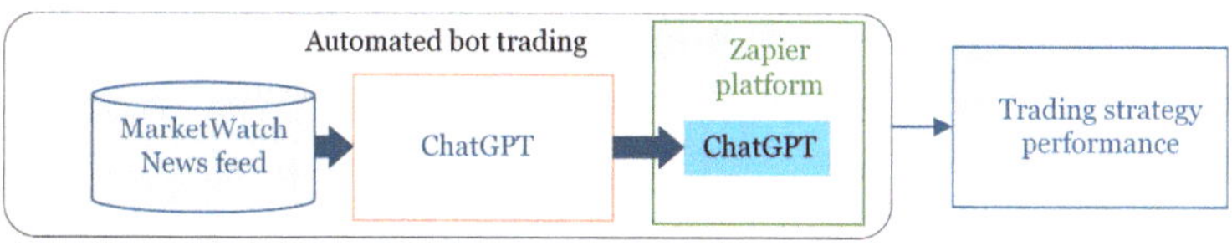

Fig. 6 Workflow of ChatGPT-enhanced automated trading

However, a critical limitation emerged: Zapier did not support integration with Alpaca's paper trading accounts, only live accounts. Consequently, the integration with Alpaca was partially abandoned, and all trade executions had to be handled manually.

The automation workflow was redesigned to integrate ChatGPT with Zapier, enabling data transfer between the AI model, financial news sources, and the trading decision engine. A trading bot was designed in Zapier, which processed financial news articles and generated trading signals through sentiment scoring. The bot accessed real-time market updates via RSS feeds, primarily sourced from MarketWatch, selected for its comprehensive coverage and timely updates. Additional feeds covering IPOs, commodity markets, and macroeconomic indicators were added to enhance analytical coverage. All sources were unrestricted by paywalls, ensuring continuous data availability.

The decision-making flow followed a structured scheme. The bot:

- Extracted the relevant financial instrument from the article title.
- Performed sentiment analysis on the article content.
- Assigned a sentiment score between 0 and 100, with low scores indicating negative sentiment and high scores reflecting strong positive outlooks.
- Interpreted scores to generate buy/sell signals.

Sentiment analysis relied on key phrases, tonal analysis, and historical market responses to news content. Signals were interpreted as follows:

- 0–20: strong negative sentiment → sell signal (or short entry if no open position)
- 81–100: strong positive sentiment → buy signal (or long entry)

To ensure transparency, the system included Telegram-based notifications. A chatbot was created to deliver each signal to a private Telegram channel, enabling real-time monitoring and manual trade execution.

Due to Alpaca's API restrictions, fully automated trade execution was not possible. As a result, all trades were executed manually. Despite this, the AI-generated signals served as a powerful decision-support tool. In practice, some positions were closed prematurely when profitable, while others were exited early to limit losses, deviating from the original systematic logic. This confirms that the system functioned as a semi-autonomous decision aid, combining automation with human oversight. The architecture of the Zapier-based trading bot is illustrated in Fig. 7.

The strategy was tested over a two-week period, from January 23 to February 6, 2025. This short test horizon was imposed by Zapier's free-tier limitations, which allow script execution for only 14 days. As a result, long-term performance under varying market conditions, such as high volatility events or illiquid periods, could not be evaluated. The strategy was tested on a sample of 40 trades, including both long and short positions. The total profit achieved was $155.70. Notably, short positions outperformed long positions significantly:

Fig. 7 Architecture of the Zapier-based trading bot

- Short trades: 16 wins, 8 losses → 66.67% win rate
- Long trades: 7 wins, 9 losses → 43.75% win rate
- Overall win rate: 57.5%

Profit factor was 2.11, indicating that profitable trades significantly outweighed losses. A moderate value of maximum drawdown suggests acceptable risk, although position sizing and stop-loss mechanisms could be improved. The average profit per trade of $3.89 confirms the overall profitability of the strategy. These results are summarized in Table 11.

The strategy demonstrated solid performance, especially in bearish markets, where short signals proved most effective. The profit factor of 2.11 and RRR of 1.56 suggest a viable model for decision-making in volatile environments. However, long positions underperformed, indicating the need for further refinement, possibly via more nuanced sentiment analysis or hybrid strategies incorporating technical indicators.

Table 11 Trading performance of the Zapier-based trading bot

Metric	Value
Net profit (USD)	155.70
Total Closed Trades	40
Short Trade Win Rate	66.67%
Long Trade Win Rate	43.75%
Overall Win Rate	57.5%
Profit Factor	2.11
Risk-Reward Ratio (RRR)	1.56
Maximum Drawdown (USD)	85
Average trade profit (USD)	3.89

5 Discussion

The findings of this study confirm that AI can meaningfully improve the process of making investment decisions, particularly with regard to analyzing market sentiment, developing strategies, and automating processes. However, the effectiveness of these approaches depends heavily on the quality of implementation, the technical infrastructure and the role of human oversight. These results align with the broader literature emphasizing the conditional success of AI systems in finance, where technological advantages must be complemented by sound domain expertise and a clear understanding of model limitations [25, 26].

In sentiment analysis, comparing the custom Google Sheets–Claude setup with the AltIndex platform revealed clear trade-offs. While the custom system offered flexibility and low cost, it suffered from limited predictive stability, particularly in short-term forecasts. This reflects the broader challenge of extracting actionable signals from noisy, context-dependent textual data [17]. AltIndex, benefiting from curated historical sentiment datasets and pre-tuned analytics, delivered more robust results, albeit at the expense of accessibility for individual investors. While these findings align with current market trends, where commercial AI platforms are increasingly shaping decision support tools in retail investing [27], they also raise questions about affordability and customizability for non-institutional users.

Backtesting of trading strategies using AI-generated models in TradingView revealed additional evidence. The DeepSeek model produced the most stable and profitable strategy, particularly for TSLA and AAPL stocks, providing a good balance between profit and drawdown. In contrast, the SuperTrend strategy with ADX filtering—generated by a GPT model—achieved higher profitability but with elevated risk, making it more volatile. This suggests that DeepSeek-based strategies are more appropriate for risk-averse investors seeking stability.

The final experiment using Zapier for trading automation could not achieve full autonomy due to the inability to connect the Alpaca API with a paper trading account, requiring manual trade execution. Interestingly, manual intervention often produced better results than fully automated systems. Short-selling strategies outperformed long positions, suggesting that under certain market conditions, AI-assisted short-term speculation may offer greater potential. A profit factor of 2.11 and a risk-reward ratio of 1.56 confirmed the effectiveness of this approach, though further performance improvements could be achieved through parameter optimization and better risk management.

6 Conclusion

The aim of this study was to explore selected AI tools that can assist both company management and non-professional investors in making informed investment decisions, including AI-based recommendations on whether to buy, sell, or hold assets. The research focused on three key areas: sentiment analysis, strategy generation, and trading automation. The results yielded valuable evidence of the effectiveness and real-world applicability of these methods.

In practice, two sentiment analysis approaches were compared: a custom-built solution using Google Cloud and Claude, and a commercial tool, AltIndex. Findings indicated that while sentiment alone is not a reliable short-term price predictor, it can enrich broader market analysis. Subsequent testing on the TradingView platform showed that AI-generated strategies—especially those built using the DeepSeek model—delivered high profitability and stability, validating AI's potential in strategy development. However, the need for thorough backtesting and parameter tuning remains critical. The experiment using Zapier and Alpaca for semi-automated trading showed promising results, especially in short-selling strategies. While automation was limited, the findings revealed that human oversight can sometimes outperform fully mechanical systems.

In summary, AI can substantially aid investment decision-making, particularly in data analysis, signal generation, and process automation. However, full autonomy is still constrained by technological barriers and the necessity of human judgment. AI should therefore be seen not as a replacement for traditional methods, but as a powerful supplement when used thoughtfully and tested rigorously. The development of AI-assisted strategies is a continuous learning process that, when properly managed, can lead to more effective and profitable investment decisions.

Several limitations emerged during the study that are important for understanding the results:

- Exclusion of transaction costs and dividends: Return calculations excluded brokerage fees and dividend payouts, which could have overestimated the strategies' net profitability.
- Short experiment duration in certain cases: The Zapier-based experiment was limited to two weeks due to the constraints of the free trial period, which restricted long-term evaluation.

- Computational constraints on cloud infrastructure: Implementing LSTM-based forecasting models on Google Cloud Platform encountered memory limitations, curtailing the complexity and depth of model tuning and evaluation.
- Integration barriers with trading APIs: Technical integration challenges with the Alpaca API prevented full automation and required manual trade execution. This limitation introduced potential human error and reduced experimental realism.
- Dependence on historical data: All models were trained and evaluated using historical market and sentiment data. However, this data may not generalize to future market conditions due to regime shifts or black swan events.

Based on the identified limitations, several promising avenues for future research emerge. Firstly, subsequent studies should incorporate transaction costs, slippage and dividend payouts to reflect real-world trading conditions more accurately and improve the reliability of profitability estimates. Secondly, longer-term experiments spanning multiple market cycles are essential for evaluating the robustness of AI-driven strategies under diverse economic conditions and structural shifts. Thirdly, it is crucial to address computational scalability; future research could leverage more powerful cloud infrastructures or efficient model architectures (e.g., transformers or lightweight LSTM variants) to enable deeper tuning and real-time processing. Fourthly, overcoming API integration challenges remains a practical hurdle. Researchers should explore standardized financial middleware or sandboxed environments that support end-to-end automation without compromising security or usability. Lastly, there is a need to develop forward-looking, adaptive models that can adjust to regime changes, unexpected events or concept drift. One approach could be to develop hybrid models that integrate real-time sentiment analysis, macroeconomic signals and probabilistic forecasting techniques to improve generalizability in volatile markets.

Acknowledgements This chapter was supported by the scientific research project of the Czech Sciences Foundation Grant No. 25-15405S.

References

1. Cao, L.: Ai in finance: challenges, techniques, and opportunities. ACM Comput. Surv. (CSUR) **55**(3), 1–38 (2022)
2. Ahmed, S., Alshater, M.M., El Ammari, A., Hammami, H.: Artificial intelligence and machine learning in finance: a bibliometric review. Res. Int. Bus. Financ. **61**, 101646 (2022)
3. Hussain, S.M., Ahmad, N., Ahmed, S.: Applications of high-frequency data in finance: a bibliometric literature review. Int. Rev. Financ. Anal. **89**, 102790 (2023)
4. Zaremba, A., Demir, E.: Chatgpt: unlocking the future of nlp in finance. Mod. Financ. **1**(1), 93–98 (2023)
5. Addy, W.A., Ajayi-Nifise, A.O., Bello, B.G., Tula, S.T., Odeyem, O., Falaiye, T.: Algorithmic trading and AI: a review of strategies and market impact. World J. Adv. Eng. Technol. Sci. **11**(1), 258–267 (2024)

6. Yadava, A.: The impact of AI-driven algorithmic trading on market efficiency and volatility: evidence from global financial markets. Inf. Sci. **36**(3), 102015 (2024)
7. Theate, T., Ernst, D.: An application of deep reinforcement learning to algorithmic trading. Expert Syst. Appl. **173**, 114632 (2021)
8. Majidi, N., Shamsi, M., Marvasti, F.: Algorithmic trading using continuous action space deep reinforcement learning. Expert Syst. Appl. **235**, 121245 (2024)
9. Martins Pereira, C.: Unregulated algorithmic trading: testing the boundaries of the european union algorithmic trading regime. J. Financ. Regul. **6**(2), 270–305 (2020)
10. Prabakar, S., Kumar, A., Jayakarthik, R., Venkatesh, D., RS, P., Khan, B.: Empirical evaluation of stock market prediction system using intelligent learning scheme with data processing logic. In: 2024 5th International Conference on Intelligent Communication Technologies and Virtual Mobile Networks (ICICV), pp. 290–297. IEEE (2024)
11. Hajek, P., Novotny, J.: Beyond sentiment in stock price prediction: integrating news sentiment and investor attention with temporal fusion transformer. In: IFIP International Conference on Artificial Intelligence Applications and Innovations, pp. 30–43, Cham (2024). Springer Nature Switzerland
12. Byrum, J.: AI in financial portfolio management: practical considerations and use cases. Innov. Technol. Interface Financ. Oper. **I**, 249–270 (2022)
13. Ali, H., Aysan, A.F.: What will chatgpt revolutionize in the financial industry? Mod. Financ. **1**(1), 116–129 (2023)
14. Adegboye, A., Kampouridis, M., Otero, F.: Algorithmic trading with directional changes. Artif. Intell. Rev. **56**(6), 5619–5644 (2023)
15. Hajek, P., Novotny, J., Kovarnik, J.: Predicting exchange rate with finbert-based sentiment analysis of online news. In: Proceedings of the 2022 6th International Conference on E-Business and Internet, pp. 133–138 (2022)
16. Sahut, J.M., Hajek, P., Olej, V., Hikkerova, L.: The role of news-based sentiment in forecasting crude oil price during the covid-19 pandemic. Ann. Oper. Res. **345**(2), 861–884 (2025)
17. Hajek, P., Novotny, J.: Fuzzy rule-based prediction of gold prices using news affect. Expert Syst. Appl. **193**, 116487 (2022)
18. Mohammed, S., Krishna, S.H., Mudalkar, P.K., Verma, N., Karthikeyan, P., Yadav, A.S.: Stock market price prediction using machine learning. In: 2023 5th International Conference on Smart Systems and Inventive Technology (ICSSIT), pp. 823–828. IEEE (2023)
19. Li, Z., Tam, V.: Developing an attention-based ensemble learning framework for financial portfolio optimisation. In: 2024 International Joint Conference on Neural Networks (IJCNN), pp. 1–8. IEEE (2024)
20. Huang, Z., Gong, W., Duan, J.: Tbdqn: a novel two-branch deep q-network for crude oil and natural gas futures trading. Appl. Energy **347**, 121321 (2023)
21. Sattar, A., Sarwar, A., Gillani, S., Bukhari, M., Rho, S., Faseeh, M.: A novel rms-driven deep reinforcement learning for optimized portfolio management in stock trading. IEEE Access **13**, 42813–42835 (2025)
22. Morelli, G.: Responsible investing and portfolio selection: a shapley-cvar approach. Ann. Oper. Res. **342**(3), 1991–2019 (2024)
23. Murugan, V.G., Nanda, S.S., Jatty, G.N., Kumar, N.R., Chamundeeswari, V.: Integrating federated learning and blockchain for enhanced financial portfolio management. In: 2025 3rd International Conference on Integrated Circuits and Communication Systems (ICICACS), pp. 1–6. IEEE (2025)
24. Hacker, P., Engel, A., Mauer, M.: Regulating chatgpt and other large generative AI models. In: Proceedings of the 2023 ACM Conference on Fairness, Accountability, and Transparency, pp. 1112–1123 (2023)
25. Dixon Roman, E., Nichols, T.P., Nyame-Mensah, A.: The racializing forces of/in AI educational technologies. Learn. Media Technol. **45**(3), 236–250 (2020)
26. Kovalerchuk, B.: Visual Knowledge Discovery and Machine Learning, vol. 144. Springer, Cham (2018)
27. Chishti, S., Barberis, J.: The Fintech book: The Financial Technology Handbook for Investors, Entrepreneurs and Visionaries. Wiley, New York (2016)

Explainable AI in Finance: Enhancing Transparency and Interpretability of AI Models in Financial Decision-Making

Anh Hoang, Hien Phan, and Van-Doan Nguyen

Abstract The integration of Artificial Intelligence (AI) into the finance sector is reshaping decision-making, risk management, and financial analysis, driving both efficiency and innovation. However, the "black box" nature of many AI models presents critical challenges regarding transparency and interpretability, especially in high-stakes financial decisions. The inability to explain AI-generated results can lead to legal, ethical, and financial risks, including regulatory penalties and loss of stakeholder trust. Explainable AI (XAI) addresses these challenges by offering methodologies that enhance the transparency and interpretability of AI models. This chapter examines the role of XAI in financial decision making, providing a comprehensive review of XAI and its applications in areas such as risk management, portfolio optimization, lending services, fraud detection, and stock market analysis. It also explores the key challenges of implementing XAI, assesses the trade-offs between performance and interpretability, and outlines future directions to advance XAI to meet the ethical, regulatory, and operational needs of the rapidly evolving financial landscape.

1 Introduction

The rapid advancement of artificial intelligence (AI) has precipitated a paradigm shift across multiple industries, with the financial sector emerging as a primary beneficiary of this technological revolution. AI-driven models, using machine learning (ML), deep learning, and natural language processing (NLP), now play a critical role in key financial applications, including predictive analytics, risk assessment, fraud

A. Hoang (✉) · H. Phan
University of Economics, Ho Chi Minh City, Vietnam
e-mail: anhh@ueh.edu.vn

H. Phan
e-mail: hienphan@ueh.edu.vn

V.-D. Nguyen
Holmes Institute, Melbourne, Australia

H. T. T. Nguyen et al. (eds.), *Data Science in Finance and Accounting*, Studies in Big Data 181, https://doi.org/10.1007/978-3-032-06179-9_9

detection, and algorithmic trading [22]. Using these technologies, financial institutions and accounting professionals can process large datasets with high dimensionality, identify nonlinear correlations, and generate real-time data-driven insights with enhanced accuracy.

Rapid integration of AI in fintech is evidenced by notable market growth: valued at $22.5 billion in 2023, the global AI fintech market reached $27 billion in 2024 (Statista, 2024). Projected to grow at a 19.7% CAGR through 2030, the sector is poised to exceed $79 billion by 2030, highlighting AI's disruptive role in reshaping financial services (Fig. 1).

However, the widespread adoption of AI in these sectors has introduced significant challenges, particularly in relation to the transparency and interpretability of AI models. The "black box" nature of many AI systems, particularly deep learning algorithms, which often operate without providing clear explanations for their decision-making processes, may pose serious risks [12]. The finance sector comprises a wide range of industries, institutions, and activities focused on managing, generating, and investing capital, where compliance, stakeholder trust, and accountability are critical. In this environment, lack of transparency or the inability to explain AI-driven decisions can lead to significant legal, ethical, and financial consequences, including regulatory penalties, revenue loss, reputational damage, diminished investor confidence, and potential legal action or sanctions. This is where XAI comes into play.

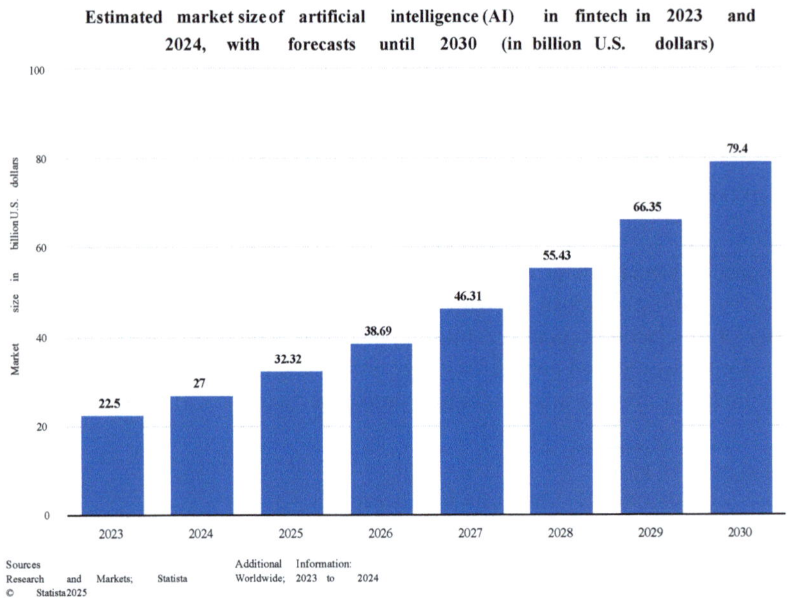

Fig. 1 Market size of AI in fintech 2023–2024, with forecasts until 2030

XAI methodologies aim to make AI models more transparent, interpretable, and trustworthy by providing insight into how these models arrive at their conclusions without compromising their predictive accuracy [13].

This chapter provides a comprehensive review of XAI in the financial sector, focusing on its role in enhancing transparency and interpretability in AI-driven decision making. In addition, it addresses the inherent challenges in the implementation of XAI within complex financial systems, such as balancing model performance with interpretability and ensuring compliance with stringent regulatory standards. The chapter also highlights future research directions and opportunities to advance XAI to better serve the evolving needs of the financial industry.

The structure of this chapter is as follows. It begins with an overview of XAI in the context of finance, followed by a comprehensive examination of current XAI methodologies, including both model-agnostic and model-specific approaches, feature importance analysis, and counterfactual explanations. The chapter then explores the application of XAI in various financial decision-making processes, including risk management, portfolio optimization, fraud detection, lending services, and stock market analysis. Subsequently, it addresses approaches for evaluating the explainability of AI models. The chapter also discusses key challenges and future research directions in the field, offering insights into emerging opportunities. Conclusions summarize key findings and highlight their implications for the future development of AI in finance.

2 Explainable AI in Finance: An Overview

The financial sector is increasingly using AI to revolutionize decision-making processes in various domains, including risk management, fraud detection, portfolio optimization, and customer service. These advancements are driven by the unparalleled capabilities of AI models, particularly those based on deep learning, which have demonstrated exceptional accuracy and efficiency in processing large amounts of data and identifying complex patterns [15, 20, 27]. However, despite these benefits, AI models often function as "black boxes", making it difficult for stakeholders to understand the reasoning behind their predictions. This lack of transparency can lead to mistrust, especially in high-stakes environments such as finance, where decisions must be not only accurate but also transparent, interpretable, and in full compliance with regulatory requirements.

This challenge is further exacerbated by the inherent complexity of financial systems, where decisions can have far-reaching consequences. The inability to explain how AI models arrive at certain predictions can hinder their adoption and raise ethical and legal concerns. This is where XAI comes into play, offering a critical solution to bridge the gap between AI's predictive power and the need for transparency and accountability [8].

XAI refers to a suite of techniques and methodologies designed to make AI models more interpretable without significantly compromising their performance. In the context of finance, XAI is not just a technical requirement but a strategic imperative. It serves several key purposes that are crucial for the sustainable integration of AI in the financial industry:

- Improved Decision-Making: The complexity of financial markets requires decision makers to rely on accurate and interpretable data. XAI empowers financial professionals by providing interpretable outputs that improve decision-making processes [1]. For example, in portfolio management, XAI can identify and highlight the characteristics that most influence investment recommendations, enabling managers to align their strategies with their risk preferences and investment goals. By offering actionable and understandable insights, XAI improves the quality of decisions, ultimately leading to better financial outcomes.
- Customer Communication: In customer-facing applications, such as robo-advisors or automated customer support systems [11], XAI plays a vital role in improving the user experience. By providing clear and understandable explanations for AI-generated advice or decisions, XAI helps bridge the gap between complex algorithms and everyday users. This transparency can significantly improve customer trust and satisfaction, as customers are more likely to engage with services that they understand. In addition, in scenarios where customers are required to make informed financial decisions, the ability to explain the rationale behind recommendations is crucial to building long-term relationships and ensuring customer loyalty.
- Trust and Accountability: Trust is the cornerstone of financial services, and the adoption of AI technologies depends on the ability to build and maintain this trust. XAI facilitates trust by unveiling the inner workings of AI models, allowing financial professionals to verify and validate the decisions made by these models [24]. This transparency fosters a culture of accountability, where organizations can trace decisions back to specific data points and model behaviors, ensuring that they align with ethical standards and business objectives. Moreover, in an era where data-driven decisions are increasingly scrutinized, the ability to explain AI outputs is essential to defend those decisions against potential challenges.
- Risk Management: Managing risk is a fundamental aspect of financial operations. XAI contributes to more robust risk management strategies by providing clear explanations for model predictions. This allows financial experts to identify and mitigate potential biases or errors in the models, reducing the probability of unexpected outcomes [30]. For example, in trading algorithms, understanding the factors driving buy or sell recommendations can help traders assess the associated risks more effectively, leading to better-informed decisions that align with the organization's risk appetite.
- Regulatory compliance: Financial institutions operate under stringent regulatory frameworks that require transparency in decision-making processes. XAI ensures that AI-driven decisions can be articulated in a way that is understandable to regulators, auditors, and other stakeholders [8]. This is particularly critical in areas such as credit scoring and loan approvals, where the rationale behind decisions must be clearly communicated to avoid regulatory penalties and to maintain pub-

lic trust. By providing explanations that satisfy regulatory scrutiny, XAI helps financial institutions navigate the complex landscape of compliance with greater confidence.

Despite its significant potential, the implementation of XAI in finance is not without challenges. One of the primary issues is balancing the trade-off between interpretability and model performance. As AI models become more complex, their interpretability tends to decrease, making it challenging to explain their predictions in a way that is both accurate and accessible. Moreover, developing XAI techniques that are not only effective, but also scalable and applicable to complex financial models is an ongoing area of research. Financial models often involve intricate data relationships and market dynamics, requiring XAI solutions that can handle this complexity without oversimplifying the underlying processes.

However, explainable AI holds the promise of transforming the financial industry by making AI-driven decisions more transparent, interpretable, and aligned with regulatory standards. As the field of XAI continues to evolve, financial institutions must actively explore and integrate these methodologies to harness the full potential of AI while maintaining trust, accountability, and compliance. The future of finance will likely be shaped by the successful integration of XAI, ensuring that AI-driven innovations are powerful and responsible [8].

3 Current Advancements in XAI Methodologies for Finance

As the use of artificial intelligence (AI) in the financial sector continues to expand, ensuring transparency and interpretability of AI-driven decisions has become increasingly critical. Many traditional AI models, such as neural networks, support vector machines, and random forests—often referred to as "black-box" systems—lack the necessary explainability to meet stringent regulatory requirements and foster stakeholder trust in high-stakes environments like finance. XAI has emerged as a solution to this challenge, offering a variety of methodologies that improve the interpretability of AI models without compromising performance. This section reviews recent advances in XAI methodologies, focusing on approaches that are particularly relevant to financial applications. These methods provide diverse techniques to interpret, explain, and validate AI predictions, ensuring that financial institutions can maintain transparency, accountability, and compliance while harnessing the power of AI. The following subsections will explore specific methodologies, including model-agnostic and model-specific approaches, feature importance analysis, and counterfactual explanations, each of which plays a crucial role in enhancing the interpretability of AI systems used in financial decision-making processes.

3.1 Model-Agnostic Methods

Model-agnostic XAI methods, such as Local Interpretable Model-Agnostic Explanations (LIME) and SHapley Additive Explanations (SHAP), are versatile and applicable across various AI models, including neural networks and ensemble methods. These methods focus on explaining individual predictions and model behavior without requiring access to or modification of the underlying model architecture [20, 22, 27]. This flexibility makes them particularly valuable in the financial sector, where various predictive models are used for tasks such as credit scoring and risk assessment [9, 10, 13].

LIME approximates the decision boundaries of complex models with simpler and interpretable models, providing information on how local changes in input data affect predictions. This approach is especially valuable in scenarios where decision justification is critical, such as loan approvals and risk assessments [27]. Following an analysis of current XAI methodologies, LIME has been extensively applied in financial applications, helping stakeholders understand the factors that influence specific credit decisions and identifying potential concerns in fraud detection [22].

LIME's mathematical formulation minimizes a loss function that balances fidelity to the original model and interpretability of the explanation. The goal is to identify an interpretable model $g \in G$ (where G is a class of interpretable models) that best approximates the black-box model f within a local neighborhood π_x. The optimization problem can be expressed as:

$$\arg\min_{g \in G} L(f, g, \pi_x) + \Omega(g) \tag{1}$$

where $L(f, g, \pi_x)$ is the fidelity function that measures how well the interpretable model g approximates f within the neighborhood π_x, and $\Omega(g)$ is the complexity term that ensures the explanation remains simple and interpretable. The locally interpretable model in LIME is typically a linear function:

$$\hat{f}(x) = \beta_0 + \sum_{i=1}^{n} \beta_i x_i \tag{2}$$

where $\hat{f}(x)$ is the locally interpretable model that approximates the complex model, β_0 is the intercept, and β_i represents the importance of the characteristic x_i in influencing the prediction. To improve interpretability and robustness, LIME assigns different weights to perturbed samples based on their proximity to the original instance. The weighting function is often chosen to be an exponential kernel:

$$\pi_x(z) = \exp\left(-\frac{D(x, z)^2}{\sigma^2}\right) \tag{3}$$

where $D(x, z)$ represents the distance between the original instance x and the perturbed sample z, and σ is a kernel width parameter that controls how much importance is given to samples near x.

On the other hand, SHAP (Shapley Additive Explanations) quantifies feature contributions to model predictions using principles from cooperative game theory. In finance, SHAP improves the interpretability of AI-driven applications, such as credit scoring, portfolio management, and financial forecasting, by providing transparent explanations that build stakeholder trust in model outputs [20]. As demonstrated in [22], SHAP's dual capability to provide global (model-wide) and local (instance-specific) explanations has established it as a leading method for deciphering complex model behavior.

The SHAP value quantifies how much a particular feature contributes to a specific prediction by computing the marginal contributions of that feature across different subsets of input features. The SHAP value for a feature i is defined as:

$$\phi_i = \sum_{S \subseteq N \setminus i} \frac{|S|!(|N| - |S| - 1)!}{|N|!} (v(S \cup i) - v(S)) \tag{4}$$

where S is a subset of all features except i, N is the set of all features, and $v(S)$ is the model prediction when using only the subset of features S.

3.2 Model-Specific Methods

Model-specific XAI methods are designed to take advantage of the unique characteristics of specific AI models, such as neural networks and deep learning architectures. These methods aim to demystify the internal workings of complex models and explain their predictions in a way that is comprehensible to humans. Neural networks employ techniques such as attention mechanisms and layer-wise relevance propagation to identify the most influential features driving the model predictions.

One of the most widely studied model-specific XAI methods for neural networks is Layer-wise Relevance Propagation (LRP). As described by Bach et al. [7], LRP traces the contribution of each input feature through the various layers of a neural network, providing a detailed understanding of which features have the greatest impact on model output. In finance, LRP has proven particularly effective in applications such as fraud detection and market analysis, where the need to understand the reasoning behind AI-driven decisions is critical. The use of LRP in these contexts provides stakeholders with valuable insights into the decision-making process, improving transparency and accountability in AI-driven financial systems.

Another prominent model-specific XAI method is the use of saliency maps [33], which provide a visual representation of the input features that most influence the model predictions. Saliency maps have been widely applied in financial settings to help analysts identify key factors driving investment strategies and risk assessments.

By visually highlighting the most influential features, saliency maps contribute to more transparent and informed decision-making processes. This is especially crucial in high-stakes financial environments, where regulatory frameworks demand clear explanations for AI-driven decisions.

3.3 Feature Importance Analysis

This process involves analyzing and ranking the variables that have the greatest influence on the model's outcomes, such as predicting credit risk, stock prices, loan defaults, or customer churn. Financial data sets typically consist of a wide range of variables, including macroeconomic indicators such as interest rates and inflation indices, as well as individual customer data points such as transactional history and credit scores. In permutation importance methods [2], the values of each feature are randomly shuffled to evaluate its impact on predictive outcomes. This helps measure the importance of each feature in the decision-making process. These techniques are particularly valuable in finance, where explainability and transparency are crucial to regulatory compliance, risk management, and building trust with stakeholders. For example, understanding which factors contribute the most to a loan approval decision or portfolio optimization can help financial institutions adjust their strategies, reduce model bias, and enhance decision-making accuracy. By quantifying the impact of each feature, these methods not only ensure transparency but also allow more targeted data collection, improved risk assessment, and continuous model refinement in dynamic financial environments.

3.4 Counterfactual Explanations

In the context of XAI for decision making, generating hypothetical scenarios can provide critical insights into the decision-making processes of AI models. For example, consider a loan approval model that denies credit to a potential borrower. Using counterfactual explanations, financial institutions can investigate how changes in input characteristics, such as income, credit score, or debt-to-income ratio, could have resulted in a different outcome, such as loan approval. This approach enables stakeholders to better understand the underlying logic and rationale behind the prediction of the model, offering actionable information on what adjustments would be necessary to achieve a positive outcome in the future [14].

This method is particularly valuable in finance because of the industry's demand for transparency and accountability, especially in regulatory environments. Financial institutions are often required to provide clear reasons for decisions, such as denying credit or making high-risk investment choices. Counterfactual explanations help fulfill these regulatory requirements by demonstrating which financial variables had the most influence on the decision and how adjustments to these variables could have altered the outcome. As Dastile and Celik highlight, counterfactual explanations can provide specific actionable changes that would make credit approval more likely, making these explanations intuitive and useful in real-world applications.

In addition, by identifying these critical decision points, financial institutions can leverage counterfactual explanations to refine their risk assessment models, develop more equitable decision-making processes, and optimize strategies to improve portfolio performance or mitigate default risk. These insights are crucial for detecting model biases, ensuring fairness in automated decision systems, and fostering trust with stakeholders by offering transparent and understandable outcomes for significant financial decisions. Just as counterfactual explanations have been applied effectively in risk-based decision making in construction, where they helped identify key factors such as bubble chamber pressure, cutter head speed, and equipment inclination in managing ground settlement risks, similar principles can be applied in finance. Counterfactuals can help financial institutions understand how adjustments in variables such as credit scores or income levels influence outcomes, enabling more informed, transparent, and actionable decisions [32].

In summary, counterfactual explanations provide a powerful mechanism for improving both the transparency and accountability of AI-driven decisions in finance. By shedding light on how different input variables impact outcomes, they offer financial institutions clear insights into the decision-making processes that underpin their models. This capability makes them indispensable not only for meeting regulatory requirements, but also for improving decision-making processes in areas such as credit scoring, fraud detection, and portfolio management. Counterfactual explanations not only help financial institutions comply with regulations, but also enable them to refine their models, ensuring fairness, transparency, and accuracy in various financial applications.

4 Applications of XAI in Financial Decision-Making

As AI systems continue to permeate the financial industry, their ability to predict outcomes, assess risks, and make autonomous decisions has become increasingly valuable. However, the complexity of AI models often leads to challenges in understanding how these decisions are made. This opacity, commonly referred to as the "black box problem", raises concerns about transparency, accountability, and fairness in AI-driven financial processes. XAI addresses these challenges by providing interpretability tools that shed light on the internal workings of AI models, helping stakeholders understand the rationale behind AI predictions and decisions. In finance, where compliance with regulatory standards and the need for clear justification of decisions are paramount, XAI plays a crucial role in improving transparency. By making AI models more interpretable, XAI empowers financial institutions to justify decisions in areas such as credit risk assessment, fraud detection, portfolio management, and lending services. It also enables institutions to detect biases, ensure fairness in automated decision-making processes, and build trust with customers.

The following subsections explore key applications of XAI in various financial domains. From predictive modeling and risk management to portfolio optimization and lending services, XAI techniques are being used to improve the precision and fairness of decision making, while ensuring that AI-driven processes remain transparent and accountable.

4.1 Predictive Modeling and Risk Management

Predictive modeling plays an important role in finance, facilitating the forecasting of financial performance, the assessment of creditworthiness, and the management of various risks. AI models have demonstrated significant capabilities in these areas by leveraging historical data to identify patterns and trends. However, the inherent opacity of AI models often impedes the transparency of their decision-making processes, thereby limiting stakeholders' ability to fully understand and trust the outcomes of these models. Lack of transparency is particularly challenging in the context of risk assessment methodologies and justification of decisions [21].

Artificial intelligence models are extensively used in credit scoring systems to assess the creditworthiness of loan applicants. XAI techniques, such as LIME, provide valuable insight into the factors that influence credit decisions. These techniques not only enable lenders to justify their decisions, but also help borrowers comprehend the rationale behind credit approvals, fostering greater transparency and trust in the lending process [17, 27].

Predicting market fluctuations and evaluating their potential impact on investment portfolios require effective and robust risk management strategies. XAI methods contribute by offering interpretability to AI-driven predictions, allowing investors to understand the reasoning behind model output. This enhanced understanding enables investors to make informed decisions and implement strategies to mitigate risks associated with market volatility [9, 19].

Moreover, the FinTech sector requires comprehensive risk management frameworks that address not only traditional financial risks but also emerging threats, such as cyber risks. The integration of XAI methods within FinTech applications improves the interpretability of risk models, allowing stakeholders to better understand, anticipate, and mitigate these evolving risks [23, 34].

4.2 Portfolio Optimization

Portfolio optimization is a critical strategy in finance that aims to achieve the best possible returns while minimizing risks through careful asset allocation [5, 36]. AI models play a pivotal role in this process by analyzing vast amounts of historical data, market trends, and financial indicators to recommend optimal portfolios. These models can assess a variety of factors, including asset performance, market volatility, and correlations between different asset classes, to generate data-driven recommendations. However, while AI models excel in identifying patterns and making predictions, their complexity often leads to a lack of transparency, making it difficult for investors to fully understand the rationale behind their recommendations. This opacity can reduce trust in AI-driven suggestions, particularly when making high-stakes financial decisions. Therefore, ensuring that investors can understand the underlying logic behind these AI-generated portfolios is crucial for informed decision-making and effective risk management.

To address this challenge, XAI techniques such as SHAP are increasingly being integrated into portfolio optimization tools [16]. SHAP is particularly effective because it quantifies the contribution of each feature, such as asset volatility, expected returns, or market conditions, to the overall portfolio recommendation. By providing a detailed breakdown of how each factor influences the final decision, SHAP allows investors to gain a deeper understanding of the reasoning behind the outputs of the AI model. This interpretability is especially valuable for aligning investment decisions with personal or institutional goals. For example, by visualizing the importance of different characteristics, investors can adjust their portfolios to better reflect their risk tolerance, time horizons, or specific investment objectives. Thus, XAI techniques not only improve portfolio performance, but also help investors make more strategic and personalized financial decisions, ultimately increasing their confidence in AI-driven tools [20].

4.3 *Fraud Detection*

AI-driven fraud detection systems are critical in the financial sector, where they help monitor and analyze large volumes of transactional and financial data to detect suspicious activities. These systems protect the integrity of financial reporting and prevent financial losses due to fraudulent transactions. However, while AI models are highly effective in identifying anomalies, they often lack transparency, making it difficult for fraud investigators and compliance officers to understand why specific transactions have been flagged. This limitation can hinder the validation of alerts and compliance with stringent regulatory requirements [27].

Enhancing the interpretability of these models is essential for overcoming these limitations. The integration of XAI methods brings much-needed transparency to fraud detection systems [35]. Techniques such as feature importance analysis and decision rule extraction allow investigators to understand the factors influencing flagged transactions. XAI methods offer clear, interpretable explanations, enabling investigators to assess whether a flagged transaction truly indicates fraudulent activity. This is crucial not only for improving fraud detection accuracy, but also for ensuring regulatory compliance and maintaining trust with stakeholders [20]. Additionally, as Awosika et al. highlight, the combination of XAI with advanced models such as Federated Learning ensures that financial institutions can collaboratively train fraud detection models while preserving data privacy. The integration of XAI provides transparency in decision-making, which is vital for both regulatory compliance and operational effectiveness [4].

4.4 Lending Services

XAI is revolutionizing the lending industry by fostering innovation with greater transparency, fairness, and accountability in automated decision-making systems. Lenders can take advantage of XAI to create more transparent loan approval processes, which are crucial for maintaining the trust of both borrowers and regulators [26, 31]. In conventional loan approval processes, decisions are often unclear, leaving borrowers uncertain of why they were denied loans. XAI handles this problem by providing clear and understandable explanations of the factors that influenced the decision, such as income, credit history, or the debt-to-income ratio. By providing these explanations, XAI allows borrowers to better understand their financial situation and what steps they can take to improve their eligibility to meet loan conditions [29]. This not only increases the trust of the borrower, but also empowers customers with actionable insights.

Moreover, by improving transparency, XAI plays a key role in mitigating biases within lending models [10]. Automated lending systems can unintentionally incorporate biases based on demographic factors such as age, race, or gender. With XAI, lenders can analyze the decision-making patterns of their models to ensure fairness, detect potential biases, and address them. This proactive approach not only helps protect against discrimination, but also enhances the institution's reputation for ethical lending practices.

In addition to compliance and fairness, XAI helps in risk management by allowing lenders to better understand the predictive variables in their risk assessment models [18]. Through explainability techniques, lenders can refine their models to improve accuracy, identifying which borrower characteristics are most indicative of creditworthiness. This helps institutions fine-tune their risk models, potentially reducing default rates and improving the overall health of their loan portfolios.

By improving transparency, fairness, and compliance, XAI is revolutionizing the lending process, creating a mutually beneficial situation for both financial institutions and borrowers [6]. For financial institutions, XAI strengthens customer relationships by providing clear and understandable explanations for loan decisions, fostering greater trust and satisfaction. Borrowers are more likely to engage with a lender they perceive as fair and transparent, improving long-term customer retention. In addition, XAI ensures compliance with increasingly stringent regulations, allowing institutions to meet legal requirements related to fairness and non-discrimination in lending practices. This not only helps avoid potential legal issues, but also improves the institution's reputation for ethical decision-making. Furthermore, XAI reduces the risks associated with lending decisions by improving the precision of risk assessments, leading to more reliable results and better overall portfolio management.

4.5 Stock Market Analysis

Artificial intelligence models have become crucial in stock market analysis, where they are employed to predict stock prices, identify profitable trading opportunities, and optimize trading strategies. These models analyze vast datasets, including historical prices, technical indicators, and market trends, to generate predictions that help traders and investors make data-driven decisions. However, despite their powerful capabilities, the complexity of AI models, particularly deep learning architectures and neural networks, often obscures the rationale behind their predictions [25]. This "black box" nature presents significant challenges for stakeholders, such as traders, analysts, and financial institutions, who struggle to fully trust or understand the logic behind the outputs. The lack of transparency in AI-generated predictions can lead to hesitation in decision making, especially in high-risk scenarios where the financial stakes are considerable. Thus, enhancing the interpretability of these models has become essential to foster trust and promote more informed, data-driven decisions throughout the financial industry [27].

XAI techniques, such as LIME and SHAP, are essential to improve the interpretability of these predictive models. By elucidating the factors driving stock price predictions, XAI enhances transparency, allowing traders and analysts to validate AI-driven insights and make informed decisions under dynamic market conditions. Recent advances in XAI have further demonstrated its applicability in more specialized contexts, such as candlestick pattern prediction. The research by Euko, Santos and Novais integrated XAI techniques such as SHAP and GRAD-CAM with Conv2D models to interpret candlestick images, a key tool in predicting market trends. This approach not only improves the accuracy of AI models, but also enhances their interpretability, making the decision-making process clearer for financial stakeholders who rely on candlestick patterns to anticipate market movements [15].

Incorporating explainability into AI models for stock market prediction is critical to maintaining transparency and fostering confidence in AI-driven decision-making. This is especially important in volatile and fast-paced financial environments, where fast and accurate decisions are crucial to capitalizing on market opportunities or avoiding losses. Using XAI techniques, traders and analysts can gain deeper insights into the factors driving model predictions, enabling them to assess risks more effectively, validate AI-generated outputs, and make more strategic adjustments to their trading strategies. Furthermore, as market conditions change rapidly, XAI allows stakeholders to adapt their decisions based on clear and interpretable insights, reducing uncertainty and increasing confidence in the use of AI technologies. In the long term, the integration of XAI not only enhances the accuracy and reliability of stock market predictions, but also builds stronger trust between financial professionals and the advanced AI systems they rely on for critical decision making.

5 Evaluating Explainability of AI Models

Assessing the explainability of AI models is essential for building trust and ensuring responsible use in financial applications [3]. As AI systems become increasingly central to decision-making processes in finance, stakeholders demand transparency in how these models generate predictions and make decisions. Without sufficient explainability, AI models may be met with skepticism due to their "black box" nature, which can obscure the reasoning behind their outputs and hinder stakeholder confidence. This lack of clarity can lead to diminished trust and slower adoption of AI technologies. Therefore, a comprehensive evaluation framework for explainability must address several key aspects, including the model's interpretability, transparency, and the clarity of its explanations. Such a framework ensures that AI systems are not only technically sound but also accessible and comprehensible to both technical and non-technical stakeholders, fostering broader acceptance and trust in their use.

Several methods can be used to evaluate explainability, each targeting a different facet of transparency and clarity in AI models. One approach is to measure the accuracy of the explanations provided, ensuring that the model's justifications for its predictions align with actual outcomes. Coherence and consistency are also essential; explanations should remain stable in similar scenarios, reflecting a reliable and systematic approach. Another critical method is to assess how well the explanations align with human intuition. In financial contexts, where experts rely on established principles and rules, it is important that AI-driven insights complement human decision-making processes. Evaluating explainability may also involve conducting user studies, where experts, regulators, and stakeholders assess the quality, comprehensibility, and usefulness of the model's explanations. These studies can provide valuable feedback for improving the alignment between AI outputs and human understanding.

It is important to recognize that the evaluation of explainability is not a one-size-fits-all process. The specific methods used will depend on various factors, including the complexity of the AI model, the financial application, and the distinct needs of the stakeholders involved. More sophisticated models, such as deep neural networks, often require advanced techniques to explain their output, while simpler models, such as decision trees, are inherently more interpretable. Furthermore, different financial applications demand varying levels of explainability due to their unique risk profiles and regulatory requirements. In addition, stakeholder expectations vary considerably. Although regulators may prioritize compliance and legal transparency, users, such as traders or analysts, may require concise and actionable explanations that align with their decision-making processes. Therefore, developing a flexible and adaptive evaluation framework is essential to ensure that the explainability of AI models meets the needs of diverse stakeholders and remains effective in multiple financial contexts.

6 Challenges and Future Directions

The integration of XAI into the financial sector, although promising, presents significant challenges that must be overcome to fully exploit its potential. Despite advances in XAI methodologies, the widespread adoption and implementation of XAI in the finance industry is hampered by issues such as scalability, trade-offs between interpretability and model performance, regulatory compliance, and the need for interdisciplinary collaboration between AI researchers and financial experts [28].

One of the main issues is the trade-off between interpretability and model performance. Highly interpretable models, such as decision trees or linear regression, often lack the predictive power of more complex models like deep neural networks. However, these complex models are typically viewed as "black boxes", offering little to no insight into their decision-making processes. The tension between maintaining high model performance and ensuring interpretability is a significant hurdle that researchers and practitioners must overcome. As Rudin argues, in high-stakes decisions, interpretable models should be favored over "black-box" models to ensure that the decision-making process is transparent and accountable [28].

Another significant challenge is the scalability of XAI techniques. As financial models become more complex, the need for scalable XAI solutions becomes increasingly critical. Current XAI methods may struggle to keep pace with the growing complexity of financial models, particularly when dealing with vast amounts of data and high-dimensional spaces. Developing XAI techniques that are not only effective, but also scalable and applicable to large-scale financial models is an ongoing area of research [8].

The financial industry operates under stringent regulatory frameworks that demand transparency and accountability. As AI models are increasingly used in decision making, ensuring that these models comply with regulatory standards is paramount. XAI plays a crucial role in this context by providing the necessary transparency to satisfy regulatory requirements. However, there is growing concern about the ethical implications of AI, including issues related to fairness, bias, and accountability. Ensuring that XAI models adhere to ethical standards while maintaining their interpretability and performance is a delicate balance that financial institutions must achieve [8].

Bias in AI models is a well-documented issue that can lead to unfair or discriminatory outcomes. In finance, this could result in biased lending practices, investment recommendations, or customer profiling. XAI techniques offer a way to identify and mitigate such biases by making the decision-making process more transparent. However, detecting and addressing bias in complex AI models remains a significant challenge, particularly since these models are often trained on large datasets that may inherently contain biases [8].

As financial institutions increasingly rely on AI-driven insights, concerns about data privacy and security have come to the fore. The ability to explain AI models could potentially expose sensitive information, leading to privacy breaches. In contrast, the lack of transparency in AI models could obscure security vulnerabilities, making it difficult to identify and mitigate potential threats. Research highlights the need for XAI techniques that can balance transparency with privacy and data security, ensuring that AI models are both interpretable and secure [8].

Looking ahead, the development of XAI in finance is likely to focus on several key areas. New methodologies that can simultaneously achieve high performance and interpretability are needed. Advances in hybrid models, which combine interpretable models with complex AI techniques, may offer a solution. Furthermore, as the financial industry continues to evolve, there will be a growing demand for XAI tools that can be easily integrated into existing workflows and systems, offering real-time insights without compromising performance.

In addition, the ethical and regulatory landscape will continue to shape the development of XAI in finance. As regulatory bodies increasingly scrutinize the use of AI, financial institutions must ensure that their XAI models are not only effective but also compliant with emerging standards. This will require ongoing collaboration between AI researchers, financial professionals, and regulators to develop XAI frameworks that are robust and adaptable.

In general, while XAI offers immense potential to revolutionize the financial industry, its successful implementation requires overcoming several significant challenges. Developing scalable and interpretable models that strike a balance between performance and transparency remains a key focus of ongoing research. Furthermore, ensuring that these models meet ethical standards and comply with stringent regulatory frameworks is crucial for their broader adoption. Achieving the full potential of XAI in finance will require close collaboration among AI researchers, financial professionals, and regulators. By emphasizing solutions that improve the interpretability of the model without sacrificing performance and fostering interdisciplinary efforts, the financial sector can effectively leverage XAI. This will improve transparent and accountable AI decisions while reinforcing trust and compliance in an industry where ethics and regulations are crucial [8].

7 Conclusion

XAI has become indispensable for fostering transparency, regulatory compliance, and trust in AI-driven decision-making across the financial sector. As AI continues to play a critical role in functions such as credit scoring, fraud detection, and portfolio management, the need for interpretability and clarity is paramount. Without sufficient transparency, the results of complex AI models can lead to mistrust and regulatory challenges, undermining the benefits they offer. By providing clear insights into how AI models generate their predictions, XAI empowers financial institutions to manage risks more effectively, optimize decision-making, and ensure that AI systems operate fairly and ethically. The integration of XAI into financial workflows not only strengthens stakeholder confidence but also enhances the accountability and reliability of AI technologies.

In the future, advances in XAI methodologies will be essential to address the ongoing challenges of balancing model complexity with interpretability. As AI models grow in sophistication and the volume of data increases, scalable and robust XAI techniques will be critical to maintaining trust and transparency. Moreover, the regulatory landscape will increasingly demand explainability in AI applications, driving the need for more adaptable and transparent models. By integrating XAI, financial institutions can unlock the full potential of AI while ensuring accountability, reducing risks, and complying with evolving regulatory standards. Ultimately, XAI will serve as a key enabler of innovation and trust in the AI-powered financial sector, driving long-term success and sustainability.

References

1. Abusitta, A., Li, M.Q., Fung, B.C.M.: Survey on explainable AI: techniques, challenges and open issues. Expert Syst. Appl. **255**, 124710 (2024)
2. Altmann, A., Toloşi, L., Sander, O., Lengauer, T.: Permutation importance: a corrected feature importance measure. Bioinformatics **26**(10), 1340–1347 (2010)
3. Arrieta, A.B., Díaz-Rodríguez, N., Del Ser, J., Bennetot, A., Tabik, S., Barbado, A., García, S., Gil-López, S., Molina, D., Benjamins, R., et al.: Explainable artificial intelligence (xai): concepts, taxonomies, opportunities and challenges toward responsible AI. Inf. Fusion **58**, 82–115 (2020)
4. Awosika, T., Shukla, R.M., Pranggono, B.: Transparency and privacy: the role of explainable AI and federated learning in financial fraud detection. IEEE Access **12**, 64551–64560 (2024)
5. Babaei, G., Giudici, P., Raffinetti, E.: Explainable artificial intelligence for crypto asset allocation. Financ. Res. Lett. **47**, 102941 (2022)
6. Babaei, G., Giudici, P., Raffinetti, E.: Explainable fintech lending. J. Econ. Bus. **125**, 106126 (2023)
7. Bach, S., Binder, A., Montavon, G., Klauschen, F., Müller, K.-R., Samek, W.: On pixel-wise explanations for non-linear classifier decisions by layer-wise relevance propagation. PLOS ONE **10**(7), 1–46 (2015)
8. Arrieta, A.B., Díaz-Rodríguez, N., Del Ser, J., Bennetot, A., Tabik, S., Barbado, A., Garcia, S., Gil-Lopez, S., Molina, D., Benjamins, R., Chatila, R., Herrera, F.: Explainable artificial intelligence (XAI): concepts, taxonomies, opportunities and challenges toward responsible AI. Inf. Fusion **58**, 82–115 (2020)
9. Bermúdez, L., Anaya, D., Belles-Sampera, J.: Explainable AI for paid-up risk management in life insurance products. Financ. Res. Lett. **57**, 104242 (2023)
10. Bertrand, A., Belloum, R., Eagan, J.R., Maxwell, W.: How cognitive biases affect xai-assisted decision-making: a systematic review. In: Proceedings of the 2022 AAAI/ACM Conference on AI, Ethics, and Society, pp. 78–91 (2022)
11. Bhatt, U., Xiang, A., Sharma, S., Weller, A., Taly, A., Jia, Y., Ghosh, J., Puri, R., Moura, J.M.F., Eckersley, P.: Explainable machine learning in deployment. In: Hildebrandt, M., Castillo, C., Celis, L.E., Ruggieri, S., Taylor, L., Zanfir-Fortuna, G., (eds.), FAT* '20: Conference on Fairness, Accountability, and Transparency, Barcelona, Spain, Jan 27–30, 2020, pp. 648–657. ACM (2020)
12. Carabantes, M.: Black-box artificial intelligence: an epistemological and critical analysis. AI Soc. **35**(2), 309–317 (2020)
13. Cerneviciene, J., Kabasinskas, A.: Explainable artificial intelligence (XAI) in finance: a systematic literature review. Artif. Intell. Rev. **57**(8), 216 (2024)

14. Dastile, X., Çelik, T.: Counterfactual explanations with multiple properties in credit scoring. IEEE Access **12**, 110713–110728 (2024)
15. Euko, J.P., Santos, F.A.O., Novais, P.: Stock market prediction: integrating explainable AI with conv2d models for candlestick image analysis. In: Rocha, Á., Adeli, H., Dzemyda, G., Moreira, F., Poniszewska-Maranda, A., (eds.), Good Practices and New Perspectives in Information Systems and Technologies - WorldCIST 2024, vol. 2, Lodz, Poland, 26–28 March 2024, Lecture Notes in Networks and Systems, vol. 986, pp. 13–23. Springer (2024)
16. Jaeger, M., Krügel, S., Marinelli, D., Papenbrock, J., Schwendner, P.: Interpretable machine learning for diversified portfolio construction. J. Financ. Data Sci. (2021)
17. Kovvuri, V.R.R., Hsuan, F., Fan, X., Seisenberger, M.: Fund performance evaluation with explainable artificial intelligence. Financ. Res. Lett. **58**, 104419 (2023)
18. Li, L.-H., Sharma, A.K., Cheng, S.-T.: Explainable AI based lightGBM prediction model to predict default borrower in social lending platform. Intell. Syst. Appl. **26**, 200514 (2025)
19. Lipton, Z.C.: The mythos of model interpretability: in machine learning, the concept of interpretability is both important and slippery. Queue **16**(3), 31–57 (2018)
20. Lundberg, S.M., Lee, S.-I.: A unified approach to interpreting model predictions. In: Guyon, I., Von Luxburg, U., Bengio, S., Wallach, H., Fergus, R., Vishwanathan, S., Garnett, R., (eds.) Advances in Neural Information Processing Systems, vol. 30. Curran Associates, Inc. (2017)
21. Ma, Y., Zhang, P., Duan, S., Zhang, T.: Credit default prediction of chinese real estate listed companies based on explainable machine learning. Financ. Res. Lett. **58**, 104305 (2023)
22. Martins, T., de Almeida, A.M., Cardoso, E., Nunes, L.: Explainable artificial intelligence (XAI): a systematic literature review on taxonomies and applications in finance. IEEE Access **12**, 618–629 (2024)
23. Mehrban, S., Nadeem, M.W., Hussain, M., Ahmed, M.M., Hakeem, O., Saqib, S., Mat Kiah, M.L., Abbas, F., Hassan, M., Khan, M.A.: Towards secure fintech: a survey, taxonomy, and open research challenges. IEEE Access **8**, 23391–23406 (2020)
24. Miller, T.: Explanation in artificial intelligence: insights from the social sciences. Artif. Intell. **267**, 1–38 (2019)
25. Muhammad, D., Ahmed, I., Naveed, K., Bendechache, M.: An explainable deep learning approach for stock market trend prediction. Heliyon **10**(21) (2024)
26. Purificato, E., Lorenzo, F., Fallucchi, F., Luca, E.W.D.: The use of responsible artificial intelligence techniques in the context of loan approval processes. Int. J. Hum.-Comput. Interact. **39**(7), 1543–1562 (2023)
27. Ribeiro, M.T., Singh, S., Guestrin, C.: Why should i trust you?: explaining the predictions of any classifier. In: Proceedings of the 22nd ACM SIGKDD International Conference on Knowledge Discovery and Data Mining, KDD '16, pp. 1135–1144, New York, NY, USA (2016). Association for Computing Machinery
28. Rudin, C.: Stop explaining black box machine learning models for high stakes decisions and use interpretable models instead. Nat. Mach. Intell. **1**(5), 206–215 (2019). May
29. Sachan, S., Yang, J.-B., Xu, D.-L., Benavides, D.E., Li, Y.: An explainable AI decision-support-system to automate loan underwriting. Expert. Syst. Appl. **144**, 113100 (2020)
30. Samek, W., Müller, K.-R.: Towards explainable artificial intelligence. In: Samek, W., Montavon, G., Vedaldi, A., Hansen, L.K., Müller, K.-R., (eds.), Explainable AI: Interpreting, Explaining and Visualizing Deep Learning, Lecture Notes in Computer Science, vol. 11700, pp. 5–22. Springer (2019)
31. Stevens, A., Deruyck, P., Van Veldhoven, Z., Vanthienen, J.: Explainability and fairness in machine learning: improve fair end-to-end lending for kiva. In: 2020 IEEE Symposium Series on Computational Intelligence (SSCI), pp. 1241–1248. IEEE (2020)
32. Zhan, J., Fang, W., Love, P.E.D., Luo, H.: Explainable artificial intelligence: counterfactual explanations for risk-based decision-making in construction. IEEE Trans. Eng. Manag. **71**, 10667–10685 (2024)
33. Zhang, H., Torres, F., Sicre, R., Avrithis, Y., Ayache, S.: Opti-cam: optimizing saliency maps for interpretability. Comput. Vis. Image Underst. **248**, 104101 (2024)

34. Zhang, T., Zhu, W., Yong, W., Zihao, W., Zhang, C., Xue, H.: An explainable financial risk early warning model based on the ds-xgboost model. Financ. Res. Lett. **56**, 104045 (2023)
35. Zhou, Y., Li, H., Xiao, Z., Qiu, J.: A user-centered explainable artificial intelligence approach for financial fraud detection. Financ. Res. Lett. **58**, 104309 (2023)
36. Zhu, M., Wang, Y., Fei, W., Yang, M., Chen, C., Liang, Q., Zheng, X.: Wise: wavelet based interpretable stock embedding for risk-averse portfolio management. In: Companion Proceedings of the Web Conference, 2022, pp. 1–11 (2022)

On Shared Directors and Liquidation: Evidence from UK SMEs

Ba Hung Nguyen and Dang Xuan Thanh Duong

Abstract Accurate credit risk assessment for Small and Medium-sized Enterprises (SMEs) is crucial for financial institutions, yet traditional methods often fail to capture the complex interconnections within the SME ecosystem. However, working with SME data presents significant challenges, including the limited availability of high-quality data, inconsistencies in director records, and variability in data formats. In this study, we address these challenges by leveraging graph-based approaches to model the intricate relationships between SMEs and their directors, focusing on how these relationships impact SMEs' liquidation likelihood. By representing SMEs as nodes and shared directors as edges, we adopt the "Entity-Attribute Relationship" paradigm to construct a graph that encapsulates both individual and relational attributes. Using a Graph Convolutional Network (GCN), we evaluate the predictive power of graph-based methods compared to traditional logistic regression. Our results demonstrate the superiority of GCN over the logistic regression baseline as an industry standard. This highlights the value of incorporating graph-structured data, such as shared director relationships, into credit risk modeling. This research contributes to the growing body of literature on AI-driven credit risk assessment, particularly for SMEs, by showcasing the effectiveness of graph-based features in capturing critical relational dynamics while addressing the inherent challenges of working with SME data.

B. H. Nguyen (✉)
School of Finance and Accounting, Industrial University of Ho Chi Minh City, Ho Chi Minh City, Vietnam
e-mail: nguyenbahung@iuh.edu.vn

D. X. Thanh Duong
Faculty of Economics and Management, Thai Binh Duong University, Nha Trang, Khanh Hoa, Vietnam
e-mail: thanh.duong@tbd.edu.vn

H. T. T. Nguyen et al. (eds.), *Data Science in Finance and Accounting*, Studies in Big Data 181, https://doi.org/10.1007/978-3-032-06179-9_10

1 Introduction

Small and Medium-sized Enterprises (SMEs) play a pivotal role in the global economy, yet their credit risk assessment remains a challenging task for financial institutions. Traditional credit scoring models often rely on limited financial data and struggle to capture the complex relationships between SMEs, their directors, and the broader business environment. This paper explores the potential of graph-based approaches to address these limitations by modeling the intricate connections within the SME ecosystem. Our focus is on the "Entity-Attribute Relationship" paradigm, where shared directors act as the bridge between SMEs. We hypothesize that the network of directors, especially those linked to liquidated SMEs, can provide valuable insights into the creditworthiness of an SME.

Graph data is increasingly common in many fields. Graph neural networks (GNNs) are a powerful machine learning method for modeling this data, used in tasks like node classification and edge prediction. GNNs are particularly useful in finance, where complex relationships and multiple data sources are common. However, the unique nature of financial data and relationships presents challenges for graph construction and GNN modeling. This paper discusses these challenges and summarizes GNN methods developed for financial tasks [15].

In this paper, we present an analysis of director networks for SMEs on how network data could help predict liquidation. We addresses the three main challenges in analyzing director networks using Graph Neural Networks (GNNs). To tackle data availability, we construct a high-quality director network, ensuring comprehensive representation of relationships between companies through shared directors. For data preprocessing, we implement a robust preprocessing pipeline that transforms raw director network data into a graph-compatible format, including cleaning, feature engineering, and generating edge indices to represent connections. The data is further standardized and encoded to prepare it for GNN input. We further employ a two-layer Graph Convolutional Network (GCN) tailored to the dataset and optimize its performance through hyperparameter tuning and dropout regularization. By addressing these challenges systematically, we demonstrate the feasibility and effectiveness of using GNNs for SME liquidation prediction, providing a scalable and reproducible framework for analyzing director networks.

The rest of the paper is organized as follows. Section 2 presents related works in network data for risk analysis and the application of GNNs. Section 3 introduces our data and the feature processing process, including the graph construction process. Section 4 presents two models we used in this study. Section 5 provides a collection of results. And Sect. 6 concludes and proposes some challenges that could be future directions of research.

2 Related Works

To frame our exploration of director networks and their predictive power in assessing SME liquidation risks, we first review related work in utilizing graph-based representations and Graph Neural Networks (GNNs) for financial and risk-related analysis. Graph structures, derived from transactional, social, or corporate data, have been extensively studied for their ability to uncover relationships and patterns that traditional models may overlook. These studies demonstrate the potential of leveraging interconnected data for tasks such as credit scoring, fraud detection, and corporate risk assessment. Building on these advancements, we focus on director networks, where GNNs offer a promising approach to analyze the structural and relational aspects of the data, enabling improved predictions of liquidation and financial risk. Below, we outline key studies that inform our work, categorized by the type of graph structure employed and its application to financial risk analysis.

2.1 Transactional Data Networks

Sukharev et al. [12] employed transactional data resulted from purchases and money transfers to improve the quality of credit scoring. Wang et al. [13] used social relationships including relatives, friends, colleagues between node pairs and [5] used shared-card network between merchants for fraud detection. Some studies utilized social behaviors from user data to identify micro-credit frauds, such as mobile network from call logs [17], embedding network [17] from residential location. In addition, guarantee network can provide useful features for risk prediction [2]. Another work in the supply-chain graph for risk analysis [16] employed a semi-supervised link prediction to extract supply chain relationship and a supervised node classification to predict loan default.

2.2 Relation and Interlock Networks

Director networks refer to the connections between directors and between directors and companies, which can be represented as a graph with directors and companies as nodes and connections between them as edges. Director networks can provide valuable insights into the relationships and dynamics not only within a company but also within a sector as a whole, and have been shown to be related to various outcomes such as corporate performance and risk. Benson et al. [1] showed that larger director networks are more beneficial during recessions, and times of increased financial uncertainty for credit ratings. Gopalan et al. [3] showed that firms take more risks when one of their directors experiences corporate bankruptcy at another firm.

GNNs are particularly well-suited for analyzing data that is structured as a graph, such as director networks, and can learn to make predictions or classify data based on the patterns and relationships within the graph structure. There have been several studies that have used graph neural networks (GNNs) to analyze director network data in the context of predicting corporate bankruptcy or other outcomes related to corporate performance and risk.

2.3 Information Networks

Lee et al. [7] used the data from an online peer-to-peer lending platform and showed that GCN outperforms existing classification models and identifies the relative contribution of input attributes. As unbalanced data could heavily affect GNN-based algorithms, which is common in sensitive domains such as finance, [8] proposed a Pick and Choose Graph Neural Network (PC-GNN for short) and experimented on both benchmark and real-world graph-based fraud detection tasks, their work demonstrated that PC-GNN outperforms state-of-the-art baselines. Another work of [14] focused on the dynamic nature of credit risk prediction and proposed temporal graphs to make default prediction on Alipay data which resulted in state-of-the-art performance.

However, to the best of our knowledge, there are little to no GNN based approaches that utilized director network data to assist corporate liquidation prediction, and most risk assessments are designed for personal finance and interest financial scenarios, such as mobile credit payment, internet loan, E-Commerce consumer lending service. Few works refer to default analysis on SMEs, the risk characteristics of which have not been extensively studied. There are several challenges that may be encountered when using graph neural networks (GNNs) to analyze director network data:

- C1: Data availability: One challenge is the availability of high-quality director network data. Director networks can be constructed from various sources, such as company records or publicly available data, but the availability and quality of these data may vary.
- C2: Data preprocessing: Another challenge is the preprocessing of director network data to prepare it for analysis by GNNs. This may involve tasks such as cleaning and formatting the data, or transforming the data into a suitable format or representation for GNNs.
- C3: Model selection and optimization: Choosing the appropriate GNN model and optimizing its performance can also be a challenge. This may involve selecting the appropriate architecture, tuning the hyperparameters, and choosing the appropriate loss function and optimization algorithm.

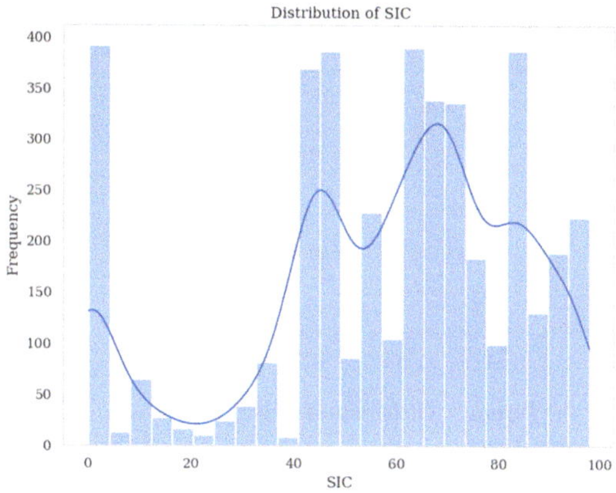

Fig. 1 Distribution of SIC

3 Data and Feature Processing

3.1 Data

In this research, we analyze a dataset of 4,146 Small and Medium Enterprises (SMEs) in Scotland, UK, provided by FAME. SMEs are defined based on the criteria outlined by [9] and the Basel Accord, as well as the European Union's SME definition.[1] Specifically, SMEs are enterprises with an annual turnover of less than 50 million euros, total assets under 43 million euros, and fewer than 250 employees. Additionally, they are limited to a maximum of six subsidiaries and ten directors. The dataset includes information on company liquidation status, aligning with the definition of SMEs in the Scotland, UK.

For data exploration, first, we present in Fig. 1 SIC code distribution graph. This graph provides insights into the industry composition of the SMEs in the dataset, based on the two-digit Standard Industrial Classification (SIC) codes as defined by Companies House.[2] These codes categorize businesses into broader industry sectors, offering an overview of the economic activities represented. Peaks in the 40s and 60s highlight a significant concentration of companies in sectors such as electricity, gas, steam, and air conditioning supply (40–49) and professional, scientific, and technical activities (60–69), reflecting the prominence of utilities and consultancy services among SMEs. Similarly, the notable frequency in the 80s corresponds to industries related to education and healthcare (80–89), encompassing companies engaged in education, medical, and social services. Conversely, lower frequencies in the 20s and 30s suggest relatively fewer SMEs in manufacturing and construction sectors, typically classified under manufacturing (20–29) and construction (30–39). These patterns underscore the dominance of service-oriented industries and professional activities in the SME landscape, with limited representation from traditional resource-

[1] http://ec.europa.eu/enterprise/policies/sme/facts-figures-analysis/sme-definition/index.htm.

[2] https://resources.companieshouse.gov.uk/sic/.

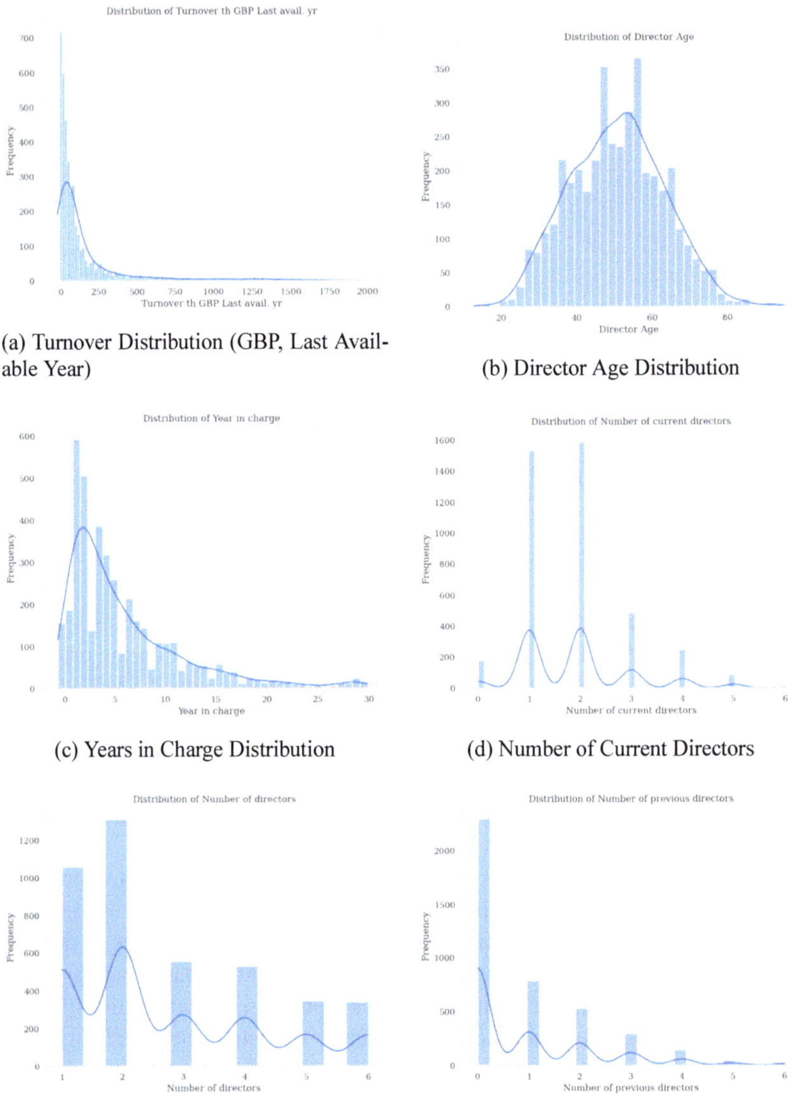

(a) Turnover Distribution (GBP, Last Available Year)

(b) Director Age Distribution

(c) Years in Charge Distribution

(d) Number of Current Directors

(e) Total Number of Directors

(f) Number of Previous Directors

Fig. 2 Distribution of key features in the SME dataset

based and manufacturing sectors. This distribution aligns with the broader trend of SMEs being more prevalent in knowledge-intensive and service-driven economic activities.

Next, we illustrate key characteristics of the SME dataset, focusing on financial and governance attributes in Fig. 2. The turnover distribution is heavily right-

Table 1 Company status

Code	Status	Count
1	In Liquidation	251
0	Active	3895

skewed, indicating that the majority of SMEs have low annual turnover, consistent with their classification as small enterprises. The director age distribution shows a near-normal pattern, with a peak around 50–60 years, highlighting that most directors are middle-aged, aligning with typical trends in SME leadership demographics. The years in charge distribution reveals a highly skewed pattern, with most directors serving fewer than five years, suggesting either a high turnover of directors or the inclusion of younger companies in the dataset. Regarding governance structures, the number of current directors is concentrated around one or two directors, reflecting the streamlined organizational structures characteristic of SMEs. The total number of directors, which includes both current and previous directors, also peaks at one or two but has a broader spread, indicating a more dynamic governance history for some companies. Finally, the number of previous directors shows that most companies have few or no previous directors, suggesting stable leadership or a relatively recent establishment. These plots collectively provide a comprehensive view of SME characteristics, emphasizing the small-scale and dynamic nature of their financial and governance structures.

Finally, we present the number of liquidation SMEs in Table 1.

3.2 *Feature Processing*

We analyze the influence of directors on SME liquidation by calculating key indicators based on their tenure and association with liquidated companies. First, we compute the tenure, or "years in charge," for both current and previous directors by evaluating the difference between relevant dates, such as the appointment, resignation, or last reported transaction dates, correcting negative values caused by data inconsistencies. For previous directors, we calculate the "incumbent duration," which measures the time between their resignation and the company's liquidation, capturing the residual impact of their leadership. We then assign Blacklist Indicators (BL) to directors, marking current directors of liquidated SMEs with BL = 1, and categorizing previous directors into five risk intervals (BL1 to BL5) based on their incumbent duration. This systematic approach enables targeted profiling of directors and longitudinal analysis of their potential roles in SME liquidation.

We update the Blacklist Indicator (BL) metrics to consolidate director-related risks across all companies associated with individual directors. Using director IDs as the grouping key, we compute the maximum values of BL and its variations (BL1 to BL5) and reassign them to ensure that each director's highest observed risk is consistently reflected throughout the dataset. We validate this process by comparing the original and updated metrics, confirming the accuracy of the aggregation. To gain deeper insights, we perform detailed crosstabulations to examine the relationship between directors' BL categories and company statuses, identifying the proportion of liquidated companies linked to high-risk directors. Additionally, we aggregate firm-level metrics by incorporating director demographics, such as age and gender, along with BL indicators, enabling a comprehensive and scalable assessment of systemic risks. This approach provides a robust framework for identifying director-specific risk patterns and their impact on SME performance.

To evaluate the impact of director-specific risk indicators and company attributes on SME liquidation outcomes, we employ logistic regression (LR) as a statistical modeling approach. Logistic regression is particularly well-suited for this analysis because it models the probability of a binary outcome—in this case, liquidation status—based on a set of predictor variables. The selected predictors include director-related metrics, such as maximum blacklist indicators (UBL, UBL1 to UBL5), demographic factors (MaxC, MaxP, Nbmc, Nbmp, Nbfc, Nbfp), and firm-level financial attributes, such as total assets, number of employees, and turnover in the last available year. Additionally, we include sector-specific features (SIC2 codes) to account for industry-level variations. Logistic regression allows us to quantify the relationship between these variables and the likelihood of liquidation, identify statistically significant predictors, and interpret the results through odds ratios.

However, the correlation heatmap in Fig. 3 reveals that features related to the number of directors (including total, current, and previous directors) are highly correlated with other features, which suggests a potential issue of multicollinearity in the logistic regression model. In addition, average age of current and previous directors are highly correlated with their corresponding max age. To address potential multicollinearity, we remove these features from our models.

4 Methods

4.1 Logistic Regression

Logistic regression (LR) is a fundamental statistical method for examining the relationship between predictor variables and a binary outcome, making it an essential tool for feature analysis in our study. By modeling the probability of an event, such as SME liquidation, based on multiple predictors, LR provides interpretable coefficients that quantify the strength and direction of these relationships. This helps identify which features significantly contribute to the outcome and their relative importance. Additionally, LR allows us to test hypotheses about individual predictors, offering insights into their statistical significance and practical relevance. LR serves as a foundational step before applying Graph Neural Networks (GNNs) by

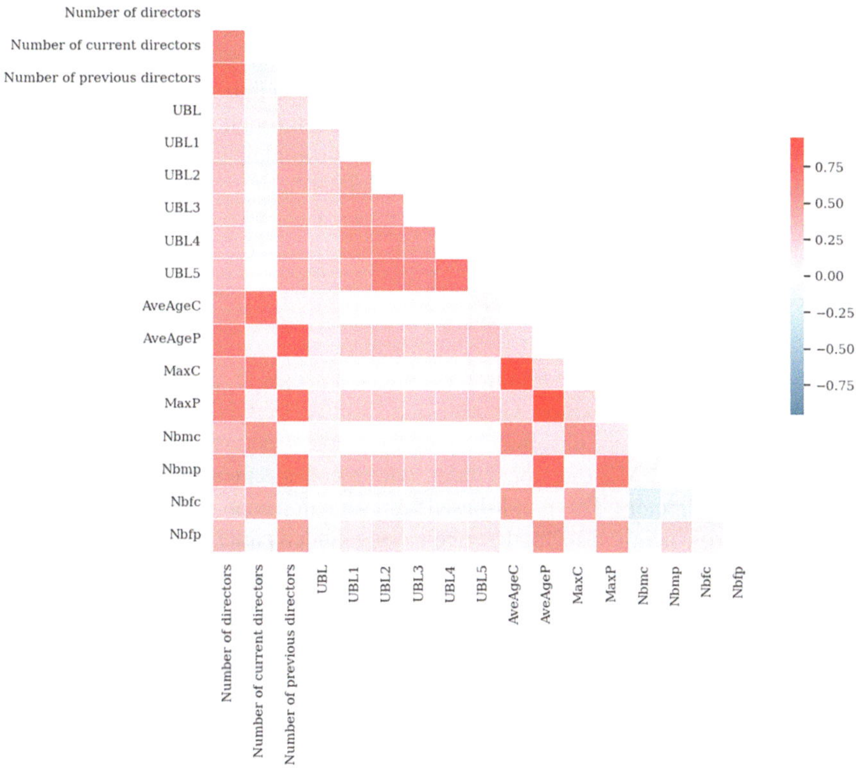

Fig. 3 Correlation heat map of features

providing a baseline understanding of the relationships between features and the target variable. LR allows for direct interpretability of feature importance, helping to identify statistically significant predictors and assess the overall predictive power of the dataset. This initial analysis ensures that the data is well-structured, relevant features are included, and multicollinearity is addressed, which are critical for constructing an effective graph-based model. By validating the dataset through LR, we establish a reference point to compare and evaluate the enhanced performance of GNNs, which can capture the relational and structural information inherent in the data.

Thus, to evaluate the impact of director-specific risk indicators and company attributes on SME liquidation outcomes, we employ logistic regression (LR) as a statistical modeling approach. Logistic regression is particularly well-suited for this analysis because it models the probability of a binary outcome–in this case, liquidation status–based on a set of predictor variables. The selected predictors include director-related metrics, such as maximum blacklist indicators (UBL, UBL1 to UBL5), demographic factors (MaxC, MaxP, Nbmc, Nbmp, Nbfc, Nbfp), and firm-level financial attributes, such as total assets, number of employees, and turnover in the last available year. Additionally, we include sector-specific features (SIC2 codes) to account for industry-level variations. Logistic regression allows us to quantify the relationship between these variables and the likelihood of liquidation, identify statistically significant predictors, and interpret the results through odds ratios. By leveraging this approach, we aim to build an interpretable model that provides actionable insights into the factors contributing to SME liquidation.

4.2 Graph Neural Networks

Building on the limitations of traditional models like logistic regression in capturing relational dynamics, we explore Graph Neural Networks (GNNs) as a more powerful alternative. GNNs are designed to process graph-structured data, making them well-suited for our SME-director dataset, where nodes represent SMEs and edges capture shared directors. This graph-based representation enables GNNs to aggregate information from neighboring nodes, uncovering hidden patterns in the network. We configure the GNN model to predict SME liquidation risk, leveraging node features such as financial indicators, blacklist metrics, and industry codes. The following sections detail the graph construction process, the chosen GNN architecture, and the evaluation strategy to benchmark its performance.

GNNs are a class of machine learning models designed to operate on graph-structured data, where entities (nodes) are connected by relationships (edges). GNNs extend traditional neural networks by incorporating relational information, allowing the model to learn representations that capture both the features of individual nodes and the structure of the graph. A key component of GNNs is the message-passing mechanism, where each node updates its representation by aggregating information from its neighbors.

Gori et al. [4] defined a GNN as a neural network model capable of processing graphs. Unlike network embedding methods, whose major aim is to generate a vector to represent each node, GNNs are designed for a variety of tasks, including node classification, edge prediction, and graph classification. Due to its wide application and superior performance, graph neural network models have drawn great attention. In this research, we focus on the homogeneous graph convolutional neural network. Formally, proposed by [6], graph convolutional network (GCN) is a widely-used graph neural network model and could encode both local graph structure and node features. Extending the convolution concepts to graphs, graph convolution could be seen as message passing and information propagation. Aggregating neighbors' feature information, graph convolutional networks could be represented with the following layer-wise propagation rule:

$$H^{l+1} = \sigma(\tilde{D}^{-\frac{1}{2}}\tilde{A}\tilde{D}^{-\frac{1}{2}}H^{l}W^{l}),$$

where $\tilde{A}$ is the adjacency matrix with added self-connections, $\tilde{D} = \text{diag}(\sum_j \tilde{A}_{ij})$ W^l is trainable weight matrix of lth layer, H^l is the node hidden feature matrix in the lth layer and $\sigma(\cdot)$ is the activation function. With its relatively simple model structure and great performance, the GCN model is often used as a benchmark method to compare with. Among the reviewed literature, over half of them have applied the GCN method as a benchmark method.

To construct the graph for our analysis, we define nodes as SMEs and edges as connections between SMEs that share one or more directors. We build the graph by iterating over all SME pairs, leveraging set operations to efficiently identify shared directors. An undirected edge is established between nodes for every identified connection. The resulting edges are then converted to a PyTorch [10] compatible tensor format for integration with Graph Neural Network (GNN) frameworks.

5 Results

The results from fitting logistic regression (LR) model in Table 2 provide valuable insights into the factors influencing SME liquidation. The model achieves a **Pseudo R-squared** value of 0.643, indicating that a substantial proportion of the variance in the dependent variable (liquidation status) is explained by the predictors. The **log-likelihood** improves significantly from −947.16 in the null model to −338.19 in the fitted model, further supported by a highly significant **likelihood ratio test**, demonstrating the overall effectiveness of the predictors in explaining liquidation outcomes.

Several predictors emerge as statistically significant contributors to the likelihood of SME liquidation. The **UBL** variable exhibits a strong positive effect, indicating that higher blacklist values substantially increase the probability of liquidation. Among the UBL intervals, only **UBL1** shows a positive and significant association, while **UBL5** demonstrates a protective effect, reducing the likelihood of liquidation. The number of current directors (Nbmc and Nbfc) has a significant negative impact, suggesting that larger current boards are associated with lower liquidation risks. Conversely, the number of previous directors (Nbmp and Nbfp) positively influences liquidation, indicating that a history of larger boards may contribute to risk.

Firm-level financial attributes also exhibit notable effects. **Total assets** significantly reduce the likelihood of liquidation, while turnover and the number of employees are not statistically significant predictors. Industry-specific variables, as captured by **SIC2 codes**, reveal significant negative coefficients for several categories, such as **SIC2_J**, suggesting that firms in these sectors (Information and communication[3]) are less likely to face liquidation compared to the reference group.

Some variables, including **UBL2**, **UBL3**, and **UBL4**, do not show significant associations with liquidation, indicating limited predictive relevance. Similarly, the maximum age of current and previous directors (MaxC, MaxP) and operational

[3] https://resources.companieshouse.gov.uk/sic/.

Table 2 Results: logit

Model:	Logit	Method:	MLE			
Dependent Variable:	Status	Pseudo R-squared:	0.643			
Date:	2024-12-26 00:59	AIC:	728.3819			
No. Observations:	4146	BIC:	892.9593			
Df Model:	25	Log-Likelihood:	−338.19			
Df Residuals:	4120	LL-Null:	−947.16			
Converged:	1.0000	LLR p-value:	2.6482e-241			
No. Iterations:	10.0000	Scale:	1.0000			
	Coef.	Std.Err.	z	P> |z|	[0.025	0.975]
UBL	3.5887	0.2548	14.0828	0.0000	3.0892	4.0881
UBL1	0.5747	0.2093	2.7457	0.0060	0.1645	0.9850
UBL2	−0.0378	0.1742	−0.2169	0.8283	−0.3792	0.3036
UBL3	0.0049	0.1721	0.0287	0.9771	−0.3324	0.3423
UBL4	0.2545	0.1892	1.3449	0.1787	−0.1164	0.6254
UBL5	−0.7341	0.1873	−3.9184	0.0001	−1.1013	−0.3669
MaxC	−0.0117	0.0074	−1.5917	0.1115	−0.0262	0.0027
MaxP	0.0107	0.0055	1.9442	0.0519	−0.0001	0.0215
Nbmc	−3.8765	0.2784	−13.9265	0.0000	−4.4220	−3.3309
Nbmp	0.5083	0.1354	3.7527	0.0002	0.2428	0.7737
Nbfc	−3.1730	0.3379	−9.3905	0.0000	−3.8353	−2.5107
Nbfp	0.4474	0.1746	2.5633	0.0104	0.1053	0.7895
Total Assets th GBP Last avail. yr	−0.0002	0.0001	−2.8287	0.0047	−0.0003	−0.0001
Number of employees Last avail. yr	0.0012	0.0013	0.8928	0.3719	−0.0014	0.0037
Turnover th GBP Last avail. yr	0.0003	0.0003	0.8528	0.3938	−0.0003	0.0009
SIC2_C	−0.6468	0.4742	−1.3641	0.1725	−1.5762	0.2825
SIC2_F	−0.3564	0.3105	−1.1480	0.2510	−0.9650	0.2521
SIC2_G	−1.3597	0.3813	−3.5656	0.0004	−2.1072	−0.6123
SIC2_I	−1.3547	0.4627	−2.9279	0.0034	−2.2616	−0.4479
SIC2_J	−2.1544	0.4111	−5.2409	0.0000	−2.9601	−1.3487
SIC2_L	−1.3072	0.6328	−2.0658	0.0388	−2.5474	−0.0670
SIC2_M	−1.5560	0.3070	−5.0689	0.0000	−2.1577	−0.9544
SIC2_N	−1.1795	0.3245	−3.6348	0.0003	−1.8154	−0.5435
SIC2_OTHERS	−1.4259	0.3075	−4.6367	0.0000	−2.0286	−0.8231
SIC2_S	−2.2990	0.6632	−3.4668	0.0005	−3.5988	−0.9993
SIC2_UNK	−3.1816	0.3918	−8.1207	0.0000	−3.9494	−2.4137

Table 3 Grid search results for logistic regression pipeline

Index	Mean test score	Std test score	Imputer strategy	Classifier C
8	0.718140	0.006718	Mean	10.000
9	0.718140	0.006718	Median	10.000
6	0.718131	0.006718	Mean	1.000
7	0.718131	0.006718	Median	1.000
4	0.718056	0.006710	Mean	0.100
5	0.718056	0.006710	Median	0.100
2	0.717334	0.006731	Mean	0.010
3	0.717334	0.006731	Median	0.010
0	0.712538	0.006805	Mean	0.001
1	0.712538	0.006805	Median	0.001

metrics like turnover and employee count also fail to achieve statistical significance, suggesting that their impact on liquidation is negligible within the current model framework.

The direction and magnitude of the coefficients provide insights into the relationships between predictors and liquidation likelihood. Positive coefficients, such as for **UBL** and Nbmp, indicate increased risks, while negative coefficients, such as for Nbmc and total assets, reflect protective factors. Confidence intervals for significant predictors exclude zero, affirming the robustness of these relationships.

Next, we construct a machine learning pipeline to predict SME liquidation by integrating preprocessing, classification, and hyperparameter tuning for LR into a unified framework. We focus on numeric features such as Total Assets, Number of Employees, and Director Attributes (MaxC and MaxP), applying transformations to handle missing values using a median or mean imputation strategy and standardize the data via StandardScaler. These preprocessing steps are managed using a ColumnTransformer and combined with a LR classifier within a pipeline. To optimize the model's performance, we employ a grid search with cross-validation to tune key hyperparameters, including the imputation strategy and the regularization strength of the LR model. We evaluate the model using the ROC-AUC metric:

Using the optimized pipeline from our grid search, we evaluate the best logistic regression model on the test set. The model achieves an AUC of 0.722, indicating strong predictive performance in distinguishing between liquidated and non-liquidated SMEs. This score reflects the effectiveness of the grid search process in selecting appropriate preprocessing steps and tuning the hyperparameters, particularly the regularization strength. By leveraging the combined strengths of systematic hyperparameter optimization and robust evaluation, we demonstrate that the chosen logistic regression model is well-suited for the task, providing a reliable baseline for further comparisons with the advanced model—GNNs (Table 3). For GNNs, we construct a graph representation of the SME dataset to enable the application of GNNs. The graph structure is built using SMEs as nodes and shared directors as

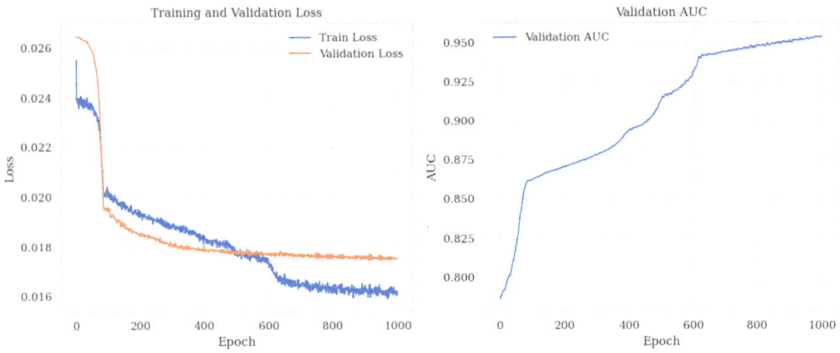

Fig. 4 Training and evaluation of GCN

the basis for creating edges. We prepare the SME dataset for Graph Neural Network (GNN) training by transforming node features, defining labels, and splitting the data into training, validation, and test sets. Node features are extracted from the dataset as a numerical tensor, where each node (SME) is represented by its feature vector. These features, along with the previously computed edge index, are combined into a PyTorch Geometric Data object to form the graph structure.

To create training, validation, and test datasets, using Scikit-learn package [11], we apply the RandomLinkSplit transformation, which splits the edges of the graph into disjoint subsets for training and evaluation. This transformation ensures that the graph remains undirected and includes negative samples for link prediction tasks, with 80% of the edges allocated to the training set. Node features are then preprocessed using a ColumnTransformer, which scales numerical features with StandardScaler to standardize their values. This preprocessing pipeline is fitted to the training data and subsequently applied to the validation and test datasets to ensure consistency across splits. The transformed features are converted back into PyTorch tensors and reassigned to the respective datasets.

Specifically, we implemented a two-layer Graph Convolutional Network (GCN) using PyTorch Geometric to predict SME liquidation based on node features and graph connectivity. The GCN architecture consists of an input graph convolutional layer that transforms node features into a 16-dimensional hidden representation, followed by a ReLU activation and dropout regularization to introduce non-linearity and prevent overfitting. A second graph convolutional layer outputs a single probability per node, which is converted into a class probability using a sigmoid activation function. The model was trained using the Mean Squared Error (MSE) loss and optimized with the Adam optimizer over 1,000 epochs. During training, we tracked the loss values, which were visualized as a learning curve to monitor convergence.

The two plots in Fig. 4 illustrate how we monitored the performance of our Graph Convolutional Network (GCN) during training and validation over 1,000 epochs. In the first plot, we show the training and validation loss curves. We observe that the training loss decreases steadily, indicating that our optimization process is effective.

Table 4 Comparison of model performance

Model	AUC (test set)	Description
LR	0.722	Baseline model using node features only
GCN	0.953	Leverages node features and graph structure

The validation loss decreases initially, then stabilizes, demonstrating that our model generalizes well to unseen data without significant overfitting. This stability reflects the effectiveness of our regularization techniques, such as dropout, in controlling overfitting.

In the second plot, we present the validation AUC (Area Under the Curve) over the training epochs. We see a rapid increase in AUC during the early epochs, indicating that the model quickly learns to distinguish between classes. The AUC continues to improve gradually and reaches approximately 0.95 by the end of training, showing excellent classification performance. These figures confirm that our GCN is well-trained and achieves both strong generalization and robust predictive accuracy, making it highly effective for SME liquidation prediction.

At the final step, we compare the performance of LR and GCN in the cross validation test set in Table 4 as follows:

The results of our experiments highlight the superior performance of the Graph Convolutional Network (GCN) compared to the logistic regression (LR) model in cross-validation. The best logistic regression model, optimized through grid search, achieved an AUC of 0.722 on the test set, indicating moderate discriminatory power in predicting SME liquidation. In contrast, the GCN model achieved a significantly higher AUC of 0.953 on the test set, demonstrating exceptional performance. This substantial improvement underscores the effectiveness of GCNs in leveraging both node features and relational information encoded in the graph structure, which traditional models like LR cannot fully utilize. These results validate the importance of incorporating graph-based models for tasks involving interconnected data, such as predicting SME liquidation through director networks.

6 Conclusion

The logistic regression model highlights key factors influencing SME liquidation, particularly blacklist indicators, board composition, and industry classifications. However, the inclusion of non-significant predictors, such as UBL intervals and some financial metrics, suggests potential multicollinearity or overlapping effects. These findings provide a strong basis for understanding liquidation risks, though future refinements using regularization techniques or feature selection may improve the model's parsimony and predictive performance. The results underscore the importance of director-related metrics and financial attributes in assessing SME viability.

In this study, we demonstrated the effectiveness of leveraging Graph Neural Networks (GNNs), specifically a Graph Convolutional Network (GCN), to predict SME liquidation by utilizing director network data. Our approach systematically addressed key challenges, including data preprocessing, network construction, and model optimization, to ensure a robust and scalable framework. The GCN significantly outperformed the baseline logistic regression model, achieving a test AUC of 0.953 compared to 0.722 for logistic regression, underscoring the value of incorporating relational information into predictive models. Furthermore, the detailed evaluation of the training process and validation metrics highlighted the GCN's ability to generalize effectively while maintaining strong classification performance. This work provides a practical and interpretable methodology for financial risk prediction, showcasing the potential of GNNs in analyzing interconnected datasets such as director networks.

While this study established a strong foundation for utilizing GNNs in SME liquidation prediction, several avenues remain for future exploration. First, incorporating temporal aspects of director networks, such as evolving relationships or time-dependent attributes, could enhance the model's predictive accuracy. Second, extending the framework to heterogeneous graphs, where nodes and edges represent different entity types and relationships, may provide deeper insights into complex corporate structures. Additionally, experimenting with advanced GNN architectures, such as attention-based or dynamic models, could further improve performance and interpretability. Finally, applying this approach to other financial risk domains, such as fraud detection or credit scoring, would validate its versatility and broader applicability in financial analytics.

References

1. Benson, B.W., Iyer, S.R., Kemper, K.J., Zhao, J.: Director networks and credit ratings. Financ. Rev. **53**(2), 301–336 (2018). _eprint: https://onlinelibrary.wiley.com/doi/pdf/10.1111/fire.12157
2. Cheng, D., Tu, Y., Ma, Z.W., Niu, Z., Zhang, L.: Risk assessment for networked-guarantee loans using high-order graph attention representation. In: IJCAI, pp. 5822–5828 (2019)
3. Gopalan, R., Gormley, T.A., Kalda, A.: It's not so bad: director bankruptcy experience and corporate risk-taking. J. Financ. Econ. S0304405X21001677 (2021)
4. Gori, M., Monfardini, G., Scarselli, F.: A new model for learning in graph domains. In: Proceedings. 2005 IEEE International Joint Conference on Neural Networks, 2005, vol. 2, pp. 729–734. IEEE (2005)
5. Khazane, A., Rider, J., Serpe, M., Gogoglou, A., Hines, K., Bruss, C.B., et al.: Deeptrax: embedding graphs of financial transactions. In: ICMLA, pp. 126–133 (2019)
6. Kipf, T.N., Welling, M.: Semi-supervised classification with graph convolutional networks. In: International Conference on Learning Representations (ICLR) (2017)
7. Lee, J.W., Lee, W.K., Sohn, S.Y.: Graph convolutional network-based credit default prediction utilizing three types of virtual distances among borrowers. Expert Syst. Appl. **168**, 114411 (2021)

8. Liu, Y., Ao, X., Qin, Z., Chi, J., Feng, J., Yang, H., et al.: Pick and choose: a GNN-based imbalanced learning approach for fraud detection. In: Proceedings of the Web Conference 2021, pp. 3168–3177. ACM, Ljubljana Slovenia (2021)
9. Orton, P., Ansell, J., Andreeva, G.: Exploring the performance of small- and medium-sized enterprises through the credit crunch. J. Oper. Res. Soc. **66**(4), 657–663 (2015)
10. Paszke, A., Gross, S., Chintala, S., Chanan, G., Yang, E., DeVito, Z., et al.: Automatic differentiation in pytorch. In: NIPS-W (2017)
11. Pedregosa, F., Varoquaux, G., Gramfort, A., Michel, V., Thirion, B., Grisel, O., et al.: Scikit-learn: machine learning in python. J. Mach. Learn. Res. **12**(Oct), 2825–2830 (2011)
12. Sukharev, I., Shumovskaia, V., Fedyanin, K., Panov, M., Berestnev, D.: Ews-gcn: Edge weight-shared graph convolutional network for transactional banking data. In: ICDM, pp. 1268–1273 (2020)
13. Wang, D., Zhang, Z., Zhou, J., Cui, P., Fang, J., Jia, Q., et al.: Temporal-aware graph neural network for credit risk prediction. In: SDM, pp. 702–710 (2021)
14. Wang, D., Zhang, Z., Zhou, J., Cui, P., Fang, J., Jia, Q., et al.: Temporal-aware graph neural network for credit risk prediction. In: Proceedings of the 2021 SIAM International Conference on Data Mining (SDM), pp. 702–710. SIAM (2021)
15. Wu, Z., Pan, S., Chen, F., Long, G., Zhang, C., Philip, S.Y.: A comprehensive survey on graph neural networks. IEEE Trans. Neural Netw. Learn. Syst. (2020)
16. Yang, S., Zhang, Z., Zhou, J., Wang, Y., Sun, W., Zhong, X., et al.: Financial risk analysis for smes with graph-based supply chain mining. In: IJCAI, pp. 4661–4667 (2020)
17. Yang, Y., Xu, Y., Wang, C., Sun, Y., Wu, F., Zhuang, Y., et al.: Understanding default behavior in online lending. In: CIKM, pp. 2043–2052 (2019)

Optimizing Portfolio and Asset Allocation Strategies

Huynh Tan Dung, Doan Van Dinh, and Do Ha Vinh

Abstract The primary objective of the research is to offer investors and readers a comprehensive guide in selecting a portfolio that maximizes profits and minimizes risks during the investment process. The portfolio is a collection of various investments held by an investor, including stocks, bonds, securities, real estate, fund certificates, bank deposits, gold, foreign currency, and other investment products. The design of a portfolio depends on an investor's risk appetite and financial goals. However, the research focuses mainly on stock portfolios. The study utilizes multiple methods to analyze data to assess risk. It is essential to consider various statistical measures, including matrix functions, variance, standard deviation, and variation coefficient. Additionally, an empirical study is conducted using the return on equity (ROEA) dataset collected from three large companies listed in Ho Chi Minh city stock exchange over the time series from 2005 to 2023. This study provides a practical understanding of how investors and readers can apply the methods used to measure risk in real investment situations. The research offers guidance to investors and readers on how to choose portfolio optimization that ensures the appropriate allocation of investments, achieves the desired level of profitability, and minimizes risks. To build a portfolio, investors should focus on three key components: investment types, allocation ratios, and risk tolerance. By considering these factors, they can align portfolios with their investment goals. The chapter discusses four basic portfolio types: risky, safe, income-providing, and balanced, detailing their advantages and disadvantages. This information helps investors make informed decisions when selecting a portfolio that suits their goals and risk appetite.

H. Tan Dung · D. Van Dinh (✉) · D. H. Vinh
School of Finance and Accounting, Industrial University of Ho Chi Minh City, Ho Chi Minh City, Vietnam
e-mail: citydinhninh@yahoo.com; doanvandinh@iuh.edu.vn

H. T. T. Nguyen et al. (eds.), *Data Science in Finance and Accounting*, Studies in Big Data 181, https://doi.org/10.1007/978-3-032-06179-9_11

1 Introduction

Investors who aim to earn profits often create a list of business projects, known as a portfolio, in which they invest. However, investing carries risks, and it's important to consider them before making investment decisions. According to economic principles, the higher the potential benefit, the higher the risk involved and vice versa. As such, investors must choose the projects that are most suitable for them, while striving for portfolio optimization.

Portfolio optimization refers to the process of selecting the best portfolio based on the quantity and type of securities that offer the highest expected profits. To minimize investment risks, investors must diversify their investments across different types of securities. To achieve this, they must study and compare various types of securities to determine which ones offer the highest expected profits. So, creating the portfolio optimization requires a careful assessment of the potential benefits and risks associated with different investment options.

Diversification is key to minimizing risks and maximizing profits. Investors must continuously monitor their portfolios and adjust as necessary to ensure they achieve their investment objectives. The process of defining and selecting new projects is one of the most crucial and intricate tasks for any developing company. Research and Development (R&D) project appraisal and portfolio selection activities have become increasingly important to keep up with the rapid evolution of technology.

However, this process becomes more challenging when it comes to selecting the most suitable combination of securities to include in a portfolio. In a study conducted by [1], investors can strategically select a project portfolio to minimize risk and enhance its overall value. The paper has created a model to identify the best project portfolio based on investment capital, risk, and rate of return. This model helps investors carefully choose a portfolio to reduce risk and increase its total worth.

In other words, the study emphasizes the importance of selecting project portfolios that balance the risk and return aspects of each project to increase the likelihood of achieving the desired outcome. By doing so, investors can achieve a higher degree of certainty in their investments, leading to better returns and a more stable financial future. Moreover, the study by [2] showed that it is possible to determine the individual probability distributions of the components that make up a portfolio.

However, it may not be possible to ascertain the correlation or dependency between the risks of these components. In such cases, the limits for the portfolio's risk measures may turn out to be impractically large and may not provide much assistance in real-world applications [3]. The study utilized the Solvency II or Risk-Based Capital (RBC) capital aggregation methodology to determine the degree to which the overall capital may be underestimated. This research highlights the importance of accurately assessing risk to make informed decisions regarding project selection and portfolio management.

The chapter focuses on the relationship between expected return and risk in the context of investment. The author employs various tests, using metrics like covariance, correlation, and variation coefficient to gauge and quantify the degree of investment risk. The objective is to determine the optimization level of anticipated return that must be sustained to make the highest return possible. The higher the risk level of the investment, the higher the possibility of capital loss.

In investment risk management, it is essential to employ standard measurement methods that encompass various variables, including return on equity (ROE), to effectively minimize and control investment risks. The chapter emphasizes that investors must consider the projected return and accept the associated investment risks to meet a high return objective when making portfolio investments. According to empirical research, expected return and investment risk have a close relationship that, depending on the degree of risk involved, can be positive, negative, linear, or non-linear.

Furthermore, the findings of earlier studies were only applicable to the unique economic circumstances of each nation. The chapter examines the impact of investment risk on the expected return of an investment. It employs commonly used metrics for managing high-risk investments to assess the relationship between investment risk and expected return. The results suggest that changes in the level of investment risk have a significant effect on the expected return of an investment.

The reason behind this is that higher investment risk levels increase the expected return but also increase the risks associated with it. Therefore, investors need to take a balanced approach to investment risk management to minimize the risks and achieve the highest expected return possible [4, 5].

2 Literature Review of the Theory of Portfolios

Modern Portfolio Theory (MPT) is a widely accepted mathematical framework in finance that provides a systematic approach for assembling a portfolio of assets to maximize the expected return for a given level of risk. The theory, which was introduced by the American economist Harry Markowitz in his 1952 paper "Portfolio Selection," later earned him a Nobel Prize [6]. The MPT revolves around several key concepts, including diversification, risk and return, efficient frontier, risk aversion, and asset correlation. Diversification is a critical aspect of MPT that emphasizes the importance of investing in multiple asset classes to reduce risk. The theory suggests that diversification can yield greater returns without necessarily increasing risk. The concept of risk and return in MPT argues that an investment's risk and return should not be viewed in isolation. Instead, they should be evaluated by how the investment affects the overall portfolio's risk and return. The efficient frontier is a set of portfolio optimizations that offers the highest expected return for a defined level of risk, and it is a key component of MPT. The theory assumes that investors are risk-averse, meaning they prefer a less risky portfolio to a riskier one for a given level of return. Further, MPT considers the statistical measures of each pair of assets in a portfolio, such as variance and correlation. The MPT has revolutionized finance and has become a key tool for asset managers, typically applied with a buy-and-hold strategy. However, it is worth noting that, like any theory, MPT has its limitations and critics. Understanding these limitations is essential to implementing MPT effectively [7]. Other literature has used return-related risk assessments in addition to conventional methods. The analysis of risk and finance has also been done using this method, mainly regarding risk investment and portfolio selection. The reason for this is that Lorenz curves can be estimated from the returns of assets, which makes it easy to apply the method in practical settings. This innovative approach was highlighted by one research [8]. In another study [9], the pre-project stage was examined to determine its impact on the success of project portfolios under specific circumstances. The researchers discovered a strong positive correlation between the success of the front-end stage and the success of the project portfolio. The impact of this correlation was found to be particularly significant for larger project portfolios that exhibit interdependence between individual projects. To evaluate the project portfolios, the market beta coefficient was utilized by the writers to calculate investment risk, so the recommendation is that incorporating low-volatile market betas into a portfolio could play a significant role in its construction. This means that portfolios with a lower investment risk may have a higher chance of success. Overall, the study highlights the importance of careful planning and analysis in the pre-project stage when it comes to the success of a project portfolio.

Risk is an important concept in the world of finance and investment. It refers to the deviation of the actual rate of return compared to the expected rate of return. When an investment has greater potential for deviation, it is considered to have greater risk. There are two types of risk that every investment can encounter: systematic risk and unsystematic risk [10, 11]. The chapter examines the impact of investment risk on the expected return of an investment. It employs commonly used metrics for managing high-risk investments to assess the relationship between investment risk and expected return. The results suggest that changes in the level of risk of investment have a significant effect on the expected return of an investment [12].

Unsystematic risk: The risk that solely impacts one or a small number of different asset or security types is referred to as unique risk or unsystematic risk. This kind of risk is frequently brought on by the company itself, for example, by poor management practices or poor choices regarding capital sources and asset allocation (using both financial and business leverage). Credit risk and default risk are two instances of non-systematic risk. However, by using a diversified investment plan, this kind of risk may be minimized. To put it another way, you can spread out your portfolio and reduce the chance that any one asset or security will have a major negative influence on your overall return by investing in several assets [13, 14]. Investors are frequently urged to diversify their stock portfolios to generate long-term profits from numerous sources of investment. However, some research shows that the degree of risk and possible rewards varies by geographic place. In completely competitive marketplaces, the relationship between risk and return may be negative. To overcome this issue, experts advocate for risk estimation to choose the best portfolio [15]. The challenge of portfolio selection under random and fuzzy uncertainty can be partially addressed by estimating the average yield and risk of the portfolio using "hybrid entropy and variance". A "medium variance entropy model" (MVHEM) was consequently developed. Much research has utilized standard assessment methodologies to assess investment risk. Much research has utilized standard assessment methodologies to assess investment risk [16]. Other research has investigated the dangers, benefits, and portfolio's prospects for diversity in the domestic financial market. While there may be differing views on the risks and profitability that must be addressed and analyzed by the portfolio, the primary goal is to investigate risk, return, and diversity. Overall, the research emphasizes the necessity of using the appropriate portfolio selection models to predict stock returns and maximize portfolio performance [17].

3 Methodology for Characteristic Models in Portfolio Optimization

Choosing the portfolio is a crucial decision in both personal and business financial management. One key reason for this choice is to diversify risks. By allocating capital to a variety of asset types, investors can minimize risks associated with negative fluctuations in any single asset. A balanced portfolio can lead to more stable profits over time and protect against market volatility. When market conditions change, a diversified portfolio can help mitigate negative impacts. Additionally, a well-structured portfolio can align with individual financial goals. Each investor has unique objectives, such as long-term growth, generating passive income, or preserving capital. To provide a clearer understanding of the research processes involved in the study, the study has developed a comprehensive diagram summarizing the key steps and methodologies. This diagram outlines the various stages of the research, including the formulation of research questions, data collection methods, analysis techniques, and interpretation of results. By presenting this information visually, the study aims to facilitate a better grasp of the overall research structure and its essential components (Fig. 1).

Modern portfolio theory, developed by Harry Markowitz, addresses investment allocation through mean-variance optimization [6]. It assumes investors seek to maximize expected returns while minimizing risk, represented by the variance in returns. By using statistical measures of expected returns and covariances among assets, investors can create an efficient frontier of portfolio optimization that maximizes returns for a given risk level. The success of this approach depends on accurately estimating expected returns and asset relationships, allowing for strategic asset allocation that aligns with individual risk tolerance and investment goals. To build portfolio optimization, follow these steps:

Step 1 *Determine Variance:* Calculate the variance by taking the weighted average of the squared deviations of each security's return from the portfolio's average return. This reflects the dispersion of returns.

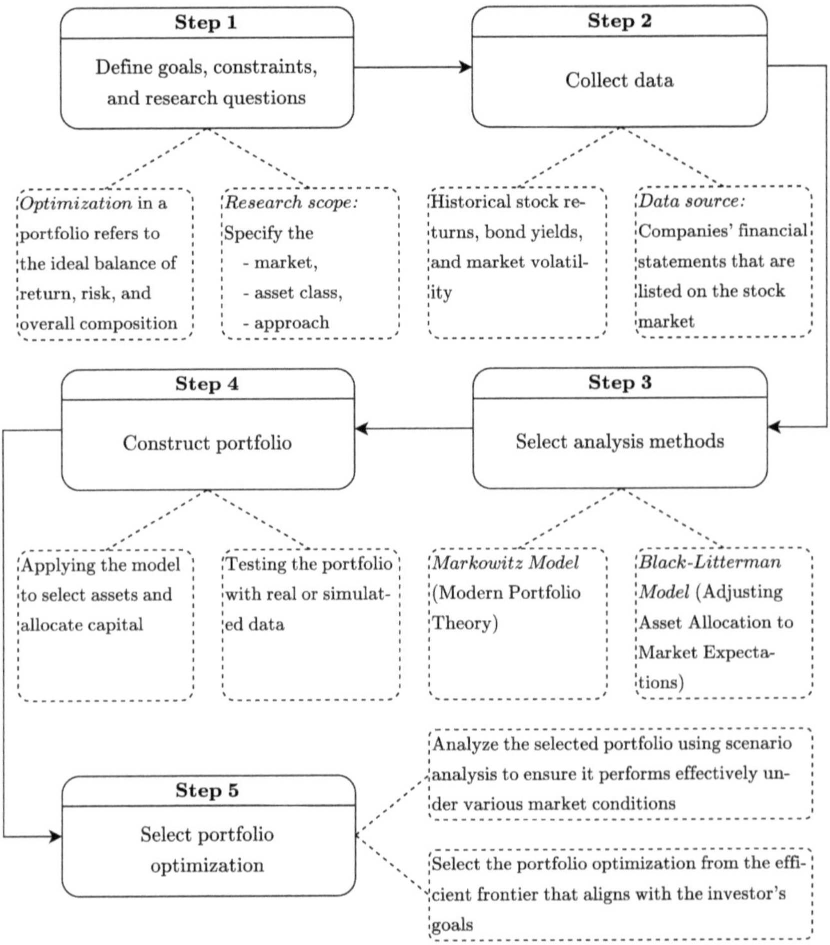

Fig. 1 Summary of the methodology

Step 2 *Calculate Standard Deviation:* Find the square root of the variance to obtain the standard deviation, a key measure of risk. Use the coefficient of variation to compare investments with different expected returns.

Step 3 *Calculate Expected Return:* Compute the expected return by taking the weighted average of each security's return based on its proportion in the portfolio.

Step 4 *Analyze Covariance:* Examine the covariance between pairs of securities to understand their movements together. Positive covariance indicates they move in the same direction, while negative covariance means they move inversely.

Step 5 To determine the expected value or rate of return, the study calculates the average value using the weighted average method based on the possible return rates in various scenarios. The study assesses risk by examining the probability distribution of these return rates. A wider dispersion of return rates indicates greater volatility, which corresponds to higher risk. Therefore, it is essential to select assets with more varied returns; lower volatility can help reduce risk in the portfolio. This careful selection is key to constructing portfolio optimization.

3.1 Data Collection

A portfolio is defined as a collection of financial assets owned by an individual, organization, or investment fund, aimed at achieving specific investment goals. The portfolio helps to diversify risks by including various types of assets. A well-managed portfolio balances the expected return with an appropriate level of risk for the investor. In this study, the focus is on a portfolio comprised of three large companies that significantly influence the economy. This situation serves as a model that can be applied to many different portfolios, each with various types of assets. The decision to select only three companies in this portfolio is intentional, as it simplifies the analysis while illustrating how similar principles can be applied to different investment cases and objectives. This study has collected data on the Return on Equity (ROE) of three joint-stock companies: Vietnam Dairy Products JSC (VNM's stock code), Hoa Phat Group JSC (HPG's stock code), and Ben Tre Seafood Import-Export Joint Stock Company (ABT's stock code) and the period for which the dataset is gathered is from 2005 to 2023. After originally selecting seven joint-stock companies for our dataset, the study encountered missing data on crucial variables for some of them. As a result, the study had to revise its selection and ultimately collect data from only three joint-stock companies (Table 1).

It is essential to note that the dataset the study collected is based on the financial statements' ROE of the three listed companies [18]. The ROE's function is to determine the profitability of a company by showing how efficiently it is using its equity to generate profits. By analyzing the empirical results, the study can gain insights into the profitability of the joint-stock companies and make informed decisions accordingly.

3.2 The Portfolio Optimization Characteristic Models

Different objectives and approaches significantly influence various research directions in portfolio optimization selection. This study focuses on three main research questions:

RQ1. How can the study construct the portfolio optimization in the stock market?
RQ2. What analytical methods can assist in optimizing the portfolio?
RQ3. How can the study minimize unsystematic risk within the portfolio?

Table 1 Data of listed companies' rate of return

Year	VNM_ROE	HPG_ROE	ABT_ROE
2005	0.31	0.02	0.29
2006	0.27	0.24	0.52
2007	0.28	0.37	0.23
2008	0.28	0.24	0.07
2009	0.43	0.29	0.23
2010	0.68	0.24	0.21
2011	0.41	0.18	0.24
2012	0.42	0.13	0.2
2013	0.4	0.22	0.19
2014	0.33	0.3	0.19
2015	0.38	0.27	0.17
2016	0.43	0.38	0.12
2017	0.44	0.31	0.07
2018	0.41	0.23	0.16
2019	0.38	0.17	0.12
2020	0.35	0.25	0.03
2021	0.3	0.46	0.08
2022	0.25	0.09	0.14
2023	0.26	0.07	0.13

To address these research questions, this study employs different methodologies. There are several methods and perspectives for calculating investment risk. Common techniques used to manage investment risk include the Capital Asset Pricing Model (CAPM), Sharpe Ratio, Beta, Standard Deviation, Value at Risk (VaR), and R-squared. These statistical techniques are essential tools for representing real-valued random variables in both the social and natural sciences. They are frequently utilized to evaluate and quantify the risk of a portfolio. By employing these techniques, investors can make informed decisions about the level of risk they are willing to accept and the potential rewards of their investments [19, 20]. When it comes to investment, there is always some degree of uncertainty involved. This uncertainty is commonly referred to as risk. To measure the risk associated with each investment, people often use probability distributions with two parameters - expected value and standard deviation. The predicted value, also referred to as the anticipated rate of return, is the mean value calculated by the weighted mean method of the rate of return that can occur in different situations. It provides an estimate of the return that an investor can expect to receive from their investment. On the other hand, standard deviation measures how much the actual rate of return can deviate from the expected rate of return. By using both these parameters, investors can assess the risk associated with each investment and make informed decisions [21, 22].

The financial market is a complex environment that generates a wide range of different types of securities returns. To resolve the challenge of selecting the appropriate securities to include in a portfolio, literature has proposed a hybrid approach that considers both random and uncertain returns. This approach applies uncertainty theory to calculate the variance of portfolio returns, and it has been shown to be effective in practice. Additionally, mean-variance models are commonly used to analyze the performance of portfolios. The primary goal of portfolio selection is to determine the optimal allocation of funds among a set of risky securities candidates to maximize profits while minimizing exposure to risks. By carefully considering the characteristics of each security and balancing the potential risks and rewards, investors can build a diversified portfolio that meets their investment objectives [23]. Investors are always searching for a strategy that can yield high profits while minimizing potential risks. However, it is important to note that there is no such thing as a perfect investment. To achieve this goal, investors must set up different portfolios, and this is where the covariance matrix model is applied. This model is used to explore the relationships between a portfolio's expected return and its level of risk. The application of this model is a novel approach that is explored in the research. The effectiveness of this project prioritization plan is then analyzed, and it is compared to the project portfolio performance. This analysis is essential for investors who want to make informed decisions that can lead to the creation of a successful portfolio. By exploring the interplay between investment risk and return and by using a well-defined set of evaluation criteria, investors can optimize their portfolios and achieve their financial goals.

The primary objective of the research is to conduct an empirical analysis that delves into the relationship between expected return and portfolio risk. The portfolio hypothesis serves as the basis for this exploration, where the study aims to understand whether there is a correlation between higher expected returns and higher risks or higher returns and lower risks. It is worth noting that "modern portfolio theory" (MPT) is a framework that offers insights into how risk-averse investors can construct portfolios that maximize expected return based on a given level of market risk [24]. Broadly speaking, a portfolio refers to as a collection of two or more investments in cash equivalents, bonds, stocks, commodities, real estate, or other assets. Investors or individuals own these assets with the intention of reducing the risk associated with investing by diversifying their investments. Furthermore, divided by the entire value of equity after a given period (as determined by the balance sheet and financial statements), the ROE is a crucial financial indicator. This metric is used to gauge the efficiency of a company in generating profits from its shareholders' investments. The ROE ratio is a financial metric that provides insight into the profitability of a company by measuring the return rate that ordinary shareholders receive on their investments. Essentially, ROE evaluates how effectively a company generates return on invested capital that its shareholders have invested in it. This ratio is an important indicator of a company's performance and is calculated by dividing the net income of a company by its shareholders' equity. In summary, the higher the ROE, the more efficient the company is in using shareholder funds to generate profits [25].

$$\text{Return on Equity (ROE)} = \frac{\text{Profits or Net Income}}{\text{Equity held by shareholders}} \tag{1}$$

ROE is a financial metric used to evaluate the profitability of a joint-stock company in terms of its share capital. It is a crucial indicator of a business's profit margin on its capital or investment.

$$f(\text{ROE}) = (\text{ROE}_{\text{VNM}},\ \text{ROE}_{\text{HPG}},\ \text{ROE}_{\text{ABT}}) \tag{2}$$

In the analysis of portfolios, the variable of interest is often the ROE, which can be used to compare the profitability of different companies. Specifically, in the case of this analysis, the study is looking at three companies: the VNM's ROE, the HPG's ROE, and the ABT's ROE. One commonly used framework for analyzing portfolios is the mean-variance approach developed by Markowitz in 1952 [6]. This approach focuses on measuring both the return on investments and the associated risk, considering the portfolio's anticipated worth and profitability. The key insight of the Markowitz model is that by combining perfectly uncorrelated assets, it is feasible to construct a portfolio in which assets' rates of return move independently of one another, so this allows investors to reduce their projected return without raising their level of investment risk. In theory, an asset's portfolio can be constructed to offer the lowest feasible risk for a given return, so this approach has important implications for investors, as it provides a way to optimize their investment strategies, balancing risk and reward in a way that aligns with their investment goals and preferences. When assessing an investment opportunity, it is essential to consider the expected return and the corresponding level of risk. In finance, an expected return is the anticipated profit or loss that an investor can expect to receive from an investment, based on various factors such as past performance, market trends, and economic conditions.

Mean-variance analysis is a method commonly used by investors to evaluate the risk and return of individual stocks. This technique determines each stock's mean predicted return and matching standard deviation, which is a statistical measure that indicates how much a security's annual return varies from its long-term historical mean return. Investors use the standard deviation as an estimate of risk, as it provides insight into the potential change in investment outcome. The higher the standard deviation, the more volatile the investment is. In other words, a high standard deviation indicates a high level of risk, while a low standard deviation suggests a lower level of risk. To compute the standard deviation, the initial step involves calculating the variance, which is obtained by finding the average of the squared variances between each return and the mean return. Subsequently, the standard deviation is determined by taking the square root of the variance. So, the standard deviation is a crucial measure of the risk of an investment, and investors use it to determine whether a particular investment is suitable for their risk tolerance level, $\delta = \sqrt{\delta^2}$.

3.3 Measure the Risk of Each Individual Investment

Assessing the risk associated with each individual investment is a crucial step in any investment decision-making process. Risk in investments is generally defined as the uncertainty or deviation of the actual rate of return compared to the expected rate of return, which may or may not occur. To measure this risk, investors often make use of probability distributions with two parameters, namely the expected value and the standard deviation. The expected rate of return is the mean value of the rate of return that can occur in different situations. This is usually calculated by using the weighted average method, which considers the probability of each possible rate of return occurring. By calculating the expected value, investors can get an idea of the most likely outcome of their investment. However, expected value alone is not sufficient to assess the risk associated with an investment. This is where the standard deviation comes into play. The standard deviation is a measure of the amount of variation or dispersion of the possible returns from the expected value. The higher the standard deviation, the greater the risk associated with the investment. By combining the expected value and standard deviation, investors can get a more comprehensive understanding of the risk associated with each investment option.

$$\bar{\bar{r}} = \sum_{i=1}^{n} p_i \times r_i \tag{3}$$

in which,

n: Number of cases (number of situations) that can occur
p_i: Probability corresponding to situation i
r_i: Return rate of investment in situation i

Risk is considered through monitoring the probability distribution of return rates. The more dispersed the return rate and the greater the fluctuation, the higher the risk. When calculating variance, the weighted average method is used to find the average value of the squares of the deviation between the actual value and the average value. This provides a measure of the dispersion of a probability distribution.

$$\sigma^2 = \sum_{i=1}^{n} p_i (r_i - \bar{r})^2 \tag{4}$$

in which,

$\bar{r}$: Expected rate of return (Mean)

The standard deviation is a measure of the degree of dispersion of a probability distribution. It is calculated as the square root of the variance. When applied to investment returns, it indicates the level of fluctuation or dispersion of the returns around the expected rate of return. This helps assess the risk level of a portfolio.

$$\sigma = \sqrt{\sigma^2} = \sqrt{\sum_{i=1}^{n} p_i(r_i - \bar{r})^2} \tag{5}$$

If two investments have identical expected rates of return, then the investment with a higher standard deviation is considered riskier. However, if the expected rates of return are not the same, then the study cannot conclude that investment with a higher standard deviation is riskier. In such cases, the study must use the coefficient of variation to evaluate the level of risk. The coefficient of variation (C_V) is a measure of risk per unit of expected return; the higher the coefficient, the greater the risk.

$$C_V = \frac{\sigma}{\bar{r}} \tag{6}$$

The matrix of covariance is a fundamental concept in probability theory and statistics that describes the relationship between a set of random variables. It is a square $m \times m$ matrix, where the diagonal elements are the variances of each variable, and the off-diagonal elements are the covariances between each pair of variables. The covariance between two random variables X and Y is defined as the expected value of their product minus the product of their expected values, $\text{Cov}(X, Y) = E[XY] - E[X]E[Y]$. If X and Y are independent, then their covariance is zero. If X and Y are positively correlated, then their covariance is positive, while if they are negatively correlated, then their covariance is negative. The matrix of covariance is a symmetric matrix, meaning that the covariance between any two variables is the same regardless of their order. Additionally, the diagonal elements of the covariance matrix are the variances of each variable, which are always non-negative. The matrix of covariance is a matrix with the composition i, j equaling covariance if the elements of a column vector are random variables with certain variances [26], this implies that the covariance matrix is a useful tool for analyzing the relationships between multiple variables and is widely used in various fields, including finance, engineering, and machine learning.

$$\sum_{i,j} \text{Cov}(X_i, X_j) = \sum_{i,j} E\left[(X_i - \mu_i) \times (X_j - \mu_j)\right] \tag{7}$$

Diversification is a key strategy for investors looking to manage risk and maximize profits in the stock market. One of the fundamental concepts in econometrics and model estimation that plays a critical role in this strategy is the covariance matrix. It provides a measure of the correlation between the returns of two or more equities. A positive covariance indicates that the returns of two stocks rise and fall together, while a negative covariance indicates that their returns move in opposite directions. By analyzing the covariance matrix, investors can identify which equities are likely to move in the same direction and which are likely to move in opposite directions. Using this information, investors can build a well-diversified portfolio that includes a variety of equities that move in different directions. By doing so, they can reduce their exposure to risk and increase their potential for profit. Overall, the covariance matrix is an essential tool for investors looking to navigate the complex world of the stock market and build a successful investment strategy. Investors use different methods to analyze the relationships between the holdings in their portfolios. Two securities in a portfolio may have connections that are linked. To determine this connection, investors must consider the variance and covariance. The $p_{A,B}$ correlation coefficient formula is used to calculate the correlation between any two investment items in the portfolio. This coefficient measures the strength of the relationship between two securities and helps investors calculate the overall risk of their portfolios. Understanding the covariance and correlation between various assets in their portfolio helps investors make informed decisions about their investments and manage their risks effectively [27].

$$p_{A,B} = \frac{\mathrm{Cov}(A, B)}{\sigma_A \sigma_B} \quad \text{or} \quad \mathrm{Cov}(A, B) = (\sigma_A \sigma_B) \times p_{A,B} \tag{8}$$

The correlation coefficient $p_{A,B}$ is a statistical measure that quantifies the relationship between the two investments' return rates, A and B, and the range of the coefficient is -1 to $+1$, which can be interpreted in three different cases. When $p_{A,B}$ equals $+1$, and it indicates that the return rate on two investments is highly positively correlated. In other words, the performance of the two investment securities is very similar, and the risk associated with investing in both is not reduced. The level of risk associated with a portfolio is equivalent to the sum of the risks of each investment within the portfolio. On the other hand, if $p_{A,B}$ equals -1, it indicates a strong negative connection between the two assets' return rates. This implies that the two securities risks are minimal and investing in both can help reduce overall risk. Lastly, when $p_{A,B}$ equals 0, it indicates that the return rates of two investments are independent and lack any correlation. In this case, investing in both investments does not have any impact on the risk associated with each investment.

3.4 Asset Allocation Strategies

Investors often hold portfolios that consist of multiple investments, which may include investment items A and B. The expected rate of return for such a portfolio can be calculated using the capital weight of the investments. Suppose an investor has allocated f_A and f_B capital weights to investments A and B, respectively. In that case, the expected return for the whole portfolio can be calculated using the formula $r_E = f_A\bar{r}_A + f_B\bar{r}_B$, where $\bar{r}_A$ is the investment's return mean A, and $\bar{r}_B$ is the investment's return mean B [28]. However, it is not enough to base decisions solely on the return means of the portfolios. Investors must also consider the risks associated with the investments. The capital weight allocation between A and B portfolios can significantly impact the expected return. Additionally, the return rates' variance and standard deviation must be assessed to determine the level of risk in each portfolio [29, 30]. When it comes to portfolio standard deviation, σ_p represents the standard deviation of the portfolios, while σ_A and σ_B stand for the standard deviations of portfolios A and B, respectively. Therefore, the variance matrices and deviations of the two stocks A and B are expressed in the following manner:

$$\begin{bmatrix} \left|f_A^2\sigma_A^2\right| & \left|f_A^2 f_B^2 \mathrm{Cov}(A,B)\right| \\ \left|f_A^2 f_B^2 \mathrm{Cov}(A,B)\right| & \left|f_B^2\sigma_B^2\right| \end{bmatrix} \tag{9}$$

which is written as:

$$\sigma_p^2 = f_A^2\sigma_A^2 + f_B^2\sigma_B^2 + 2 f_A f_B \mathrm{Cov}(A,B) \quad \text{and} \quad \sigma_p = \sqrt{\sigma_p^2} \tag{10}$$

Therefore, to make informed investment decisions, investors must evaluate both the expected return, and the risk associated with each portfolio. By considering these factors, investors can allocate their investment capital weight to optimize their portfolio's performance. The covariance matrix of two stocks, A and B, serves as a foundation for extending the analysis to a portfolio consisting of n stocks. To assess the risk level of any combination of investments in a portfolio, it's important to consider the covariance of returns among the individual investment items. When a portfolio contains multiple investments or securities, each with its own risk, the overall risk, or standard deviation, of the portfolio can be calculated by using a formula that considers the standard deviations of each investment as well as their respective covariances. This provides a more comprehensive understanding of the risk associated with the entire portfolio. This formula helps in assessing the overall volatility and risk of the portfolio, which is crucial in making informed investment decisions.

$$\sigma_p = \sqrt{\sum_{i=1}^{n} f_i^2\sigma_i^2 + 2\sum_{i=1}^{n}\sum_{j=1;j\neq i}^{n} f_i f_j \mathrm{Cov}(i,j)}$$

Based on the analysis and evaluation, the forecast model is established with a general formula as follows:

$$r_E = \sum_{i=1}^{n} \bar{r}_i \times f_i \tag{11}$$

This material explores the model that investors may use to adjust their capital allocation for investment ratios (f_i) into different portfolios, which results in variable predicted returns. The model is then utilized to calculate the investors' return on investment. The fundamental methodologies explained above are used to analyze the risky portfolio. The analytical findings aid investors, risk managers, and scholars in diversifying their portfolios, as diversity is crucial in reducing investment risks. If a particular investment suffers a loss, but other assets gain during the same period, they can be used to compensate for each other, resulting in investment risk dispersion [31].

The chapter study is focused on an empirical hypothesis investigating the correlation between risky investments and expected returns when investors allocate their funds across three different companies. The assessment of portfolio risk involves the calculation of the portfolio's standard deviation. When several individual assets are combined within a portfolio, their returns are interconnected, and covariance is a statistical metric used to determine how closely the profit margins of two different securities are related. The degree of linkage or correlation between the investment components in the portfolio affects its standard deviation.

4 Research Findings

Using the covariance matrix model and standard deviation for n securities across m portfolios, where n is the volume of securities from the three largest Vietnamese enterprises and m is the volume of portfolios, this chapter seeks to ascertain the relationship between risk and returns in notable Vietnamese companies. Investors and scholars can use the study's findings to develop an appropriate investment model. Risk metrics were evaluated based on the given dataset, including variance, covariance, mean sample, and standard deviation [32] (see Table 2).

The normal distribution, which is the most used probability distribution in statistical mathematics, can accurately depict the nature and extent of the study data. This distribution is defined by two key parameters: the expected value (also known as the mean value of r), denoted by μ, and the standard deviation, denoted by σ. While μ represents the average value of all data, σ reflects the consistency or uniformity of the research data. The mean return on equity (ROE) for VNM is 36.89%, with a standard deviation of 9.949%. HPG has a mean value ROE of 23.47%, with an as-yet-unknown standard deviation. Lastly, ABT's mean ROE stands at 17.84%, with an also as-yet-unknown standard deviation. When evaluating the potential impact of two investments on portfolio diversification, one crucial measure is the coefficient of correlation, which assesses the relationship between their rates of return over time. The correlation coefficient ranges from -1 to 1, with values of $r < 0$ indicating a negative correlation, $r > 0$ indicating a positive correlation, and $-1 \leq r \leq 1$. A

Table 2 Descriptive statistics

Items	N	Minimum	Maximum	Mean	Std. Deviation	Variance
ROE_{VNM}	19	0.25	0.68	0.3689	0.09949	0.01
ROE_{HPG}	19	0.02	0.46	0.2347	0.10987	0.012
ROE_{ABT}	19	0.03	0.52	0.1784	0.10704	0.011
Valid N (listwise)	19					

positive correlation ($r > 0$) suggests that the returns of the two investments move in the same direction, meaning they tend to grow or decline together and thus may not contribute significantly to portfolio diversification. Conversely, a negative correlation ($r < 0$) indicates that the returns of the two investments move in opposite directions, presenting an opportunity for diversification as they may help to offset each other's performance within the portfolio. Therefore, when seeking to diversify a portfolio, investors might consider investments with a negative correlation to balance their overall risk and return profile [33].

The relationship between the return on equity of HPG and VNM was positive and weak, with a correlation coefficient of 14.9687%. As a result, investors have the option to invest in either company to potentially make a profit without having to choose one over the other. On the other hand, the correlation between the return on equity of ABT and VNM is negative and weak, with a correlation coefficient of −6.4999%. Consequently, investors can choose to invest in one of these two companies to earn a return. However, the relationship between the return on equity of HPG and ABT is negative and weak, with a correlation coefficient of −19.7097%. Therefore, investors must choose one of these two companies for investment. As per the methods for evaluating portfolio risk, which quantify the level of risk associated with investment items in a portfolio, the said investment items are assessed through covariance, which gauges the connection between the returns of the two investment items. When the covariance between investment items is positive, their returns move in tandem and conversely, as shown in Table 3. The ideal distribution of securities in the portfolio is essential for allocating investment assets to achieve the VNM's ROE, the HPG's ROE, and ABT's ROE the portfolio. If investors can allocate 40% of their investment assets to the HPG's ROE, 30% to the ABT' ROE, and the remaining 30% to the VNM's ROE, the total individual securities' standard deviation in the portfolio is a measure that indicates the extent of variation or dispersion of these assets from their mean value. It provides insight into the risk or volatility associated with the portfolio's securities, as follows (Table 4).

The portfolio's standard deviation is not simply the average standard deviation of the individual stocks within it. Instead, it is calculated by considering the covariance between the different stocks in the portfolio, which reflects the relationships between

Table 3 Covariance and correlation analysis: ordinary

Covariance	ROE_{ABT}	ROE_{HPG}	ROE_{VNM}
ROE_{ABT}	0.010787		
ROE_{HPG}	−0.00219	0.011444	
ROE_{VNM}	−0.000651	0.001545	0.009309
Correlation	ROE_{ABT}	ROE_{HPG}	ROE_{VNM}
ROE_{ABT}	1.000000		
ROE_{HPG}	−0.197097	1.000000	
ROE_{VNM}	−0.064999	0.149687	1.000000

Table 4 The risk measurement results of the three stocks

Items	ROE_{ABT}	ROE_{HPG}	ROE_{VNM}
Mean rate of return $\bar{r}$	0.1784	0.2347	0.3689
Standard deviation, σ	0.10704	0.10987	0.09949
Coefficient of variation, $C_V = \sigma/\bar{r}$	0.011	0.012	0.01
Assumption of the investment capital weight $\sum f = 1$ (%)	40	30	30
Expected return of individual stock $r_E = \sum_{i=1}^{n} \bar{r}_i \times f_i$	0.0714	0.0704	0.1107
Anticipated return on portfolio stocks	0.252		

the various stocks. When using this method, the correlation level of each stock in the portfolio is considered, providing a more comprehensive measure of the portfolio's risk and volatility [34].

When analyzing portfolios, the standard deviation is crucial in measuring the risk associated with the investments. In the given scenario, the standard deviation of the portfolio is calculated to be 4.5% based on the formula $\sigma_p = \sqrt{0.00339} \approx 0.06$ or 6.0%. This percentage indicates the level of risk that the companies' stocks carry within the portfolio. It's important to note that changes in the investment capital weight can directly impact the portfolio's standard deviation. As the weight of different assets in the portfolio changes, the portfolio's overall risk is altered, leading to a variance in the standard deviation value. Therefore, investors and financial analysts closely monitor these changes to assess the evolving risk levels within their portfolios. When analyzing portfolios, the standard deviation plays a crucial role in measuring the risk associated with the investments. In the given scenario, the standard deviation of the portfolio is calculated to be 4.5% based on the formula of σ_p. This percentage indicates the level of risk that the companies' stocks carry within the portfolio. It's important to note that changes in the investment capital weight can directly impact the portfolio's standard deviation. As the weight of different assets in the portfolio changes, the overall risk profile of the portfolio is altered, leading to a variance in the standard deviation value. Therefore, investors and financial analysts closely monitor these changes to assess the evolving risk levels within their portfolios.

5 Analysis of Portfolios Policies and Solutions

Comparing individual stocks within a portfolio involves evaluating not only their expected returns but also their standard deviations. When investors consider not only the anticipated returns but also the standard deviation of each individual stock, they are able to make well-informed choices regarding how to structure their portfolios and develop effective strategies for managing risk. Empirical results serve as the foundation for discussions on risk measurement within a portfolio. These results, based on real-world data and statistical analysis, provide valuable insights into the risk profiles of different assets and help in shaping risk management strategies within the portfolio. By relying on empirical evidence, investors can make more informed decisions regarding portfolio diversification, asset allocation, and overall risk mitigation techniques. So, the standard deviation of a portfolio is a key metric that reflects the level of risk associated with the investments it holds. Changes in investment capital weight can influence this metric, and comparing individual stocks based on expected returns and standard deviations is essential for effective portfolio management. Empirical results form the basis of discussions on risk measurement within a portfolio, guiding investors in making well-informed decisions to optimize risk-return trade-offs and achieve their financial objectives. Diversification in portfolios involves spreading investments across various assets with different risk levels. The risk measurement of these investments is typically done using standard deviations and covariance. By diversifying across assets with differing risk profiles, investors aim to establish a balanced asset structure that helps spread risk and minimize overall investment risks. While diversification cannot eliminate risks, it can reduce them to some extent, aligning with the investment principle of not putting all your eggs in one basket. This strategy helps investors manage risk exposure and enhance the resilience of their portfolios against market fluctuations and specific asset volatility [35]. Based on the given hypothesis, the portfolio's expected return can be calculated by multiplying the total mean return rate of all investments by the percentage of each investment's capital in the portfolio. This method allows for a comprehensive analysis of the expected return based on the composition of the portfolio.

Hypothesis: If investors allocate 30% to listed company VNM, 30% to listed company HPG, and 40% to listed company ABT, the expected return would be 22.5% of the portfolio's total investment assets. This calculation considers the weighted mean returns of the individual investments based on their respective capital weights, providing an estimate of the overall return expected from the portfolio. This approach allows investors to forecast potential returns based on their portfolios composition and helps make informed decisions regarding asset allocation and risk management strategies. The correlation coefficient (r) is a statistical measure that ranges from ± 1, with a coefficient of 0 indicating no relationship between two investments in assets. A positive coefficient ($r > 0$) signifies a positive correlation between the two investment assets' return rates, while a negative coefficient ($r < 0$) indicates a negative correlation. In our findings, the study observed a negative correlation in the return rate of ABT with listed company VNM and listed company HPG, indicating that the return rates of these companies are not compatible with listed company ATB's return rate. This suggests that investors who invested in both the listed company HPG and listed company ABT may experience negative rates of return.

Specifically, the relationship between listed company VNM and listed company HPG' return rate was 15%, while the correlation between listed company HPG and listed company ABT is -19.7%. The relatively close correlation between the

Table 5 The risk results of the three stocks

Items	ROE_{ABT}	ROE_{HPG}	ROE_{VNM}	Total
ROE_{ABT}	0.00182	−0.00026	−0.00008	0.00148
ROE_{HPG}	−0.00026	0.00109	0.00014	0.00096
ROE_{VNM}	−0.00008	0.00014	0.00088	0.00095
Total σ_p^2				0.00339

listed company VNM and the listed company HPG implies that their return rates are positively correlated. As a result, if investors invest in both companies, their risks increase and decrease. However, if the two assets have a strong, positive connection (>80%), they won't be able to divide the risk, as indicated in Table 5. A statistical metric is called covariance that indicates the degree to which two variables change together. In the context of finance, covariance is used to measure the relationship between the return rates of different assets. When calculating the covariance between two investment securities, several possible scenarios can occur based on the sign of the covariance. Positive Covariance: A positive covariance indicates that the two assets tend to move in the same direction. So, a covariance of 1.14441 between assets HPG and ABT suggests a positive relationship where, when the return rate of one asset increases, the return rate of the other asset also tends to increase. Negative Covariance: On the other hand, a negative covariance implies that the two assets move in opposite directions. So, a covariance of −0.651 between assets ABT and VNM indicates an inverse relationship, where when the return rate of one asset goes up, the return rate of the other asset tends to go down (see Table 5). When considering investments in companies like listed company VNM and listed company HPG, or listed company VNM and listed company ABT, it's important to understand the concept of covariance. If the return rates of these two companies move in the same direction, it suggests a positive covariance, indicating that when one company's return increases, the other company's return also tends to increase. Conversely, if the assets are invested in companies like HPG and ABT, where there is a negative covariance, an increase in ABT Company's return might lead to a decrease in HPG Company's return, and vice versa. Understanding covariance is crucial as it sheds light on the interrelationship between the return rates of different companies.

This understanding is essential for effective portfolio management as it allows investors to gauge the potential risks associated with their investment decisions portfolio risk and making informed investment decisions. The comparison between the initial full investment capital in VNM stock and the allocation of investment capital based on a portfolio with 30% VNM shares, 30% HPG stock, and 40% ABT shares reveals interesting findings. When investment capital is diversified across these three stocks, the risk level of VNM stock decreases as indicated by the decrease in its standard deviation from 9.949 to 6.0%. However, this risk reduction comes at a cost, with the expected profit also decreasing from 36.89 to 25.2%.

Similarly, the standard deviations of both HPG and ABT stocks also decrease significantly from 23.47% and 10.987 to 6% when included in the portfolio. The expected return for HPG decreases by 23.47 to 25.2%, while ABT's expected return increases from 17.84 to 25.2%. This shift in standard deviations and expected returns highlights the trade-off between risk and return in investment decisions. It's important to note that although ABT stock presents a higher risk compared to HPG and VNM due to its higher standard deviation, this higher risk is associated with the potential for higher returns. However, investing in riskier stocks does not guarantee higher profits. Investors need to carefully assess their risk tolerance, investment goals, and portfolio diversification strategy to make informed decisions that align with their financial objectives and risk preferences.

6 Conclusion

When investors choose to invest in a portfolio consisting of multiple securities items, they need to determine the expected return and risk of the portfolio to make informed decisions. Calculating the expected return and risk ratios of a portfolio is crucial for investors as it helps them assess the potential outcomes and make strategic investment choices based on their risk tolerance and investment goals. To determine the expected return of a portfolio, investors typically consider the individual securities' expected returns and their respective weights in the portfolio. By multiplying the expected return of each security by its weight in the portfolio and summing up these values, investors can calculate the overall expected return of the portfolio. This calculation considers both the return potential of each asset and how much of the portfolio is allocated to each asset, providing a comprehensive view of the expected performance of the portfolio. Risk assessment in a portfolio involves evaluating the volatility and covariance of the individual securities to understand how they interact within the portfolio. Investors commonly use measures such as standard deviation and correlation coefficients to assess the overall risk of the portfolio. By considering the risk characteristics of each security and their correlations with one another, investors can calculate the portfolio's overall risk level, considering diversification benefits and potential downside risks. The process of determining the expected return and risk of a portfolio is essential for investors to make well-informed decisions and build portfolios that align with their financial goals and risk preferences. By leveraging quantitative methods, data analysis, and portfolio management techniques, investors can optimize their portfolio construction and enhance their investment outcomes.

Diversification is a key strategy that investors can use to minimize risks in their portfolios. By combining two or more different investment items in a portfolio, investors can spread out their risk exposure and potentially reduce the impact of losses from any single asset. When investors diversify their portfolios, they aim to invest in securities that are not perfectly correlated with each other. This means that the prices of these securities do not move in perfect sync, providing a hedge against potential losses. The empirical results show that if one stock in the portfolio decreases in value, the gains from another stock can help offset the losses, leading to a more balanced overall return. Moreover, diversification can extend beyond investing in different types of securities within the same market. Investors can also diversify by investing in international stock markets instead of focusing solely on the stock market of a specific country.

This approach allows investors to spread their risk across different economies, industries, and currencies, reducing their exposure to country-specific risks and market fluctuations. By diversifying globally, investors can benefit from opportunities in various regions and sectors while potentially lowering the overall risk of their portfolio. Diversification is an effective risk management strategy that investors can use to enhance the resilience of their portfolios and potentially improve their overall returns. By spreading investments across multiple assets and markets, investors can reduce the impact of individual security or market fluctuations and create a more balanced and diversified portfolio.

References

1. Mohaghegh, V., Mousavi, S.M., Vahdani, B., Shahriari, R.M.: R&d project evaluation and project portfolio selection by a new interval type-2 fuzzy optimization approach. Neural Comput. Appl. **28**(12), 3869–3888 (2017)
2. Carole, B., Michel, D., Steven, V.: Measuring portfolio risk under partial dependence information. J. Risk Insur. **85**(3), 843–863 (2016)
3. Abate, G., Basile, I., Ferrari, P.: The integration of environmental, social and governance criteria in portfolio optimization: an empirical analysis. Corp. Soc. Responsib. Environ. Manag. **31**(3), 2054–2065 (2024)
4. Dinh, D.V.: Analyzed relationship between risks and expected returns. J. Econ. Adm. Sci. **39**(4), 749–759 (2023)
5. Kim, M., Jeong, Y.J., Jeong, J.: Two empirical studies of portfolio optimization using cryptocurrency allocation ratios. IEEE Access **12**, 63827–63838 (2024)
6. Markowitz, H.: Portfolio selection. J. Financ. **7**(1), 77–91 (1952)
7. Yu, J., Zhang, J.: A comprehensive analysis of the modern portfolio theory. BCP Bus. Manag. **38**, 2111–2114 (2023)
8. Haim, S.: Portfolio risk management using the Lorenz curve. J. Portf. Manag. **40**(3), 152–159 (2014)
9. Alexander, K., Wilderich, H., Hans, G.G.: A contingency approach on the impact of front-end success on project portfolio success. Proj. Manag. J. **47**(2), 115–129 (2016)
10. Kumar, M., Gregory, M.: An exploration of risk management in global industrial investment. Risk Manage. **15**(4), 272–300 (2013)
11. Dinh, D.V.: Forecasting domestic credit growth based on arima model: Evidence from vietnam and china. Manag. Sci. Lett. **10**(5), 1001–1010 (2020)
12. Insana, A.: Does systematic risk change when markets close? an analysis using stocks' beta. Econ. Model. **109**, 105782 (2022)
13. Virlics, A.: Investment decision making and risk. Procedia Econ. Financ. **6**, 169–177 (2013)
14. Dinh, D.V.: Determinants of commercial banks' profitability: evidence from vietnam. Adv. Appl. Stat. **63**(1), 39–58 (2020)
15. Jason, H., Tzee, C., Feifei, L., Li-Lan, K.: A study of low-volatility portfolio construction methods. J. Portf. Manag. **40**(4), 89–105 (2014)
16. Linda, K., Erik, J.H., Abbie, G.: An empirical investigation of the antecedents and outcomes of npd portfolio success. J. Prod. Innov. Manag. **31**(6), 1199–1213 (2014)

17. Rongxi, Z., Yu, Z., Ru, C., Guanqun, T.: A mean-variance hybrid-entropy model for portfolio selection with fuzzy returns. Entropy **17**(5), 3319–3331 (2015)
18. VIETSTOCK. Company financial reports, 2002. Accessed: 2025-06-20
19. Andrew, C.W., Helen, H.: Art as an investment: risk, return and portfolio diversification in major painting markets. Account. Financ. **44**(2), 257–272 (2004)
20. Dinh, D.V.: Public debt management and economic growth: Var model approach. Ind. Eng. Manag. Syst. **22**(2), 173–187 (2023)
21. Fabio, M.: Measuring bond mutual fund performance with portfolio characteristics. J. Empir. Financ. **33**(3), 223–242 (2015)
22. Dinh, D.V., Ha, N.T.: The tax policy and macro management: evidence in vietnam. Financ.: Theory Pract. **27**(5), 150–159 (2023)
23. Qin, Z.: Mean-variance model for portfolio optimization problem in the simultaneous presence of random and uncertain returns. Eur. J. Oper. Res. **245**(2), 480–488 (2015)
24. Surtee, T.G., Alagidede, I.P.: A novel approach to using modern portfolio theory. Borsa Istanbul Rev. **23**(3), 527–540 (2023)
25. The Economic Times. Return on equity, 2021. Accessed: 2025-06-20
26. Snedecor, G.W., Cochran, W.G.: Statistical Methods. Iowa State Press, Ames, IA, 7 edition (1980)
27. Weisstein, E.W.: Covariance matrix (1999). Accessed: 2025-06-20
28. Uspensky, J.V.: Introduction to Mathematical Probability. McGraw-Hill Book, New York (1938)
29. Dinh, D.V.: Applied individual investment risk measurement method to forecast expected return rate. Int. J. Appl. Math. Stat. **58**(3), 60–17 (2019)
30. Fuhre, A., Hock, T.: Uncertainty in the black–litterman model: empirical estimation of the equilibrium. J. Empir. Financ. **72**, 251–275 (2023)
31. Hongke, Z., Le, W., Qi, L., Yong, G., Enhong, C.: Investment recommendation in p2p lending: a portfolio perspective with risk management. IEEE Trans. Syst., Man, Cybern.: Syst. **48**(1), 1109–1114 (2018)
32. Kolari, J.W., Liu, W., Pynnönen, S.: Portfolio performance measures. In: Professional Investment Portfolio Management, pp. 97–119. Palgrave Macmillan, Cham, Switzerland (2024)
33. Sahu, S., Vázquez, J.H., Ramírez, A.F., Kim, J.-M.: Analyzing portfolio optimization in cryptocurrency markets: a comparative study of short-term investment strategies using hourly data approach. J. Risk Financ. Manag. **17**(3), 2–14 (2024)
34. Giunta, N., Orlando, G., Carleo, A., Ricci, J.M.: Exploring entropy-based portfolio strategies: empirical analysis and cryptocurrency impact. Risks **12**(5), 1–26 (2024)
35. Masdemont, J.J., Luis, O.-G.: Haar wavelets-based approach for quantifying credit portfolio losses. Quant. Financ. **1**(9), 1587–1595 (2011)

The Impact of Capital Structure on the Performance of Non-financial Enterprises in Vietnam

Lai Cao Mai Phuong

Abstract Between 2014 and 2019, Vietnam's economic growth rate consistently exceeded twice the global average. Even amid the COVID-19 pandemic in 2020, Vietnam remained one of the few countries that sustained growth. This raises questions regarding the policy environment supporting business performance, enabling firms to continue expanding and contributing to economic development. This study investigates the effect of capital structure on the performance of non-financial enterprises in Vietnam—a developing economy. The research focuses on firms operating in the Food and Beverage, Steel, and Energy sectors during the 2017–2020 period, as these industries represent a significant proportion of Vietnam's stock market and are classified as non-financial. Balanced panel data regression models are applied across all three sectors, using two distinct regression equations. After controlling firm-specific variables, such as asset growth, the ratio of fixed assets to total assets, revenue growth, size, and liquidity, the findings indicate that capital structure negatively affects firm performance in both the short and long term across all industries analyzed. These results suggest that Vietnamese firms should give priority to internal financing rather than relying heavily on debt. Moreover, to enhance operational efficiency, Food and Beverage firms should prioritize investment in fixed assets, while Steel and Energy firms should focus on boosting revenue and asset growth. Increasing firm size and asset accumulation are also recommended for all three sectors to improve performance outcomes.

Keywords Asset · Firm size · Leverage · Profitability · Liquidity · Revenue · Sector · Vietnam

L. C. M. Phuong (✉)
School of Finance and Accounting, Industrial University of Hochiminh City, Ho Chi Minh City, Vietnam
e-mail: laicaomaiphuong@iuh.edu.vn

H. T. T. Nguyen et al. (eds.), *Data Science in Finance and Accounting*, Studies in Big Data 181, https://doi.org/10.1007/978-3-032-06179-9_12

1 Introduction

In recent years, Vietnam has remained one of the fastest-growing economies both regionally and globally. During the period 2014–2019, Vietnam's average annual GDP growth consistently ranged between 6–7% [1], approximately double the global average growth rate [2]. Despite the economic disruptions caused by the COVID-19 pandemic in 2020, Vietnam continued to experience growth, whereas the average growth rate of ASEAN-5 was −3.7% and the global growth rate was −3.5% [3]. This remarkable economic performance was largely driven by the substantial contributions of Vietnam's business sector.

By the end of 2020, nearly 20 years after its inception on July 28, 2000, Vietnam's stock market had achieved a market capitalization equivalent to 87.68% of GDP [4]. Over this period, the market has attracted major enterprises across various industries, including efficiently operated firms seeking listing opportunities. These developments underscore the growing contribution of publicly traded companies and the stock market's role in supporting Vietnam's economic progress. Given this context, research aimed at enhancing the operational efficiency of listed firms is crucial for ensuring economic stability and fostering sustainable growth.

The choice of funding source, or capital structure decision, is one of the three fundamental aspects of corporate financial management, as it directly influences both business performance [5] and stock market reactions [6]. Empirical studies indicate that the relationship between capital structure and firm performance varies across different countries and industries [7–9]. Moreover, many existing publications on this topic focus exclusively on a single industry [9, 10], select representative samples from a single market [7, 11], or primarily examine stock markets with a long-established history [8, 12]. However, research that simultaneously analyzes multiple industries to compare the impact of capital structure on firm performance remains relatively limited.

The findings of this study offer several contributions. Firstly, analyzing three distinct industries concurrently, rather than focusing on a single industry, enables readers to more effectively discern variations in the influence of structural factors across different sectors. This study finds that both long-term and short-term capital structures exert a negative impact on the return on assets across all three industries. However, when return on equity is used as the dependent variable, capital structure influences only the steel and energy industries, while its effect on the food and beverage industry remains statistically insignificant. Secondly, this study examines non-financial firms, all of which adhere to a uniform financial reporting format. As a result, the findings regarding the effects of control variables provide deeper insights into industry-specific characteristics that may impact firm performance. This study reveals that size significantly influences performance across all three industries. Furthermore, enhancing business performance requires industry-specific strategies: increasing fixed asset value is crucial in the food and beverage sector, revenue expansion demands greater attention in the steel industry, and asset growth should be prioritized in the energy sector. Finally, this study contributes to the empirical evidence on developing economies, particularly those with stock markets at a frontier stage of development.

2 Literature Review and Hypothesis

Modigliani and Miller [13] assert that, in a perfectly efficient market, a firm's value is independent of its capital structure. However, since perfect markets are rare in practice, their assumptions have been progressively relaxed to align more closely with real-world conditions. The trade-off theory and agency cost framework are commonly employed to explain how capital structure influences firm performance and valuation [14–16].

When firms utilize debt financing, they must weigh the tax benefits against the potential costs of financial distress to optimize operational efficiency. Excessive reliance on debt can adversely impact a company's growth prospects [11], as higher debt ratios are associated with increased financial distress and a greater likelihood of bankruptcy [5, 17]. According to the trade-off theory [14, 15], in such cases, the tax advantages of debt financing may be insufficient to offset the risks and potential losses associated with a rising debt burden.

Empirical studies support this perspective, including [11, 18]. Using an unbalanced panel dataset of 167 firms listed on the Jordanian stock market from 1989 to 2003, Zeitun and Tian [11] found that a higher debt ratio is associated with lower firm performance, consistent across both accounting-based and market-based valuation methods. Their findings suggest that firms often underestimate the financial distress and bankruptcy costs incurred when excessively increasing their debt levels, ultimately leading to a decline in performance.

To investigate the relationship between financing choices and debt burden, Dezi and Giudice [18] analyzed 320 small and medium-sized textile enterprises in Italy over the extended period of 2005–2010. Their study revealed that firms favoring debt financing experienced rising unstable costs, posing a significant risk to corporate reputation. In other words, for these firms, financial distress costs exceeded the tax benefits derived from debt utilization [18].

Moreover, as a company's equity becomes more diluted, the separation between control and ownership increases [19], potentially leading to higher agency costs. These costs arise due to conflicts between managers and shareholders [15, 16]. A greater separation between control and ownership allows managers to pursue personal objectives rather than maximizing corporate value. In such cases, managers may invest in low-efficiency projects, reducing the firm's operational performance and ultimately harming shareholder interests.

Majumdar and Chhibber [20] support this argument, having examined companies listed on the Bombay Stock Exchange in India between 1988 and 1994. Their findings indicate that a higher debt ratio in the capital structure negatively impacts firm performance. They argue that this relationship exists because lending institutions during that period were government-owned, giving creditors little incentive to closely monitor or improve borrowers' operational efficiency.

One approach to mitigating agency costs is for firms to increase their debt levels, thereby exerting pressure on managers to enhance performance [10]. Since creditors impose stricter oversight and constraints on firms, managers must prioritize high-return investments to maintain sufficient cash flow for interest and principal payments. This perspective suggests that higher debt reduces agency costs, increases corporate value, and minimizes conflicts between managers and shareholders.

Empirical studies support this view across various contexts, including U.S. banks with above-average debt ratios from 1990 to 1995 [10], traditional manufacturing and growth industries on the French stock market from 2002 to 2005 [8], listed manufacturing and service firms in the U.S. from 2005 to 2007 [12], and Ghanaian firms from 1998 to 2002 [7]. These studies [7, 8, 10–12] consistently demonstrate a positive relationship between capital structure and firm performance. In other words, when debt is utilized effectively, managers can mitigate agency cost issues, allowing debt financing to enhance business performance rather than hinder it.

In emerging stock markets such as Vietnam, regulatory effectiveness—particularly from Fama's [21] perspective—and the timeliness of legal enforcement to protect shareholders remain limited [22, 23] compared to more developed markets. Consequently, the agency cost issue is likely to persist at a higher level than in more mature stock markets.

Additionally, compared to advanced economies, Vietnamese enterprises that achieve international recognition are predominantly concentrated in the financial sector, rather than non-financial industries [24, 25]. As a result, the overall management capabilities of these firms tend to be lower. From the trade-off theory perspective, the net benefit of using debt is likely to be negative, making it improbable that increased debt financing will effectively mitigate agency costs for Vietnamese enterprises [26].

A study analyzing 488 non-financial listed firms in Vietnam between 2013 and 2018 [27] found a negative correlation between capital structure and firm performance, applicable to both state-owned and privately owned enterprises listed on the Vietnamese stock market [28]. However, in terms of industry analysis, [27] only examined correlations between capital structure and financial indicators such as ROE, ROA, and EPS, without conducting separate regression analyses for each sector.

Based on these considerations, this study hypothesizes the impact of capital structure on the performance of non-financial enterprises in Vietnam as follows:

H1: Capital structure has a negative impact on the performance of non-financial enterprises.

Besides variables representing capital structure, studies on the impact of capital structure on the performance of businesses also use many different control variables. Control variables are often used such as: scale conducted in research in India and Jordan [11, 20], growth opportunities in Ghana, Jordan and India [7, 11, 20], the proportion of fixed assets in Romania and Jordan [11, 29] and accounting liquidity in India and Romania [20, 29].

3 Methods and Data

3.1 Method

The research model presented in this study is grounded in prior literature examining the relationship between capital structure and corporate performance. To estimate the proposed models, panel data regression techniques are employed. The selection of the most appropriate estimation results is based on diagnostic tests, including assessments for autocorrelation, multicollinearity [30], and the suitability of the estimation method. When issues of heteroscedasticity and autocorrelation are detected, the feasible generalized least squares (FGLS) method is applied. If, in addition to these issues, endogeneity is present, the generalized method of moments (GMM) is utilized for estimation.

To examine the impact of capital structure on the performance of non-financial enterprises, this study incorporates variables previously explored in related research [7, 11, 29]. Accordingly, the general model is structured as follows:

$$\text{Profitability} = f[\text{capital structure; growth opportunities; size; fixed assets ratio; liquidity}]$$

$$\text{ROA or ROE} = f[(\text{SDA, LDA}); (\text{AG, SG}); \text{SIZE}; \text{TANG}; \text{LIQ}] \quad (1)$$

where:

Dependent variables:

- The dependent variables in this study measure profitability, represented by return on assets (ROA) and return on equity (ROE).

Independent variables:

- Independent variables include two metrics reflecting capital structure: short-term debt ratio (SDA) and long-term debt ratio (LDA), both measured as a proportion of total assets.

Control variables consist of several factors influencing firm performance:

- **Growth opportunities,** captured through asset growth (AG) and revenue growth (SG). Asset growth is calculated as the percentage change in total asset value between the beginning and end of the year. Revenue growth is measured as the percentage change in revenue from year $(t - 1)$ to year t.
- **Firm size (SIZE),** defined as the natural logarithm of total assets at year-end.
- **Fixed asset ratio (TANG),** computed as the proportion of fixed assets to total assets.
- **Liquidity (LIQ),** assessed by the ratio of total cash and cash equivalents to total assets.

The general model is formulated through two specific regression Eqs. 2 and 3 to analyze the impact of capital structure on firm performance.

$$\begin{aligned} \mathrm{ROA}_{it} &= \alpha_i + \beta_1 \mathrm{SDA}_{it} + \beta_2 \mathrm{LDA}_{it} + \beta_3 \mathrm{AG}_{it} + \beta_4 \mathrm{SG}_{it} \\ &\quad + \beta_5 \mathrm{SIZE}_{it} + \beta_6 \mathrm{TANG}_{it} + \beta_7 \mathrm{LIQ}_{it} + \varepsilon_{it} \end{aligned} \tag{2}$$

$$\begin{aligned} \mathrm{ROE}_{it} &= \alpha_i + \beta_1 \mathrm{SDA}_{it} + \beta_2 \mathrm{LDA}_{it} + \beta_3 \mathrm{AG}_{it} + \beta_4 \mathrm{SG}_{it} \\ &\quad + \beta_5 \mathrm{SIZE}_{it} + \beta_6 \mathrm{TANG}_{it} + \beta_7 \mathrm{LIQ}_{it} + \varepsilon_{it} \end{aligned} \tag{3}$$

Each industry is analyzed using panel data regression based on Eqs. 2 and 3, employing the feasible generalized least squares (FGLS) method. FGLS offers advantages over alternative regression techniques, such as pooled ordinary least squares (POLS), fixed-effects models (FEM), and random-effects models (REM), as it effectively addresses issues of autocorrelation and heteroscedasticity within the research model.

The panel data undergoes several diagnostic tests, including the Chow test to distinguish between POLS and FEM, the Hausman test [31] to determine whether FEM or REM is more suitable, the Wald [32] and Breusch-Pagan tests [33] for heteroscedasticity, the Wooldridge test [34] for autocorrelation, and the Wu-Hausman test to assess variable endogeneity. FGLS is preferred as it mitigates autocorrelation and heteroscedasticity; however, when endogeneity is present, the generalized method of moments (GMM) regression is applied, as it resolves endogeneity concerns while addressing the limitations of FGLS.

Based on the test results, the most appropriate regression model is selected. Subsequently, the influence of capital structure on industry-specific performance is analyzed. Finally, the regression outcomes—considering both control variables and capital structure—are used to formulate recommendations for relevant stakeholders.

3.2 Data

This study utilizes annual audit report data from non-financial companies listed on the Vietnam stock market, sourced from Vietstock, to examine the impact of capital structure on firm performance. The analysis focuses on companies from three key industries—Food and Beverages, Steel, and Energy—that meet data availability criteria for the 2017–2020 period.

4 Experimental Results

Table 1 descriptive statistical analysis indicates that the Steel industry exhibits the lowest average return on assets (ROA) and return on equity (ROE), followed by the

Table 1 Descriptive statistics and variance inflation factor (VIF) for variables in the Food and Beverage, Steel, and Energy sectors

	Food and Beverage			Steel			Energy		
Variable	Obs	Mean	VIF	Obs	Mean	VIF	Obs	Mean	VIF
ROA	120	0.079		112	0.035		120	0.052	
ROE	120	0.138		112	0.082		120	0.125	
SDA	120	0.306	1.05	112	0.512	1.25	120	0.321	1.29
LDA	120	0.068	1.29	112	0.071	1.41	120	0.182	1.71
AG	120	0.078	1.4	112	0.050	1.46	120	0.061	1.17
SG	120	0.066	1.44	112	0.074	1.46	120	0.209	1.1
SIZE	120	12.556	1.3	112	12.103	1.43	120	12.623	1.3
TANG	120	0.269	1.15	112	0.199	1.54	120	0.338	1.75
LIQ	120	0.089	1.04	112	0.060	1.23	120	0.103	1.43
Mean VIF			1.24			1.40			1.39

Energy sector, while the Food and Beverage industry achieves the highest profitability metrics.

The two variables representing a firm's debt utilization are the ratio of short-term debt to total assets (SDA) and the ratio of long-term debt to total assets (LDA). Across all three industries, firms exhibit a preference for short-term debt, ranging from 1.7 times (Energy sector) to 7.2 times (Steel sector) the usage of long-term debt. Among these industries, the Steel sector has the highest average SDA ratio, whereas the Food and Beverage sector has the lowest. Conversely, the Energy sector records the highest average LDA ratio, while the Food and Beverage sector maintains the lowest.

Regarding control variables, except for asset growth (AG), the Energy sector exhibits the highest average values for four out of five variables among the three industries, including firm size (SIZE), fixed asset ratio (TANG), revenue growth rate (SG), and liquidity (LIQ). The Food and Beverage sector records the highest asset growth rate (AG). Conversely, the Steel sector has the lowest values for four out of five indicators—AG, SIZE, TANG, and LIQ—among the industries analyzed.

None of the three industries exhibits multicollinearity among variables, as the variance inflation factor for all variables remains below 2 [35]. Consequently, these variables are appropriate for inclusion in the regression analysis based on Eqs. 2 and 3 for each industry.

The correlation analysis results presented in Table 2 indicate that the correlation coefficients for the ROA and ROE variables exceed 0.7. This outcome is expected, as these are two dependent variables derived from separate regression equations.

Additionally, the highest pairwise correlation coefficients among independent variables for the three industries are as follows: (LDA; TANG) = 0.48 for the Steel industry, (AG; SG) = 0.52 for the Food and Beverage industry, and (LDA; TANG) =

Table 2 Correlation tables for the Food and Beverage, Steel, and Energy sectors

Variable	ROA	ROE	SDA	LDA	AG	SIZE	TANG	SG	LIQ
Panel A: Food and Beverage sector									
ROA	1.00								
ROE	0.84	1.00							
SDA	−0.08	0.12	1.00						
LDA	−0.31	−0.25	−0.05	1.00					
AG	0.20	0.26	0.01	0.16	1.00				
SIZE	0.27	0.18	−0.12	0.40	0.20	1.00			
TANG	0.14	0.09	−0.02	0.31	−0.01	0.22	1.00		
SG	0.08	0.08	0.12	0.13	0.52	0.22	−0.02	1.00	
LIQ	0.15	0.08	0.06	−0.11	−0.06	0.00	−0.15	0.02	1.00
Panel B: Steel sector									
ROA	1.00								
ROE	0.92	1.00							
SDA	−0.20	−0.07	1.00						
LDA	−0.10	−0.05	−0.24	1.00					
AG	0.34	0.33	0.15	−0.03	1.00				
SIZE	0.29	0.32	0.04	0.25	0.32	1.00			
TANG	−0.04	0.04	−0.18	0.48	−0.07	0.31	1.00		
SG	0.29	0.32	0.12	0.25	0.40	0.21	0.31	1.00	
LIQ	0.02	−0.02	−0.31	−0.02	−0.06	−0.29	−0.06	−0.14	1.00
Panel C: Energy sector									
ROA	1.00								
ROE	0.78	1.00							
SDA	−0.29	0.04	1.00						
LDA	−0.28	−0.10	−0.10	1.00					
AG	0.07	0.26	0.26	−0.11	1.00				
SIZE	0.01	−0.13	−0.31	0.07	−0.15	1.00			
TANG	−0.06	−0.04	−0.17	0.56	−0.28	0.29	1.00		
SG	0.18	0.09	−0.10	−0.12	0.06	−0.20	−0.09	1.00	
LIQ	0.14	−0.08	−0.23	−0.45	−0.07	0.13	−0.30	0.07	1.00

0.56 for the Energy industry. Since all values remain below the 0.70 threshold, they do not pose multicollinearity concerns.

According to the criteria established by Farrar and Glauber [35], all explanatory and control variables meet the suitability requirements for inclusion in the regression analysis based on Eqs. 2 and 3.

The regression results for the industries are presented in Tables 3, 4, and 5.

The test results in Table 3 indicate that the model with dependent variables ROA and ROE exhibits heteroscedasticity. To address this issue, the regression results of

Table 3 Regression results of Food and Beverage (F&B) sector

	FGLS		GMM	
Variable	ROA	ROE	ROA	ROE
L1.ROA			0.367**	
L1.ROE				0.304**
SDA	−0.075**	0.028	−0.101*	0.004
LDA	−0.344***	−0.484***	−0.417***	−0.905**
AG	0.068***	0.120***	0.127	0.734**
SIZE	0.042***	0.065***	0.054	−0.045
TANG	0.095**	0.135*	0.02	0.766*
SG	0.001	−0.006	0.017	−0.038
LIQ	0.016	−0.017	−0.264	0.91
Constant	−0.437***	−0.713***	−0.571	0.382
Obs	120	120	90	90
Chow test	$F(29, 83) = 11.82;\ Prob > F = 0.0000$	$F(29, 83) = 4.81;\ Prob > F = 0.0000$		
Wald test	$chi2(30) = 2.6e+05;\ Prob > chi2 = 0.0000$	$chi2(30) = 1.5e+06;\ Prob > chi2 = 0.0000$		
Hausman test	$chi2(7) = 513.70;\ Prob > chi2 = 0.0000$	$chi2(7) = 23.07;\ Prob > chi2 = 0.0017$		
Breusch-Pagan test	$chibar2(01) = 49.14;\ Prob > chibar2 = 0.0$	$chibar2(01) = 23.77;\ Prob > chibar2 = 0.0$		
Wooldridge test	$F(1, 29) = 3.995;\ Prob > F = 0.0551$	$F(1, 29) = 7.548;\ Prob > F = 0.0102$		
Wu-Hausman			< 0.05	< 0.05
AR(1)			0.009	0.066
AR(2)			0.148	0.283
Hansen J Test			0.249	0.381

* Notes: *, **, *** are statistical significance levels at the 10, 5 and 1% levels, respectively

the feasible generalized least squares (FGLS) model are presented after correcting for these shortcomings. Additionally, the Wu-Hausman test yields a $P - value < 0.05$, indicating the presence of endogeneity among variables in the model. To mitigate this problem, the generalized method of moments (GMM) estimation is employed. In Table 3, both the FGLS and GMM models indicate that capital structure plays a crucial role in the performance of Food and Beverage enterprises. The short-term debt ratio (SDA) significantly impacts ROA but has no statistically significant effect on ROE. In contrast, the long-term debt ratio (LDA) exerts a significant influence

Table 4 Steel sector regression results

Variable	ROA		ROE	
	FEM	FGLS	FEM	FGLS
SDA	−0.104***	−0.089***	−0.137**	−0.091**
LDA	−0.146***	−0.148***	−0.209*	−0.245***
AG	0.022	0.018*	0.044	0.033
SIZE	0.017**	0.023***	0.035*	0.055***
TANG	−0.031	−0.017	−0.016	−0.029
SG	0.050***	0.034***	0.095***	0.080***
LIQ	−0.008	0.016	−0.039	0.063
Constant	−0.111	−0.184***	−0.264	−0.528***
Obs	112	112	112	112
Chow test	$F(27,77) = 3.29$; $Prob > F = 0.0000$		$F(27,77) = 3.00$; $Prob > F = 0.0001$	
Wald test	$chi2(28) = 22532.97$; $Prob > chi2 = 0.0000$		$chi2(28) = 62613.90$; $Prob > chi2 = 0.0000$	
Hausman test	$chi2(7) = 11.07$; $Prob > chi2 = 0.1355$		$chi2(7) = 8.19$; $Prob > chi2 = 0.3159$	
Breusch-Pagan test	$chibar2(01) = 13.98$; $Prob > chibar2 = 0.0001$		$chibar2(01) = 12.43$; $Prob > chibar2 = 0.0002$	
Wooldridge test	$F(1,27) = 10.872$; $Prob > F = 0.0027$		$F(1,27) = 16.601$; $Prob > F = 0.0004$	
Wu-Hausman	$P - value > 0.05$		$P - value > 0.05$	

* Notes: *, **, *** are statistical significance levels at the 10, 5 and 1% levels, respectively

on both dependent variables—ROA and ROE. Compared to the FGLS regression, certain control variables in the GMM model lose statistical significance. Specifically, AG, SIZE, and TANG exhibit an insignificant impact on ROA, while SIZE has no significant effect on ROE. Moreover, firm performance from the previous year remains an important determinant of current-year outcomes for companies in the Food and Beverage sector.

Tables 4 and 5 present the regression results for the FEM and FGLS models. The Chow and Hausman tests confirm that the FEM estimation is more appropriate than the POLS and REM models. However, the Wald and Wooldridge tests reveal that the model suffers from heteroscedasticity and autocorrelation. To address these issues, the FGLS estimation method is applied. Furthermore, diagnostic tests indicate no endogeneity among independent variables, reinforcing the reliability of the FGLS results over FEM.

Table 4 indicates that the effects of variables on firm performance in the FEM and FGLS models for the steel industry exhibit considerable similarity. As observed in the Food and Beverage sector, capital structure plays a crucial role in shaping

Table 5 Energy sector regression results

	ROA		ROE	
Variable	FEM	FGLS	Variable	FEM
SDA	−0.090***	−0.074***	SDA	−0.090***
LDA	−0.092**	−0.081***	LDA	−0.092**
AG	0.019	0.017***	AG	0.019
SIZE	−0.003	−0.002	SIZE	−0.003
TANG	−0.025	0.005	TANG	−0.025
SG	0.007**	0.008***	SG	0.007**
LIQ	−0.031	−0.021	LIQ	−0.031
Constant	0.142	0.103	Constant	0.142
Obs	120	120	Obs	120
Chow test	$F(29, 83) = 8.88$; $Prob > F = 0.0000$		$F(29, 83) = 4.85$; $Prob > F = 0.0000$	
Wald test	$chi2(30) = 1.5e+05$; $Prob > chi2 = 0.0000$		$chi2(30) = 94087.34$; $Prob > chi2 = 0.0000$	
Hausman test	$chi2(7) = 5.85$; $Prob > chi2 = 0.5573$		$chi2(7) = 19.53$; $Prob > chi2 = 0.0067$	
Breusch-Pagan test	$chibar2(01) = 73.75$; $Prob > chibar2 =$ 0.0000		$chibar2(01) = 30.26$; $Prob > chibar2 =$ 0.0000	
Wooldridge test	$F(1, 29) = 8.232$; $Prob > F = 0.0076$		$F(1, 29) = 3.471$; $Prob > F = 0.0726$	
Wu-Hausman	$P - value > 0.05$		$P - value > 0.05$	

* Notes: *, **, *** are statistical significance levels at the 10, 5 and 1% levels, respectively

business performance within the steel industry. Both short-term debt ratio (SDA) and long-term debt ratio (LDA) exert a significant negative influence on firms' ROA and ROE. Additionally, firm performance in the steel industry is impacted by company size (SIZE) and revenue growth (SG). Furthermore, in the FGLS estimation, the dependent variable ROA is also affected by asset growth (AG).

The regression results in Table 5 indicate that both short-term debt ratio (SDA) and long-term debt ratio (LDA) have a significant impact on the return on assets (ROA) of the energy sector in both fixed-effects model (FEM) and feasible generalized least squares (FGLS) estimations. However, for the dependent variable return on equity (ROE), only LDA is statistically significant, highlighting its role in shaping firm performance within the energy industry. Compared to FEM estimates, the FGLS results reveal that ROA and ROE in this sector are influenced by a broader set of variables. Specifically, the FGLS estimation indicates that firm performance is affected by asset growth (AG) and revenue growth (SG), while ROE is additionally impacted by the fixed asset ratio (TANG).

Based on the regression findings across the three industries analyzed, this study discusses the most reliable estimation approach for each sector. For the Food and Beverage industry, the generalized method of moments (GMM) estimation is used, while for the Steel and Energy sectors, the FGLS approach provides the most reliable results.

4.1 Variable Representing Capital Structure

Overall, capital structure plays a significant role in determining firm performance across all three industries, as evidenced by the statistical significance of the short-term debt ratio (SDA) and long-term debt ratio (LDA) in the regression results. The LDA negatively affects both ROA and ROE across all industries, while the SDA significantly impacts ROA for all industries but only influences ROE in the Steel and Energy sectors, with no statistically significant effect observed in the Food and Beverage sector.

These findings indicate that the capital structure—encompassing both short-term and long-term debt—generally supports hypothesis **H1**. In other words, empirical results demonstrate a negative relationship between capital structure and firm performance among non-financial enterprises in Vietnam. This suggests that the net benefit of debt usage is negative, as the trade-off between benefits and costs does not mitigate the persistence of agency problems in these industries. Increasing debt ratios fail to reduce agency costs; instead, they diminish business efficiency.

Thus, the capital structure of the three industries during the study period implicitly supports both the trade-off theory [14, 17] and agency cost theory [15, 16]. This phenomenon may be attributed to corporate managers' insufficient and incomplete assessment of debt-related costs. Additionally, Vietnam's stock market remains classified at a marginal level, and its market efficiency—according to Fama [21]—is lower than that of more developed economies. This lower efficiency may explain why agency cost issues are more prevalent among Vietnamese enterprises compared to firms operating in mature stock markets.

4.2 Control Variables

Company size (SIZE): At the 1% statistical significance level, both models with dependent variables ROA and ROE confirm the influence of company size on the performance of enterprises in the steel industry. Specifically, larger firms in this sector demonstrate higher profit margins, reflecting the benefits of economies of scale, where firms leverage resources more efficiently to maximize profitability. This finding in Vietnam aligns with previous research conducted by Majumdar and Chhibber [20] in India during 1988–1994 and Zeitun and Tian [11] in Jordan, reinforcing the argument that firm size plays a crucial role in financial performance within the steel sector. However, the FGLS regression results for the Energy industry and the GMM regression results for the Food and Beverage industry do not indicate a statistically significant impact of company size on business performance. This suggests that while economies of scale contribute to profitability in the steel sector, their influence may be less pronounced or negligible in other industries examined in this study.

Growth opportunities (AG): A company's growth opportunities are reflected through the regression results of two key variables: asset growth (AG) and revenue growth (SG). The regression coefficient of AG is statistically significant across all three industries, confirming its substantial impact on firm performance. Firms that exhibit a higher average increase in assets at year-end compared to the beginning of the year tend to generate greater profits. This suggests that asset expansion plays a crucial role in boosting business profitability. These findings align with prior studies on firms listed in the Ghanaian market during 1998–2002 [7] and those trading on the Amman Stock Exchange in Jordan [11], reinforcing the broader relevance of asset growth as a determinant of financial success across different economic contexts.

Unlike asset growth, revenue growth demonstrates statistical significance in only two of the three industries. Firms in the Energy and Steel sectors with higher revenue growth tend to achieve greater profitability. However, this variable has a negligible impact on the financial performance of the Food and Beverage industry, similar to findings on listed companies in India [20]. Thus, while revenue growth serves as a reliable indicator of business performance in the Energy and Steel industries, there is insufficient empirical evidence to confirm its role in influencing profitability within the Food and Beverage sector.

Fixed assets to total assets ratio (TANG): Companies that allocate more resources to fixed asset investments tend to achieve higher profitability. However, this relationship holds true only for the Food and Beverage and Energy sectors when ROE is the dependent variable, whereas no statistically significant effect is observed for the Steel sector. In other words, the positive correlation between TANG and firm performance is confirmed for the Food and Beverage and Energy sectors, but there is insufficient evidence to generalize this conclusion to the Steel sector. This finding contrasts with results from studies on listed companies in Romania [29] and Jordan [11], suggesting that sector-specific dynamics influence the role of fixed assets in firm performance. Moreover, in the context of Vietnam, the observed relationship highlights that investments in modern assets and equipment contribute to enhancing the performance of firms in the Food and Beverage sector.

Accounting liquidity: The liquidity ratio to total assets (LIQ), averaging between 6 and 10.3% (Table 1), does not exhibit a statistically significant impact on profitability across the three industries analyzed (Tables 3, 4, and 5). These findings contrast with previous studies, including [20], which examined firms listed on the Bombay Stock Exchange in India, and [29], which analyzed companies in Romania. This discrepancy suggests that the role of liquidity in firm performance may vary across different economic environments and market structures.

5 Conclusion

This study examines the impact of capital structure on the business performance of enterprises in three sectors—Food and Beverage, Steel, and Energy—listed on the Vietnam stock market during the period 2017–2020. The findings indicate that both short-term and long-term capital structures play a significant role in shaping firm performance, even after accounting for control variables such as firm size, revenue growth, asset growth, fixed asset ratio, and liquidity. Given these results, corporate managers in these sectors must evaluate and monitor the efficiency of capital utilization to ensure an optimal balance in capital structure, thereby enhancing overall business performance.

Food and Beverage sector: Profitability in this sector is influenced by key control variables, including asset growth, firm size, and the proportion of fixed assets to total assets. Consequently, companies seeking to enhance profits should focus on expanding production and business investments, particularly through asset acquisitions, with a strong emphasis on fixed assets. Investment capital should primarily be sourced from internal funds and retained earnings, while increasing debt utilization must be carefully evaluated, as excessive reliance on debt may adversely impact return rates.

Steel sector: To enhance profit margins, firms in the steel sector should prioritize scaling operations by expanding asset growth and increasing market share to boost revenue. However, asset growth must be carefully managed to prevent excessive reliance on debt, as higher debt ratios can negatively impact business performance.

Energy sector: Companies in the energy sector should focus on increasing revenue and asset growth while implementing stringent debt management strategies to improve profitability. Maintaining a balanced capital structure is essential to ensuring sustainable financial performance in this industry.

References

1. Epstein, M.N.P., Gornicka, L., Ha, N., Musil, K., Nalban, V.: Quarterly Projection Model for Vietnam: A Hybrid Approach for Monetary Policy Implementation. Technical Report 2022-2125, International Monetary Fund (2022)
2. Long, A., Ascent, D.: World Economic Outlook. International Monetary Fund (IMF), Washington (2020)
3. Khanh, V., Muralikumar, A.: Vietnam's 2020 GDP growth likely to slow to 2%-3%. Thomson Reuters news (2020). https://www.reuters.com/article/idUSL4N2HB0WF/, (accessed on 20 Nov 2020)
4. Vietnamplus. Stock market size reaches 88 percent of GDP (2021). https://link.gov.vn/s0SLQ1vZ, (accessed on 02 March 2021)
5. Ross, S.A.: The determination of financial structure: the incentive-signalling approach. Bell J. Econ. 23–40 (1977)

6. Machado, C., Pereira, A.E.: Optimal capital structure with stock market feedback. Rev. Financ. **27**(4), 1329–1371 (2023)
7. Abor, J.: The effect of capital structure on profitability: an empirical analysis of listed firms in Ghana. J. Risk Financ. **6**(5), 438–445 (2005)
8. Margaritis, D., Psillaki, M.: Capital structure, equity ownership and firm performance. J. Bank. Financ. **34**(3), 621–632 (2010)
9. Uddin, N.: Productivity relevance of capital structure choices. Int. J. Econ. Financ. **7**(9), 31–49 (2015)
10. Berger, A.N., Di Patti, E.B.: Capital structure and firm performance: a new approach to testing agency theory and an application to the banking industry. J. Bank. Financ. **30**(4), 1065–1102 (2006)
11. Zeitun, R., Tian, G.G.: Capital structure and corporate performance: evidence from Jordan. Australas. Account., Bus. Financ. J. **1**(4), 40–61 (2007)
12. Gill, A., Biger, N., Mathur, N.: The effect of capital structure on profitability: evidence from the United States. Int. J. Manag. **28**(4), 3–15 (2011)
13. Modigliani, F., Miller, M.H.: The cost of capital, corporation finance and the theory of investment. Am. Econ. Rev. **48**(3), 261–297 (1958)
14. Kraus, A., Litzenberger, R.H.: A state-preference model of optimal financial leverage. J. Financ. **28**(4), 911–922 (1973)
15. Meckling, W.H., Jensen, M.C.: Theory of the firm: managerial behavior, agency costs and ownership structure. J. Financ. Econ. **3**(4), 305–360 (1976)
16. Jensen, M.C.: Agency costs of free cash flow, corporate finance, and takeovers. Am. Econ. Rev. **76**(2), 323–329 (1986)
17. Myers, S.: The capital structure puzzle. J. Financ. **39**(3), 575–592 (1984)
18. Dezi, L., Del Giudice, M.: Economic performance and capital structure choices: a nonparametric analysis of the southern Italian textile-clothing industry. Int. Stud. Manag. Organ. **44**(1), 42–54 (2014)
19. Berle, A.A.: For whom corporate managers are trustees: a note. Harv. Law Rev. **45**(8), 1365–1372 (1932)
20. Majumdar, S.K., Chhibber, P.: Capital structure and performance: evidence from a transition economy on an aspect of corporate governance. Public Choice **98**(3–4), 287–305 (1999)
21. Fama, E.F.: Efficient capital markets. J. Financ. **25**(2), 383–417 (1970)
22. Giang, N.K.: Vietnam Police Seek Prosecution of 51 People as FLC Probe Ends. Bloomberg (2024). https://www.bloomberg.com/news/articles/2024-02-24/vietnam-police-seek-prosecution-of-51-people-as-flc-probe-ends
23. Phuong, L.C.M.: Stock price reactions to information about top managers. Banks Bank Syst. **16**(2), 159–169 (2021)
24. Aseanvietnam. 64 Vietnamese enterprises win at Asia Pacific Enterprise Awards, 2022. https://aseanvietnam.vn/en/post/64-vietnamese-enterprises-win-at-asia-pacific-enterprise-awards, (accessed on 01 March 2023)
25. Vietnam Report. Top 50 Doanh nghiep xuat sac Viet Nam nam 2023 - Top 50 Vietnam the Best (2023). https://vnr500.com.vn/Top-50-viet-nam-the-best-4643-1009.html, (accessed on 02 March 2023)
26. Vy, L.T.P., Nguyet, P.T.B.N.: Capital structure and firm performance: empirical evidence from a small transition country. Res. Int. Bus. Financ. **42**, 710–726 (2017)
27. Hieu, N.T., Anh, N.H.: Capital structure and firm performance of non-financial listed companies: cross-sector empirical evidences from Vietnam. Accounting **6**(2), 137–150 (2020)
28. Hieu, N.T., Anh, N.H.: The impact of capital structure on firm performance: evidence from Vietnam. J. Asian Financ., Econ. Bus. **7**(4), 97–105 (2020)
29. Vătavu, S.: The impact of capital structure on financial performance in Romanian listed companies. Procedia Econ. Financ. **32**, 1314–1322 (2015)
30. Farrar, D.E., Glauber, R.R.: Multicollinearity in regression analysis: the problem revisited. Rev. Econ. Stat. 92–107 (1967)
31. Hausman, J.A.: Specification tests in econometrics. Econ.: J. Econ. Soc. 1251–1271 (1978)

32. Wald, A.: Tests of statistical hypotheses concerning several parameters when the number of observations is large. Trans. Am. Math. Soc. **54**(3), 426–482 (1943)
33. Breusch, T.S., Pagan, A.R.: A simple test for heteroscedasticity and random coefficient variation. Econ.: J. Econ. Soc. 1287–1294 (1979)
34. Wooldridge, J.M.: Econometric Analysis of Cross Section and Panel Data. MIT Press (2010)
35. Essel, R.E.: The effect of capital structure on corporate performance: panel empirical evidence of an emerging capital market. J. Afr. Bus. **25**(2), 224–263 (2024)

Analyze Financial Data Using Benford's Law: Evidence From Vietnam Before and During the COVID-19 Pandemic

Hai Hong Phan, Ba Thu Tran, Anh Tu Pham, and Ha Thi Thu Nguyen

Abstract This study analyzes financial accounting data to investigate the extent and nature of deviations from Benford's Law. Using the dataset of Vietnamese-listed companies from 2015 to 2022, we examine the distribution of the first digits in accounts such as revenue, cost of goods sold, selling expense, administrative expenses, total assets, and total liabilities that may carry significant risks of misstatement in financial statements. The dataset is divided into two distinct periods: pre- and during the COVID-19 pandemic. Our analysis reveals differences in deviations from Benford's Law across these periods. The results suggest that the extent of financial data misstatements declined during the COVID-19 period compared to the pre-pandemic period.

Keywords Benford's law · Revenue · Selling costs · Administrative costs · Abnormalities · And vietnamese listed firms

1 Introduction

This study applies financial accounting data from Vietnamese firms to examine the extent of deviations from Benford's Law, which may signal potential manipulative practices. While prior studies have provided empirical evidence supporting the application of Benford's Law in detecting earnings management and financial misstatements for the whole financial statements across various contexts [1–4], this study applies the Benford's Law to several important financial accounting data from Vietnam, a developing country, where accounts such as revenue, cost of goods sold, selling expense, administrative expenses, total assets, and total liabilities may carry

H. H. Phan · B. T. Tran · H. T. T. Nguyen
Industrial University of Ho Chi Minh City, Ho Chi Minh City, Vietnam
e-mail: phanhonghai@iuh.edu.vn

A. T. Pham (✉)
School of Finance and Accounting, Industrial University of Ho Chi Minh City, Ho Chi Minh City, Vietnam
e-mail: phamtuanh@iuh.edu.vn

H. T. T. Nguyen et al. (eds.), *Data Science in Finance and Accounting*, Studies in Big Data 181, https://doi.org/10.1007/978-3-032-06179-9_13

significant risks of misstatement to examine whether these figures conform to Benford's expected distribution. These accounting data are selected because they are commonly employed in accounting research to quantify the level of earnings management and to identify instances of misstatements.

Benford's Law is a statistical law which reveals that in many naturally occurring datasets, significant digits do not follow a uniform distribution as one might anticipate. Instead, these digits adhere to a specific logarithmic distribution pattern. Many previous studies indicated that although a dataset's alignment with Benford's Law does not inherently confirm its authenticity, any deviation from the law should prompt a degree of scrutiny or concern [5, 6]. When manipulation affects assets and revenues, the reported figures are likely to be inflated. As a result, the observed frequency of large leading digits will exceed the expected proportions under Benford's Law, while the observed frequency of smaller leading digits will fall below expectations. In contrast, for liabilities and expenses, manipulation is typically aimed at understatement, leading to a higher occurrence of smaller leading digits than predicted by Benford's Law [7].

Our objective is to detect potential irregularities in specific financial items, especially in the context of the COVID-19 pandemic, which introduced unprecedented challenges to financial reporting. We divide the dataset into two periods: before COVID-19 (2015–2019) and during the COVID-19 period (2020–2022), in order to examine whether there are differences in the conformity of major financial statement items with Benford's Law between these two periods in Vietnam context. The first period begins in 2015, marked by the implementation of Circular 200/TT-BTC, which introduced a comprehensive overhaul of Vietnam's accounting system. The COVID-19 pandemic emerged unexpectedly and impacted every sector of the economy. Accordingly, this event leads to an increased inclination toward financial data manipulation during the COVID-19 crisis. It was observed among a group of European companies across diverse industries [8]. In contrast, Sequeira et al. [9] find no evidence of revenue manipulation among Portuguese companies in the wine and tourism sectors during the period of COVID-19. Due to opposite findings of existing literature about the impact of COVID-19 on misstatements of financial data, the conformity of the first digit in several major items presented in the financial statements to Benford's Law, especially before and during COVID-19 in developing countries, is questionable. This paper contributes to the literature by addressing a previously unexplored area.

Using a sample of 2,698 firm-year observations, our study examines digit distribution of both income statement accounts (e.g., revenue, cost of goods sold, selling expenses, administrative expenses) and statement of financial position accounts (e.g., total assets, total liabilities), which is used by prior studies, for example, Alali and Romero [7], Jordan and Clark [10], Johnson [11] and Carslaw [12]. The results of our study indicate that the conformity of the first digit of major items presented in the financial statements during the COVID-19 period to Benford's Law is higher than that in before COVID-19. This study contributes to the literature by applying Benford's Law at the account-level rather than to the entire financial statements, which is the common approach in previous studies. While prior research generally finds that Vietnamese accounting data conforms to Benford's distribution when aggregated, our study focuses on specific line items (i.e., revenues, expenses, assets, and liabilities), which are frequently examined for potential misstatements in financial statement audits. We find that, before the COVID-19 pandemic, certain key accounts deviated from Benford's Law, suggesting potential irregularities. However, during the COVID-19 period, these same accounts showed greater conformity with Benford's Law, indicating a reduction in potential manipulation or misreporting.

This study is structured as follows: Sect. 2 presents Preliminaries. Section 3 shows research methodology. Section 4 presents the experimental study. Section 5 provides the conclusion.

2 Preliminaries

2.1 *Benford's Law Distribution*

Benford's Law, also known as the first digit law, is a statistical law discovered and studied by American physicist and engineer Frank Benford. This is a very interesting law with many applications in economics and society. The first person to discover this law was Simon Newcomb [13], a mathematician and astronomer, who published the first paper on this topic in an American mathematical journal. The paper presented issues that are now known as Benford's law. He observed that the pages of a logarithm table became more worn out when they began with smaller digits compared to pages with higher digits. From this, he inferred that his colleagues, who were scientists, tended to use logarithm tables to look up numbers starting with digit 1 more often than numbers starting with digit 2, 3 ..., 9 (with decreasing frequency). He concluded that numbers beginning with digit 1 appear more frequently than numbers starting with larger digits. Newcomb calculated the probability of a number having a specific first digit using a formula.

It reveals that in many naturally occurring datasets, significant digits do not follow a uniform distribution as one might anticipate. Instead, these digits adhere to a specific logarithmic distribution pattern. In its most common application, Benford's Law applies to the first significant digit (i.e., the first non-zero decimal digit), indicating that the leading digit is not equally likely to be any of the nine possible digits (1 through 9). Specifically, digit 1 appears more than 30% of the time, while the digit 9 occurs less than 5% of the time, with the likelihood of each digit gradually decreasing in between. The exact probability distribution for the first significant digit is described more precisely by this law.

A dataset is determined to follow Benford's law when the frequency of the first digit of the numbers in the dataset corresponds to the probability or likelihood of its occurrence within that dataset. The probability is calculated as the number of occurrences of numbers with the first digit d_1 (denoted as n_1) divided by the total number of observations (denoted as N).

The probability or expected frequency of the first digit of a number is determined by the formula:

$$P(D_1 = d_1) = \lim_{N\to\infty} \frac{n_1}{N} \quad (d_1 = 1, 2, 3, \ldots, 9) \tag{1}$$

When N is large enough:

$$P\left(D_1 = d_1\right) = \log_{10}\left(1 + \frac{1}{d_1}\right) \tag{2}$$

Where:

- D_1 is the first position in the number (value).
- d_1 is the first digit (d_1 can be one of the digits 1, 2, 3, …, 9).
- P is the probability.

2.2 *Leading Digits to Follow Benford's Law*

The first digit of a number is the digit in the first position of a value and is one of the digits in the set $\{1, 2, 3, 4, 5, 6, 7, 8, 9\}$.
Notation: $d_1(d_1 = 1, 2, 3, \ldots, 9)$

The first leading digit is the first non-zero digit that appears on the leftmost side of a number. For negative numbers, the sign is ignored, and for numbers beginning with zeros, those zeros are omitted. Therefore, it is sometimes referred to as the significant first digit when analyzing a dataset expected to follow Benford's Law. For analysis based on Benford's law, we divide or segment numbers with the same starting digit for examination, leading to the following concept:

Region i is the set of all values in the dataset X that begin with the digit i.
Notation:

$$S_i = \{s_k \in X \mid \text{left}(s_k) = i\}$$

Example: Determine the first digit for a number:

53445214 → digit **5**
0.000**1**45 → digit **1**
−**8**787212 → digit **8**

Additionally, the jth digit (where $j = 1, 2, 3, \ldots, 9$) of a number is the digit in the jth position from the left of a number. If j is not the first position, then j can take values from 0 to 9.
Notation: D_j

2.3 Statistical Test

Statistical testing, also known as hypothesis testing, is a method used to evaluate hypotheses related to the probability of leading digits of a population based on an observed dataset. Quantitative analysis is carried out using basic methods, namely calculation of the z-statistic (test statistics), confidence intervals, and hypothesis testing. The purpose of this method is to check for deviations from Benford's law and estimate misstatements.

Hypotheses for testing are as follows:

H_0 : The observed proportion of set $S_i$equals Benford's proportion.

H_1 : The observed proportion of set $S_i$does not equal Benford's proportion.

Where: S_i is region i, the set of all values in the dataset that begin with the digit i, with $i = 1, 2, 3, \ldots, 9$.

Details about the z-statistic and standard deviation σ_i are as follows:

The z-statistic measures the difference between the observed frequency of a digit and the expected frequency according to Benford's law. It considers:

Absolute magnitude of the difference: The z-statistic takes into account the absolute difference between the observed value (actual frequency of a digit) and the expected value (predicted by Benford's Law).

Size of the dataset: The size of the dataset is crucial because it affects the variance. Larger datasets generally result in smaller variances, making it easier to detect significant deviations from the expected distribution.

Expected proportion: The expected proportion is derived from Benford's law and serves as the benchmark for comparison. Following Azevedo et al. [14], and Fleiss et al. [15], the z-statistic essentially quantifies how far the observed distribution is from this expected distribution as below:

$$z = |p_i - p_{ei}| \sqrt{\frac{n}{p_{ei} \times (1 - p_{ei})}} \tag{3}$$

Where:

p_i is the actual probability of occurrence of the digit i

p_{ei} is the probability of occurrence of the digit i based on Benford's law

n is the total number of observations.

With the confidence level $(1 - \alpha)$, if the z-statistic is less than or equal to the critical regions in hypothesis testing corresponding to $\frac{\alpha}{2}$ $(z_{\alpha/2})$, we have grounds to accept the null hypothesis H_0, meaning the dataset S_i follows Benford's distribution.

Conversely, if the z-statistic satisfies $z > z_{\alpha/2}$, we reject the null hypothesis H_0 and accept the alternative hypothesis H_1, indicating that the dataset S_i does not follow Benford's distribution.

In this case, the examiner must estimate the error rate for the dataset S_i based on the confidence interval for the proportion of set S_i. The allowable error threshold in the 95% confidence interval estimation method should not exceed 1.96 times the corresponding standard deviation for set S_i.

2.4 Data Types that Can Apply Benford's Law

Benford's Law is a valuable tool for detecting anomalies in datasets, but its effectiveness depends on the type of data being analyzed. However, not all datasets are suitable for Benford's analysis. Tables 1 and 2 show the data types that are suitable and not suitable for Benford's analysis, as illustrated by Durtschi et al. [16].

Table 1 Types of data suitable for Benford's analysis

Types of data suitable for Benford's analysis	Examples
Data generated from mathematical operations—results obtained from multiplying two or more numbers	Accounts payable (quantity × price), amounts paid (quantity × price)
Transaction-level data—no need for sampling	Disbursements, revenue, costs
Large datasets—the more observations, the better	Daily stock trades over a year
Data where numbers naturally occur—when the average value is not constrained by minimum or maximum thresholds, and numbers are not exclusively positive	Almost all accounting datasets

Table 2 Types of data not suitable for Benford's analysis

Types of data not suitable for Benford's analysis	**Examples**
Datasets that include assigned numbers	Inspection numbers, invoice numbers, zip codes
Numbers influenced by human thought or judgment	Nominal values, reference values, ATM withdrawals
Accounts with a fixed size or numbers that have been predetermined	A special account created to record a full refund
Accounts with artificial number boundaries, such as minimum or maximum values	Sets of assets whose recorded value does not comply with Benford's law

3 Research Methodology

To enhance the robustness and reliability of the analysis, this study employs four complementary methods including the probability distribution chart based on Benford's Law, confidence interval testing, the Chi-square test, and the probability method. The probability distribution chart based on Benford's Law provides a visual comparison between the actual distribution of leading digits and the theoretical distribution prescribed by Benford's Law, allowing for the quick identification of anomalies. Confidence intervals offer a range of values within which the observed data points are expected to fall with a confidence level of $1 - \alpha$ (i.e., the significance level is α), serving as a flexible tool for evaluating the conformity of the observed values. The Chi-square test is a widely recognized hypothesis test that rigorously assesses whether the observed data significantly deviates from Benford's Law. Finally, the probability method offers a comprehensive assessment of the overall probability distribution, helping to identify the level of error risk within the dataset. The risks of error and the expected magnitude of errors may not be evident when analyzing each digit in isolation. The integration of all four methods ensures a more comprehensive and reliable evaluation of the dataset's conformity to Benford's Law. We illustrate our research method in the Fig. 1.

3.1 Benford Probability Distribution Chart

The Fig. 2 helps visualize the research data by comparing the observed proportions of the Si sets with the expected proportions according to Benford's Law. The advantage of this method is its ability to quickly identify S_i sets whose proportions are approximately consistent with or conform to Benford's Law, as well as those that significantly deviate and do not conform to the law.

3.2 Confidence Interval Testing

The null hypothesis (H_0) and alternative hypothesis (H_1) are tested as follows:

$$H_0 : \text{The observed proportion of set } S_i = \text{Benford's expected proportion}$$
$$H_1 : \text{The observed proportion of set } S_i \neq \text{Benford's expected proportion}$$

Let F_i be the Benford expected proportion. We estimate the confidence interval for the population proportion (P_i) of set S_i at the confidence level $1 - \alpha$ (typically 95%). The margin of error (ε_i) of the estimate is calculated as:

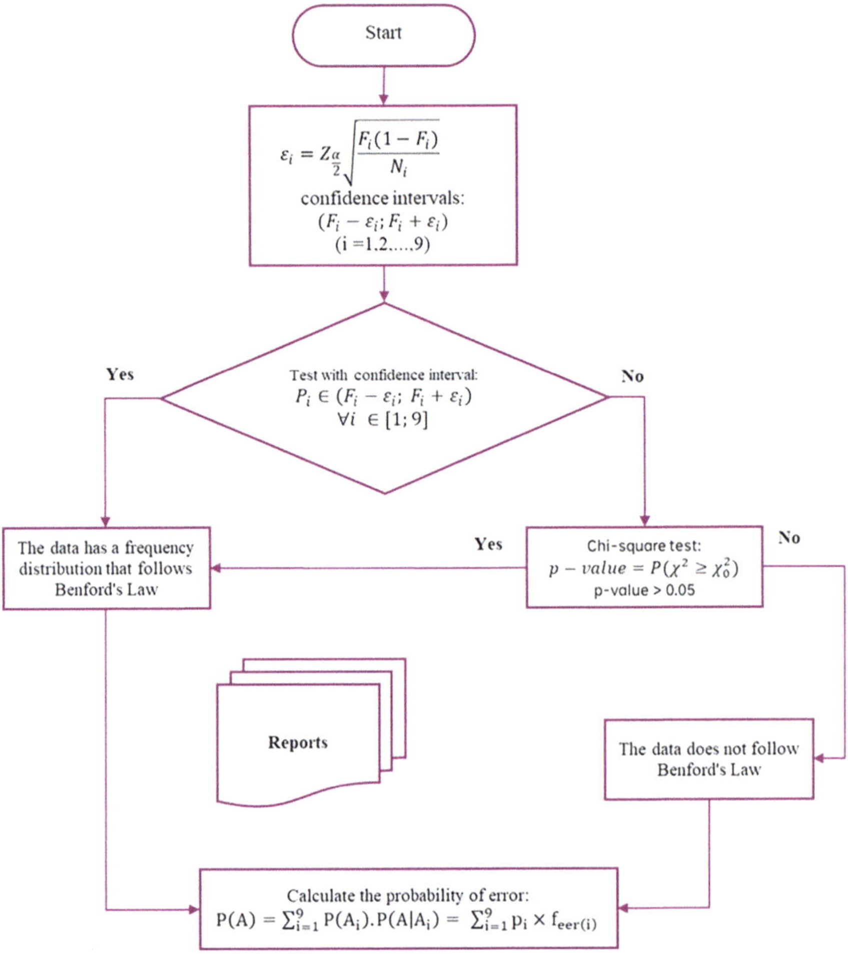

Fig. 1 Summary of research methodology

$$\varepsilon_i = Z_{\alpha/2}\sqrt{\frac{F_i(1-F_i)}{N_i}} \tag{4}$$

where: N_i is the number of observations in set S_i; $Z_{\alpha/2}$ is the critical value corresponding to the significance level α. Thus, the confidence interval for the population proportion P_i of set S_i is:

$$F_i - \varepsilon_i < P_i < F_i + \varepsilon_i \tag{5}$$

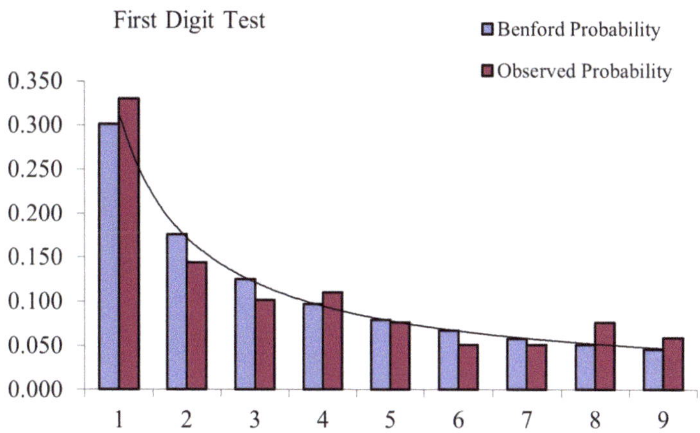

Fig. 2 Benford's law expected digital frequencies

Decision Rule

If the observed proportion of set S_i falls outside the confidence interval $(F_i - \varepsilon_i,\ F_i + \varepsilon_i)$, then we reject the null hypothesis H_0 and accept the alternative hypothesis H_1. This indicates that set S_i contains a large number of erroneous values.

If the observed proportion of set S_i falls within the confidence interval $(F_i - \varepsilon_i,\ F_i + \varepsilon_i)$, then we fail to reject the null hypothesis H_0. This means the observed proportion equals the Benford proportion, and set S_i conforms to Benford's Law.

If all sets S_i have observed proportions equal to the Benford proportions, then the dataset is said to follow Benford's Law. In other words, the data is considered natural, with no signs of manipulation or major distortion.

3.3 Chi-Square Testing

Assume the following hypotheses:

Null Hypothesis (H_0): The data (under review) follows the Benford distribution.
Alternative Hypothesis (H_1): The data (under review) does not follow the Benford distribution.

The formula for calculating the Chi-square test statistic is:

$$\chi_0^2 = \sum_{i=1}^{9} \frac{(O_i - E_i)^2}{E_i} \tag{6}$$

Where:

- O_i is the observed frequency of digit set S_i.
- E_i is the expected frequency of digit set S_i, based on the total number of observations and Benford's Law.

At a significance level $\alpha = 0.05$ (commonly used level), the p-value can be obtained using the `pchisq(`χ_0^2, df), with where df $= n - 1$. In this case $n = 9$:

$$p - value = P(\chi^2 \geq \chi_0^2) \tag{7}$$

Decision Rule:
If $p - value > 0.05$, then we fail to reject the null hypothesis H_0. This means the tested data follows Benford's distribution.

If $p - value < 0.05$, then we reject the null hypothesis H_0. We accept the alternative hypothesis H_1. This means the tested data does not follow Benford's distribution.

3.4 *Probability Testing*

Let us denote A_i $(i = 1, 2, \ldots, 9)$ as the event "a number with the first digit i is selected from the dataset of the item being tested $(A_i \in S_i)$," and let A be the event "a number selected from the dataset shows a significant deviation compared to the audit value (actual value)." Then, the set of events A_i forms a complete and mutually exclusive set of events.

The method of calculating the error probability of the tested item is based on the full probability formula, as follows:

$$P(A) = \sum_{i=1}^{9} P(A_i) \cdot P(A \mid A_i) = \sum_{i=1}^{9} p_i \cdot f_{\text{eer}(i)} \tag{8}$$

where:

$P(A_i)$ is the proportion of set S_i.
$P(A \mid A_i)$ is the error rate in set S_i.
$P(A)$ is the probability of error in the tested item.
p_i is the actual (empirical) proportion.
$f_{\text{eer}(i)}$ is the error deviation rate (i.e., the rate exceeding the allowable error threshold).

Moreover, we have the condition:

$$\sum_{i=1}^{9} p_i = 1 \tag{9}$$

4 Experimental Study

4.1 Sample

We analyze a sample of 732 firms, comprising 2,698 observations, covering the pre-COVID-19 period (2015–2019) and the COVID-19 period (2020–2022) for all Vietnamese listed companies. The reason for choosing the year 2015 is that the Ministry of Finance changed the accounting system from Decision 15/2006/QD-BTC to Circular 200/2014/TT-BTC to avoid the macro environment effect on corporate decisions. This study aims to examine the impact of COVID-19 on the misstatements of sales, expenses, total assets, and total liabilities reported in the financial statements of Vietnamese listed companies, using data obtained from FiinPro platform.

4.2 Main Results

This section presents the results of the Benford's Law analysis applied to the test variables. The findings for the pre-COVID-19 period and the during-COVID-19 period for items in income statements such as revenues, costs of goods sold, administrative expenses, and selling expenses, are shown in Table 3, and for items in balance sheets (i.e., total assets, total liabilities) are presented in Table 4.

As shown in Table 3, the data for revenue, costs of goods sold, administrative expenses, liabilities, and cost of goods sold prior to COVID-19 do not conform

Table 3 Summary results of sales, cost of goods sold, selling and administrative expenses before and during the COVID-19 pandemic

Period	χ^2 value	p-value	Misstatement probability
Panel A. Revenues			
Pre-COVID 19	18.7371	0.0163	4.4%
During COVID 19	11.8169	0.1596	6.6%
Panel B. Cost of goods sold			
Pre-COVID 19	23.7365	0.0025	4.9%
During COVID 19	9.9897	0.2657	2.9%
Panel C. Selling expenses			
Pre-COVID 19	4.1767	0.8408	1.6%
During COVID 19	6.6973	0.5696	4.8%
Panel D. Administrative expenses			
Pre-COVID 19	21.2202	0.0066	3.5%
During COVID 19	6.0790	0.6384	3.9%

Table 4 Summary results of total assets and total liabilities before and during the Covid-19 Pandemic

Period	χ^2 value	p-value	Misstatement probability
Panel A. Total assets			
Pre-COVID 19	8.8871	0.3519	4.9%
During COVID 19	10.9966	0.2019	5.3%
Panel B. Total Liabilities			
Pre-COVID 19	24.0737	0.0022	5.2%
During COVID 19	9.8754	0.2739	3.8%

to Benford's Law, suggesting potential manipulation or significant errors in these figures. In contrast, the data for selling expenses and total assets conform to Benford's Law, indicating no evidence of manipulation or major inaccuracies.

The estimated probability of misstatement that also refers to as the risk of error ranges from 1.6 to 5.2%, with an average of 4.1% per item. This probability reflects the general likelihood of errors in the data, regardless of materiality. A higher probability implies a greater risk of misstatement, while a lower probability suggests stronger data reliability.

Furthermore, based on Fig. 2 and Table 5 (Appendix), prior to the COVID-19 pandemic, several financial statement items exhibit significant deviations from Benford's expected digit distribution. Specifically, revenue data show the deviations for first digits 2, 4, and 7. For cost of goods sold, the frequencies of first digits 1 and 5 significantly differ from Benford's proportions, suggesting non-conformity. In the case of administrative expenses, the first digit 9 displays a significant deviation. Similarly, for total assets, the frequency of first digit 1 diverges from the expected pattern. Lastly, for total liabilities, the first digits 3, 5, and 8 show significant deviations from Benford's Law.

Table 6 presents the results indicating that the data for items such as revenue, selling expenses, administrative expenses, total assets, liabilities, and cost of goods sold after COVID-19 conform to Benford's Law. This suggests that these financial figures were not manipulated and do not exhibit significant irregularities.

Furthermore, the estimated probability of misstatement ranges from 2.9 to 6.6%. On average, the likelihood of error for each item during the COVID-19 period is 4.6%, slightly higher than the pre-COVID-19 average of 4.1%. These findings suggest an increased risk of financial misstatements during the pandemic, likely reflecting greater volatility and uncertainty in business operations during that time. In addition, based on Figs. 9, 10, 11, 12, 13, 14 during the COVID-19 pandemic, the first digits of most selected items in the S_i sets generally follow Benford's Law (Fig. 3).

These figures show Benford's Law of expected digital frequencies of selected items before and during COVID-19 (Figs. 4, 5, 6, 7, 8, 9, 10, 11, 12, 13 and 14).

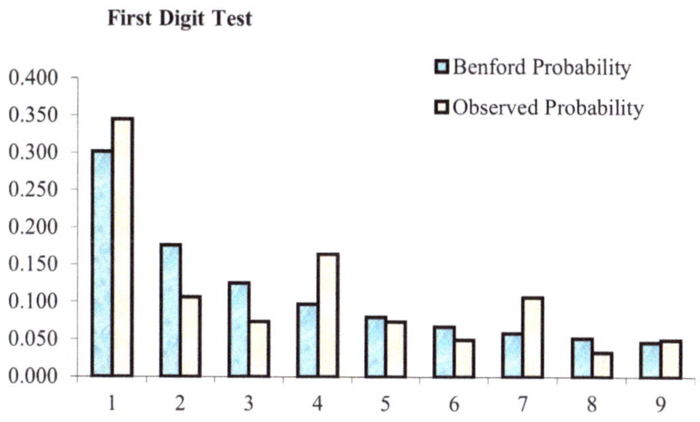

Fig. 3 Benford's Law of expected digital frequencies of revenues before COVID-19

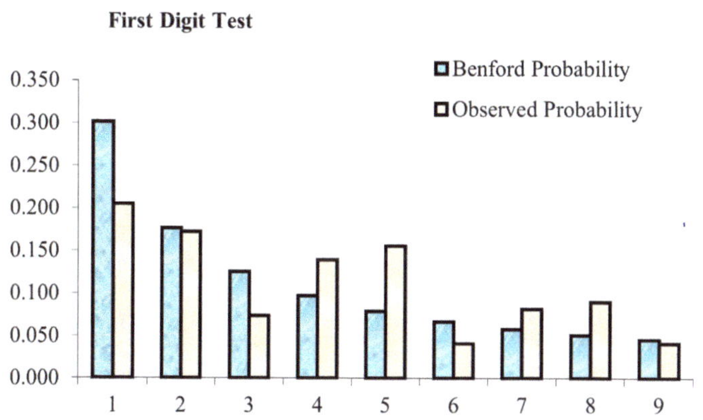

Fig. 4 Benford's Law of expected digital frequencies of costs of goods sold before COVID-19

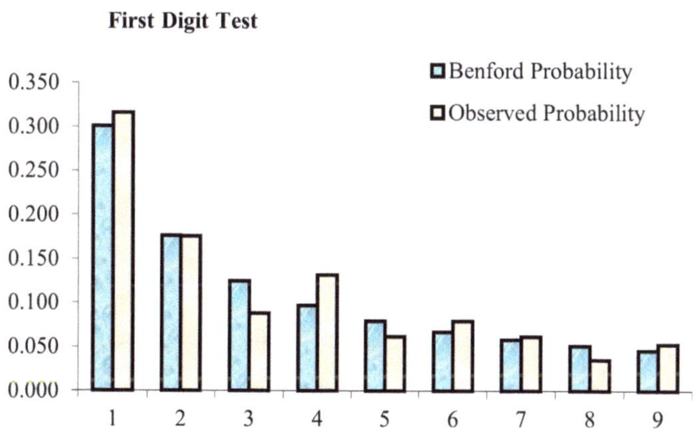

Fig. 5 Benford's Law of expected digital frequencies of selling expenses before COVID-19

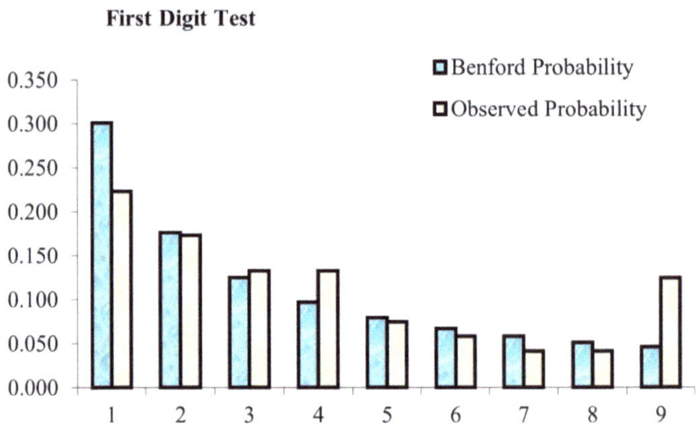

Fig. 6 Benford's Law of expected digital frequencies of administrative expenses before COVID-19

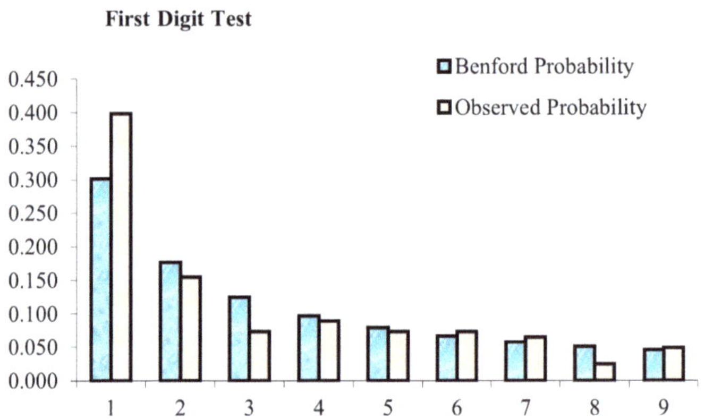

Fig. 7 Benford's Law of expected digital frequencies of total assets before COVID-19

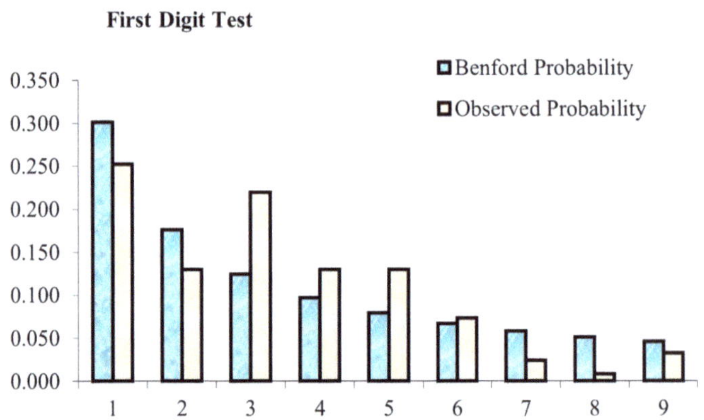

Fig. 8 Benford's Law of expected digital frequencies of total liabilities before COVID-19

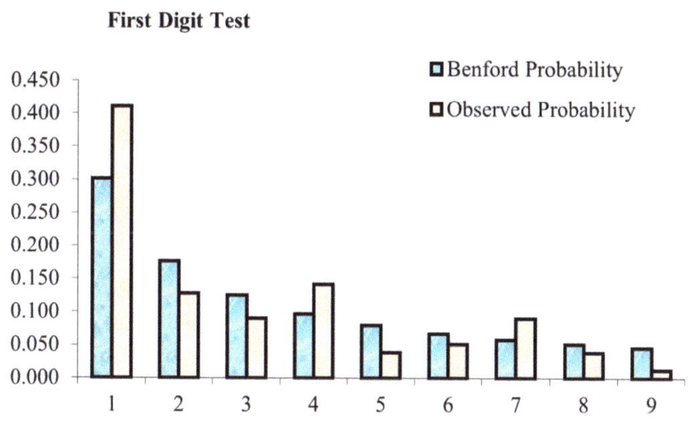

Fig. 9 Benford's Law of expected digital frequencies of revenues during COVID-19

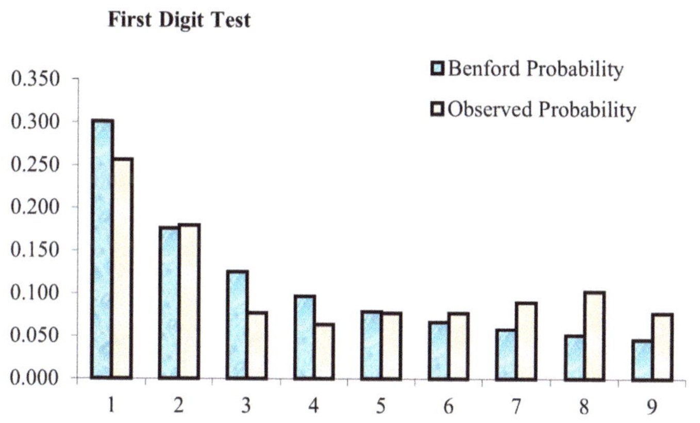

Fig. 10 Benford's Law of expected digital frequencies of costs of goods sold during COVID-19

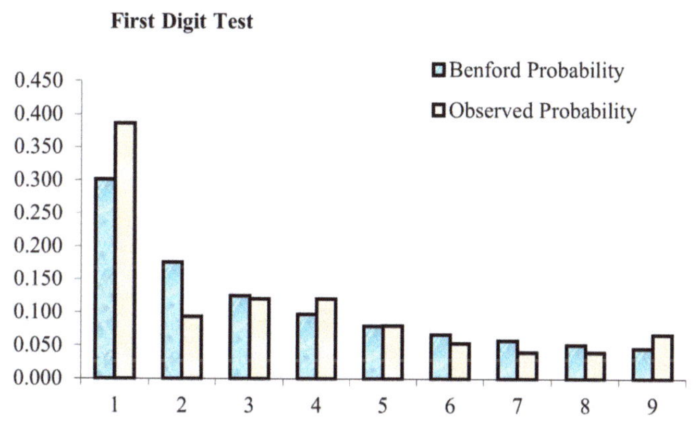

Fig. 11 Benford's Law of expected digital frequencies of selling expenses during COVID-19

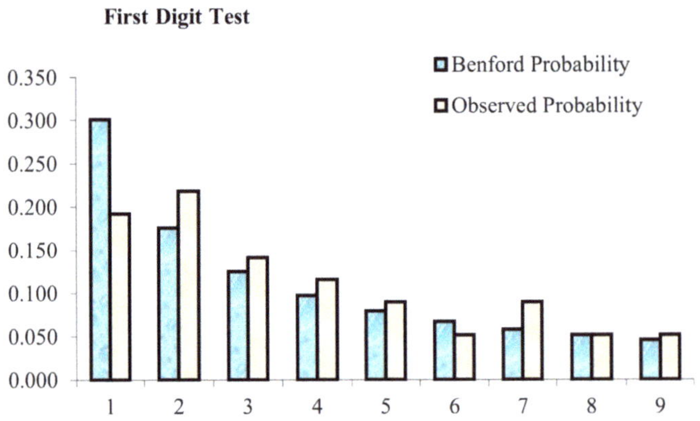

Fig. 12 Benford's Law of expected digital frequencies of administrative expenses during COVID-19

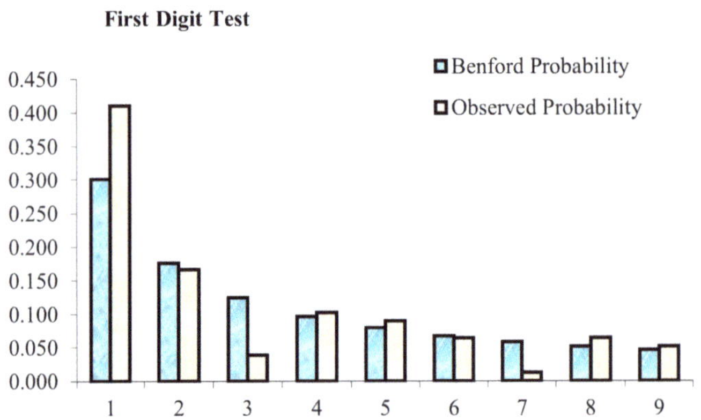

Fig. 13 Benford's Law of expected digital frequencies of total assets during COVID-19

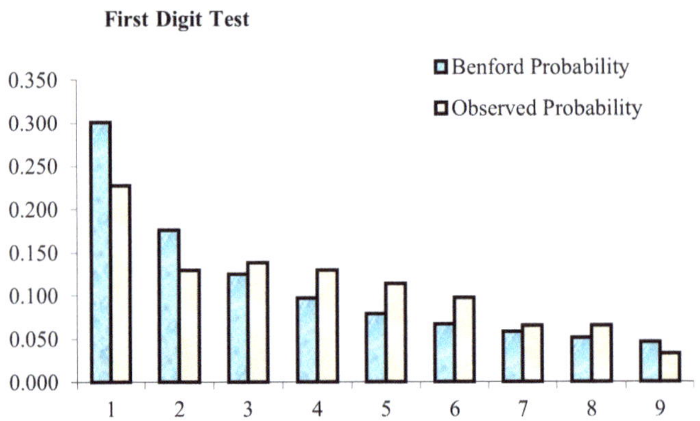

Fig. 14 Benford's Law of expected digital frequencies of total liabilities during COVID-19

5 Conclusion

The findings of this study indicate that, prior to the COVID-19 pandemic, several financial items, namely sales, cost of goods sold, administrative expenses, and total liabilities, deviated from Benford's Law, suggesting the presence of potential misstatements. In contrast, during the COVID-19 period, all selected items aligned with Benford's Law, reflecting a reduction in irregularities and an improvement in data integrity.

The findings of this study have several important implications. By applying Benford's Law at the account level, the study offers a more precise tool for auditors and regulators to detect potential misstatements in specific financial items. The improved conformity during the COVID-19 period suggests that crisis conditions may discourage aggressive reporting practices, highlighting the need for heightened oversight in normal periods.

In conclusion, the application of Benford's Law to assess potential misstatements in financial statement items of Vietnamese listed companies demonstrates its value as a robust and efficient analytical tool. By identifying anomalies and irregularities in reported figures, this method offers critical insights into potential errors or fraudulent activities. The findings can serve as a valuable resource for auditors and regulators seeking to enhance the transparency and reliability of financial reporting in the Vietnamese capital market.

Appendix

See Tables 5 and 6.

Table 5 Detailed analysis results of items before COVID-19

S_i	Lower bound	Upper bound	Actual proportion	Proportion exceeding the allowable threshold	Error probability	Decision
Panel A. Revenue						
1	0.220	0.382	0.344	0.043	0.015	Accept
2	0.109	0.244	0.107	0.070	0.007	Reject
3	0.066	0.184	0.074	0.051	0.004	Accept
4	0.044	0.149	0.164	0.067	0.011	Reject
5	0.031	0.127	0.074	0.005	0.000	Accept
6	0.023	0.111	0.049	0.018	0.001	Accept
7	0.017	0.099	0.107	0.049	0.005	Reject
8	0.012	0.090	0.033	0.018	0.001	Accept
9	0.009	0.083	0.049	0.003	0.000	Accept

(continued)

Table 5 (continued)

S_i	Lower bound	Upper bound	Actual proportion	Proportion exceeding the allowable threshold	Error probability	Decision
Panel B. Selling expenses						
1	0.217	0.385	0.316	0.015	0.005	Accept
2	0.106	0.246	0.176	0.001	0.000	Accept
3	0.064	0.186	0.088	0.037	0.003	Accept
4	0.043	0.151	0.132	0.035	0.005	Accept
5	0.030	0.129	0.061	0.018	0.001	Accept
6	0.021	0.113	0.079	0.012	0.001	Accept
7	0.015	0.101	0.061	0.003	0.000	Accept
8	0.011	0.092	0.035	0.016	0.000	Accept
9	0.007	0.084	0.053	0.007	0.000	Accept
Panel C. Administrative expenses						
1	0.219	0.383	0.223	0.078	0.017	Accept
2	0.108	0.244	0.174	0.003	0.000	Accept
3	0.066	0.184	0.132	0.007	0.001	Accept
4	0.044	0.150	0.132	0.035	0.005	Accept
5	0.031	0.127	0.074	0.005	0.000	Accept
6	0.022	0.111	0.058	0.009	0.001	Accept
7	0.016	0.100	0.041	0.017	0.001	Accept
8	0.012	0.090	0.041	0.010	0.000	Accept
9	0.009	0.083	0.124	0.078	0.010	Reject
Panel D. Total assets						
1	0.220	0.382	0.398	0.097	0.039	Reject
2	0.109	0.243	0.154	0.022	0.003	Accept
3	0.067	0.183	0.073	0.052	0.004	Accept
4	0.045	0.149	0.089	0.007	0.001	Accept
5	0.031	0.127	0.073	0.006	0.000	Accept
6	0.023	0.111	0.073	0.006	0.000	Accept
7	0.017	0.099	0.065	0.007	0.000	Accept
8	0.012	0.090	0.024	0.027	0.001	Accept
9	0.009	0.083	0.049	0.003	0.000	Accept
Panel E. Total liabilities						
1	0.220	0.382	0.252	0.049	0.012	Accept
2	0.109	0.243	0.130	0.046	0.006	Accept
3	0.067	0.183	0.220	0.095	0.021	Reject
4	0.045	0.149	0.130	0.033	0.004	Accept
5	0.031	0.127	0.130	0.051	0.007	Reject
6	0.023	0.111	0.073	0.006	0.000	Accept

(continued)

Table 5 (continued)

S_i	Lower bound	Upper bound	Actual proportion	Proportion exceeding the allowable threshold	Error probability	Decision
7	0.017	0.099	0.024	0.034	0.000	Accept
8	0.012	0.090	0.008	0.043	0.000	Reject
9	0.009	0.083	0.033	0.013	0.000	Accept
Panel F. Cost of goods sold						
1	0.220	0.382	0.205	0.096	0.020	Reject
2	0.109	0.244	0.172	0.004	0.001	Accept
3	0.066	0.184	0.074	0.051	0.004	Accept
4	0.044	0.149	0.139	0.042	0.006	Accept
5	0.031	0.127	0.156	0.077	0.012	Reject
6	0.023	0.111	0.041	0.026	0.001	Accept
7	0.017	0.099	0.082	0.024	0.002	Accept
8	0.012	0.090	0.090	0.039	0.004	Accept
9	0.009	0.083	0.041	0.005	0.000	Accept

Table 6 Detailed analysis results of items after COVID-19

S_i	Lower bound	Upper bound	Actual proportion	Proportion exceeding the allowable threshold	Error probability	Decision
Panel A. Revenues						
1	0.199	0.403	0.410	0.109	0.045	Reject
2	0.092	0.261	0.128	0.048	0.006	Accept
3	0.052	0.198	0.090	0.035	0.003	Accept
4	0.031	0.163	0.141	0.044	0.006	Accept
5	0.019	0.139	0.038	0.041	0.002	Accept
6	0.011	0.122	0.051	0.016	0.001	Accept
7	0.006	0.110	0.090	0.032	0.003	Accept
8	0.002	0.100	0.038	0.013	0.000	Accept
9	-0.001	0.092	0.013	0.033	0.000	Accept
Panel B. Selling expenses						
1	0.1972	0.4048	0.3867	0.0856	0.0331	Accept
2	0.0899	0.2623	0.0933	0.0828	0.0077	Accept
3	0.0501	0.1998	0.1200	0.0049	0.0006	Accept
4	0.0300	0.1639	0.1200	0.0231	0.0028	Accept
5	0.0181	0.1403	0.0800	0.0008	0.0001	Accept
6	0.0104	0.1235	0.0533	0.0136	0.0007	Accept
7	0.0051	0.1109	0.0400	0.0180	0.0007	Accept
8	0.0013	0.1010	0.0400	0.0112	0.0004	Accept
9	-0.0015	0.0930	0.0667	0.0209	0.0014	Accept

(continued)

Table 6 (continued)

S_i	Lower bound	Upper bound	Actual proportion	Proportion exceeding the allowable threshold	Error probability	Decision
Panel C. Administrative expense						
1	0.199	0.403	0.192	0.109	0.021	Reject
2	0.092	0.261	0.218	0.042	0.009	Accept
3	0.052	0.198	0.141	0.016	0.002	Accept
4	0.031	0.163	0.115	0.018	0.002	Accept
5	0.019	0.139	0.090	0.011	0.001	Accept
6	0.011	0.122	0.051	0.016	0.001	Accept
7	0.006	0.110	0.090	0.032	0.003	Accept
8	0.002	0.100	0.051	0.000	0.000	Accept
9	-0.001	0.092	0.051	0.006	0.000	Accept
Panel D. Total assets						
1	0.199	0.403	0.410	0.109	0.045	Reject
2	0.092	0.261	0.167	0.009	0.002	Accept
3	0.052	0.198	0.038	0.086	0.003	Reject
4	0.031	0.163	0.103	0.006	0.001	Accept
5	0.019	0.139	0.090	0.011	0.001	Accept
6	0.011	0.122	0.064	0.003	0.000	Accept
7	0.006	0.110	0.013	0.045	0.001	Accept
8	0.002	0.100	0.064	0.013	0.001	Accept
9	-0.001	0.092	0.051	0.006	0.000	Accept
Panel E. Total liabilities						
1	0.220	0.382	0.228	0.073	0.017	Accept
2	0.109	0.243	0.130	0.046	0.006	Accept
3	0.067	0.183	0.138	0.013	0.002	Accept
4	0.045	0.149	0.130	0.033	0.004	Accept
5	0.031	0.127	0.114	0.035	0.004	Accept
6	0.023	0.111	0.098	0.031	0.003	Accept
7	0.017	0.099	0.065	0.007	0.000	Accept
8	0.012	0.090	0.065	0.014	0.001	Accept
9	0.009	0.083	0.033	0.013	0.000	Accept
Panel F. Cost of goods sold						
1	0.199	0.403	0.256	0.045	0.011	Accept
2	0.092	0.261	0.179	0.003	0.001	Accept
3	0.052	0.198	0.077	0.048	0.004	Accept
4	0.031	0.163	0.064	0.033	0.002	Accept
5	0.019	0.139	0.077	0.002	0.000	Accept
6	0.011	0.122	0.077	0.010	0.001	Accept
7	0.006	0.110	0.090	0.032	0.003	Accept
8	0.002	0.100	0.103	0.051	0.005	Reject
9	-0.001	0.092	0.077	0.031	0.002	Accept

References

1. Gong, Y., Li, J., Zhenghan, X., Li, G.: Detecting financial fraud using two types of benford factors: evidence from China. Procedia Comput. Sci. **214**, 656–663 (2022)
2. Leonov, P.Y., Suyts, V.P., Norkina, A.N., Sushkov, V.M.: Integrated application of benford's law tests to detect corporate fraud. Procedia Comput. Sci. **213**, 332–337 (2022)
3. Mehdi Rad, A.A., Ranjbar, M.H., Salari, H.: Predictability of financial statements fraud-risk using benford's law. Cogent Econ. Financ. **9**(1), 1889756 (2021)
4. Nguyen, L.H.T., Nguyen, T.T., Ngoc Le, T.V., Duc Mai, N.: Applying benford's law to examine earnings management: evidence from emerging asean-5 countries. J. Financ. Report. Account. (2023)
5. Nigrini, M.J.: A taxpayer compliance application of benford's law. J. Am. Taxat. Assoc. **18**(1), 72 (1996)
6. Hill, T.P.: Random-number guessing and the first digit phenomenon. Psychol. Rep. **62**(3), 967–971 (1988)
7. Alali, F.A., Romero, S.: Benford's law: analyzing a decade of financial data. J. Emerg. Technol. Account. **10**(1), 1–39 (2013)
8. Lassoued, N., Khanchel, I.: Impact of covid-19 pandemic on earnings management: an evidence from financial reporting in European firms. Glob. Bus. Rev. page 09721509211053491 (2021)
9. Sequeira, N., Mota, M., Costa, R., Luty, P.: The influence of the covid-19 crisis on financial statements manipulations in the portuguese wine and tourism sector. In: International Conference on Tourism, Technology and Systems, pp. 483–493. Springer (2023)
10. Jordan, C.E., Clark, S.J.: Detecting cosmetic earnings management using benford's law. CPA J. **81**(2), 32 (2011)
11. Johnson, G.C.: Using benford's law to determine if selected company characteristics are red flags for earnings management. J. Forensic Stud. Account. Bus. **1**(2) (2009)
12. Carslaw, C.A.P.N.: Anomalies in income numbers: evidence of goal oriented behavior. Account. Rev. pp. 321–327 (1988)
13. Newcomb, S.: Note on the frequency of use of the different digits in natural numbers. Am. J. Math. **4**(1), 39–40 (1881)
14. da Silva Azevedo, C., Gonçalves, R.F., Gava, V.L., de Mesquita Spinola, M.: A benford's law based method for fraud detection using r library. MethodsX **8**, 101575 (2021)
15. Fleiss, J.L., Levin, B., Paik, M.C.: Statistical Methods for Rates and Proportions. Wiley (2013)
16. Durtschi, C., Hillison, W., Pacini, C., et al.: The effective use of benford's law to assist in detecting fraud in accounting data. J. Foren. Account. **5**(1), 17–34 (2004)

IT Governance and Perceived Usefulness of CAATs: An Empirical Study

Tra Lam Pham, Phuoc Bao An Nguyen, and Thi Hoang Minh Mai

Abstract Industry 4.0 has been taking place rapidly and profoundly affecting all areas of social life, especially significantly changing how organizations operate. IT tends to be increasingly implemented in organizations and is receiving large investments. Auditing practice does not go beyond this phenomenon. Computer-assisted audit techniques (CAATs) have become important for performing audit tasks. However, there is a lack of research on CAATs in emerging markets. This study investigates the impact of IT governance (ITG) on CAATs' perceived usefulness, based on the integration of technology acceptance models and agency theory. We also examine the impact of CAAT type on CAATs' perceived usefulness of internal and external auditors. PLS analysis shows that ITG mechanisms, including IT support, IT strategy, and IT committee, correlate more strongly with CAATs' perceived usefulness than top management support. CAATs' perceived ease of use plays a significant role in perceived usefulness. CAATs' perceived usefulness is heterogeneous among the three software groups: spreadsheets, experts, and in-house software. This study provides initial empirical support and confirms the relationship between the types of CAATs and CAATs' perceived usefulness. The findings also suggest that IT support, IT strategy, IT committee, and CAATs' perceived ease of use may make it easier to adopt CAATs.

Keywords CAATs · Perceived usefulness · IT governance · Vietnam

1 Introduction

In the current world, IT has enhanced productivity and promoted the value of enterprises, which enables businesses to reach themselves and grow stronger in the fierce market. It has been widely deployed in businesses [1]. IT developments have also changed all aspects of professional careers, such as auditing and accounting. The

T. L. Pham (✉) · P. B. A. Nguyen · T. H. M. Mai
Department of Accounting Information Systems, School of Accounting, University of Economics Ho Chi Minh City, Ho Chi Minh City, Vietnam
e-mail: phamtralamais@ueh.edu.vn

H. T. T. Nguyen et al. (eds.), *Data Science in Finance and Accounting*, Studies in Big Data 181, https://doi.org/10.1007/978-3-032-06179-9_14

accounting profession is at the top of the professions that are most affected by the development of technology and globalization [2]. As Frey et al. [3] noted, this career is at the top of 702 occupations with a high probability of automation and digitization in the near future. With the development of IT, accounting and auditing activities can be performed more easily, comprehensively, and rapidly [1].

The large-scale adoption of IT in enterprises, such as accounting software, enterprise resource planning systems (ERP), cloud-based accounting, blockchain technology (BT), tools in information disclosure, artificial intelligence (AI), and big data analytics, have made the use of technology in auditing a necessity [4, 5]. Technological tools, such as CAATs, enable auditors to audit complex business environments with high inherent risk and utilize professional judgment in a more structured audit process [6]. The use of technology can improve efficiency, automation, accountability, and information processing, and reduce costs, human errors, audit risk, and the level of technical information required to perform audit work [7, 8]. Inability and unwillingness to use technology in auditing can cause challenges and have negative consequences in providing auditing services [9], resulting in detrimental effects on the auditing profession [10] and producing poor audit quality services [11].

Many industrial enterprises and public audit companies have not adopted or used CAATs in internal and independent audit activities [12–15], as designers of CAATs have not been able to consider critical factors [16]. Prior studies examined the factors affecting the acceptance of technology by auditors and audit firms (e.g., [16–21]) with often inconsistent and mixed results. There is evidence of the ineffective application of CAATs [15, 22, 23]. The main challenge of audit firms face limited restrictions on the use of CAATs when conducting data analyses [24].

A meta-analysis of 88 studies showed that, while there are minor differences between professionals in developed markets and those in developing markets or smaller firms within developed markets, as well as between traditional technologies such as CAATs and more advanced technologies such as advanced data analytics, the mere provision of a tool is insufficient for its adoption. Practitioners require assurance regarding the cost-benefit ratio before integrating a technology into their routine practices [25]. The implementation of CAATs is intricately connected to client interactions, staff engagement, and overall effectiveness of audit operations. Furthermore, its impact on operational effectiveness is contingent upon the audit organization's level of preparedness. Hence, they highlight the significance of making strategic technological investments to enhance audit firms' nonfinancial performance outcomes [26].

Based on the Technology Acceptance Model (TAM), which explains significant exogenous variables in IT adoption, the perceived usefulness of IT is an important determinant of IT use behavior [27]. Tripathi and Mishra [28] confirmed that the perceived usefulness of cloud computing is a significant determinant of cloud computing for both adopters and non-adopters. The perceived usefulness of CAATs has proven to be a significant determinant of CAATs adoption by both external auditors [22, 29] and internal auditors [30]. From an organizational perspective, Zainol et al. [31] found that perceived usefulness has a significant influence on behavioral intention to adopt CAATs in Malaysia. Pedrosa et al. [16] substantiated the association between perceived utility and user behavior and desire to use technology by investigating the determinants influencing the adoption of CAATs among statutory auditors in a European country. Because of the important role of the perceived usefulness of CAATs, we explore the antecedents of CAATs' perceived usefulness by auditors.

As organizations rely on IT to some extent to conduct their business, IT governance (ITG) is considered a framework of determinants of decision rights and accountability to stimulate anticipated behavior in the use of technology [32], and has become more significant in recent years as a key aspect of corporate governance [33]. It is imperative to ensure the service of the business to accomplish business objectives through the decision-making process and design monitoring systems that align with organizational goals [34]. In order to employ effective ITG, a set of ITG mechanisms (e.g., an IT steering committee and IT organizational structure) is compulsory. These ITG mechanisms focus on several technological issues that may arise in enterprises. For example, the way significant information systems (IS)-related practices are being implemented or the way the strategies are aligned into practices [35]. Scholars such as Nawi et al. [36] and Amid et al. [37] revealed that the main factors provoking the failure of IT projects relate to inadequate management mechanisms, namely managerial support, IT strategy, and IT support.

Veerankutty et al. [38] investigate the impact of ITG on audit technology performance in Malaysia. They found that ITG mechanisms, including IT strategies and management support, significantly impact CAAT performance. In Vietnam, the current market for CAATs remains relatively small. Independent services supplied by professional audit companies have become implementing CAATs in some situations, as the accounting information systems (AIS) of their customers are computerized. Regarding internal control systems, a few enterprises use different types of CAATs to support internal audit activities. However, very little research has been conducted on CAAT adoption and performance. In addition, Black and Gerard [39] identified five priority IT application topics in auditing that need further research, including making the transition—individual professional, organization, and environment-enabling building blocks, development of new techniques, refinement of existing techniques, and enablers and barriers to adoption. Hence, this study focuses on exploring the relationship between ITG mechanisms and perceived usefulness of CAAT in both public audit companies and industrial enterprises in Vietnam.

2 Literature Review and Hypothesis Development

2.1 CAATs and Technology Acceptance Models

CAATs can be defined as any use of technology to support the completion of an audit [16]. The extended definition of CAATs includes both automated working papers and traditional word processing applications such as e-worksheets, artificial intelligence (AI) tools, and statistical analysis software that predicts violations or failures in financial statements [4, 40]. Sayana and Cisa [41] stated that CAATs are "the use of certain software that can be used by the auditor to perform audits and to achieve the goals of auditing" (p. 1). Auditors and audit firms benefit greatly from CAATs because they contribute to lower audit costs, improve audit quality and productivity, help maintain audit reports, and improve overall efficiency assessments [42].

Technology acceptance models are a group of theories that explore user acceptance in the IT context. The TAM is the most widely used theory by researchers to examine IT behavior [43]. Theories of this group are TAM [27], applied in the information system (IS) field to explain and predict technology acceptance, TAM2 [44], and TAM3 [45].

Davis [27] indicated *perceived usefulness*, *perceived ease of use*, attitude toward the use of the system, behavioral intention to use, and actual use in the TAM. Taylor and Todd (1995) used data collected from 786 potential users of a computer resource center and proved a model that factors impact usage behavior including perceived usefulness, attitude, peer influence, superior's influence, subjective norm, self-efficacy, resource facilitating conditions, technology facilitating conditions, perceived behavioral control and behavioral control. In the TAM2, Venkatesh and Davis [44] demonstrated that perceived usefulness applies to intentions in terms of social influence and cognitive instrumental processes. Constructs in the TAM2 are the subjective norm, image, job relevance, output quality, result demonstrability, perceived usefulness, perceived ease of use, intention to use, experience, voluntariness, and usage behavior. The latest TAM is TAM3, which is shown in Fig. 1 [45].

Researchers have applied the TAM to a wide variety of information systems such as communications systems, general-purpose systems, office systems, and specialized business systems (e.g., computerized models, case tools, hospital systems, decision support systems, expert support systems, and MRP) [43]. In addition, many external TAM variables have been examined [46].

This study focuses on CAATs' perceived usefulness and ease of use of auditors in public independent audit companies and internal auditors in enterprises. The perceived usefulness of IT is defined as the degree to which an individual believes that using the system will help him or her attain job performance gains [27]. Hence, CAATs' perceived usefulness is considered the degree to which an auditor believes that using CAATs will help him or her attain gains in audit performance. Based on the TAM [27, 44, 45], we suggest that CAATs' perceived ease of use, which is the degree of ease associated with the use of CAATs, impacts users' perceived usefulness.

Tripathi and Mishra [28] examined the factors of perceived ease of use, perceived usefulness, and other factors in determining behavioral intention to use cloud computing. They found that perceived ease of use of cloud computing strongly impacts perceived usefulness in adopter and non-adopter firms. Additionally, perceived ease of use significantly affects internal auditors' perceived usefulness of CAATs [47]. Curtis and Payne [48], who interviewed external auditors in the US, found that when budget pressure is high, the intention to utilize CAATs is driven by effort expectancy or perceived ease of use through perceived usefulness. Based on the TAM, in line with previous scholars, we hypothesize the link between CAATs' perceived ease of use and perceived usefulness as follows:

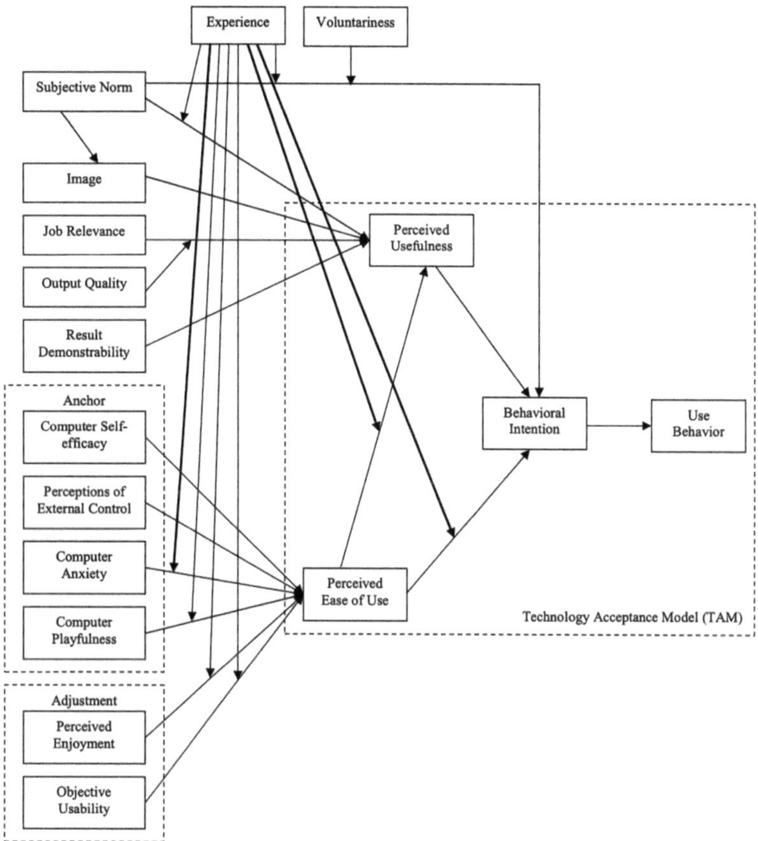

Fig. 1 Technology Acceptance Model 3 (TAM3)

Hypothesis 1. CAATs' perceived ease of use has a direct positive influence on auditors' perceived usefulness.

2.2 *CAATs and ITG*

Weill [32] defined that "ITG represents the framework for decision rights and accountabilities to encourage desirable behavior in the use of IT" [p. 3]. Effective ITG draws on corporate governance principles to manage and implement IT to achieve business performance goals [35]. An effective ITG structure is important for overcoming potential impediments to effective information security risk management [49].

According to Bradley [43], a combination of the TAM with other approaches to user acceptance suggests promising advancements similar to those that have been operated, such as the TAM with the TTF model (Dishaw et al., 2002) or the TAM with the integrated model developed by Wixom and Todd (2005). Hence, Dishaw et al. (2002) recommended that a combination of user aspects (acceptance, satisfaction, and resistance) may result in effective future research. In this study, we create a novel approach by integrating CAATs' perceived usefulness as the TAM suggests with ITG, which acts as the mechanism on behalf of principals in monitoring the behavior of management – an argument of agency theory [38].

Good ITG stimulates and influences the workforce in technology usage and warrants compliance with an organization's missions and norms [32]. If company authorities cannot operate the ITG profitably, it may result in a low performance of technology assets [50]. Therefore, in connection with CAATs, user acceptance behavior is strongly influenced by ITG in enterprises

Nfuka and Rusu [50] argued that effective ITG should include IT structures, management participation, well-disseminated IT strategies and IT policies, as well as defined performance indicators. Top management participation in IT decision-making, as well as the establishment of IT steering committees in handling IT-related issues to ensure they are in line with organizational goals, have significantly measured the effectiveness of ITG [51]. Top managers, who represent potential decision-makers in an organization [52], can stimulate change by communicating and reinforcing values through an articulated vision of the organization [53]. They can create a positive environment that facilitates the adoption of new technologies by creating an appealing vision of how adoption will benefit the firm, securing sufficient resources, and overcoming any member-level resistance to change [52, 54]. Top management support provides the strategic direction and mechanisms needed to align business goals with technology adoption [55–57]. It can therefore facilitate the organizational adoption of IT [58–65]. Internal audits and information security professionals have long argued that top management support and involvement are important for information security [66–69]. A positive relationship exists between management support and audit technology performance [38]. Senior management supports a supportive climate and allocates sufficient resources for the successful utilization of IT [56, 70]. It is a significant driver in decision-making on IT adoption in audit organizations [15, 30]. Huang et al. [47] also pointed out that organizational support influences the perceived ease of use and perceived usefulness of CAATs. Based on the above arguments, we propose the following hypothesis:

Hypothesis 2. Management support has a direct positive influence on CAATs' perceived usefulness of auditors.

Hypothesis 3. IT Committee has a direct positive influence on CAATs' perceived usefulness of auditors.

IT strategy, which refers to information processing required by organizations [71], is a construct with four dimensions: role, competencies, infrastructure, and system design and development [72]. Chan et al. [73] asserted that the strategic orientation of an IS emphasizes the organizational application portfolio, which blankets several attributes such as business strategic orientation, the rudiments of being aggressive, defensive, analytical, proactive, futuristic, innovative, and risk-averse. IT strategies include environmental scanning and the strategic use of IT [71]. Weill and Ross [32] noted that well-disseminated IT strategies and policies are designed to ensure employees' behaviors are consistent with organizational objectives.

Ahmi [15] built a research model to explain how auditors accept Generalized Audit Software (GAS) based on their motivations and limitations. The organizational influence dimension in his model comprises top management support, IT support, IT audit expertise in the organization (Unified Theory of Acceptance and Use of Technology - UTAUT facilitating conditions), and internal and external training in implementing and maintaining GAS. Scholars such as Wilkin and Chenhall [74] and Ali and Green [75] indicated that IT strategies and IT plans can ensure that IT investments have been evaluated for relevant risk and assist in business expectations. IT strategy significantly influences audit technology performance in Malaysia [38]. Regarding access to organizational data, the majority of auditors agreed with the statement that it is difficult for the internal audit function to obtain without IT support [23]. Blanton et al. [76] suggested that personalized IT support is more effective than generalized IT support in improving service quality. IS support services play a significant role in the successful implementation of the system through the availability of IT experts and IS consultants [77]. Accordingly, we propose the following hypothesis:

Hypothesis 4. IT strategy has a direct positive influence on CAATs' perceived usefulness of auditors.

Hypothesis 5. IT support has a direct positive influence on CAATs' perceived usefulness of auditors.

2.3 *Types of CAATs*

There are various types of CAATs, ranging from word processing or electronic spreadsheets to expert systems [12]. CAATs used by external and internal auditors can be grouped into electronic audit working papers, fraud detection, GAS, and continuous monitoring [30]. As the ACCA document and ACCA F8 Certificate, CAATs are divided into three main groups: testing software, test data, and other techniques. The difference in the types of CAATs is similar to the difference between a single accounting software and ERP system. While electronic spreadsheets such as Microsoft Excel, a common CAAT tool, are quite simple and easy to use in audit activities, expert systems such as ACL, IDEA, AS2, SQL, and Power BI have a wide variety of features, and they require a lot of effort from users to benefit audit activities. The acceptance of accounting software and ERP software is contradictory [78–81]. Kim et al. [14] showed that there is a negative relationship between complexity features and audit software acceptance of internal auditors in the US. Furthermore, Mahzan and Lymer [30] proved that technical complexity is a barrier to the use of CAATs in the UK and Malaysia. In the Vietnamese environment, we examine the impact of three types of CAATs: Microsoft Excel, professional software, and in-house software on CAATs' perceived usefulness of auditors.

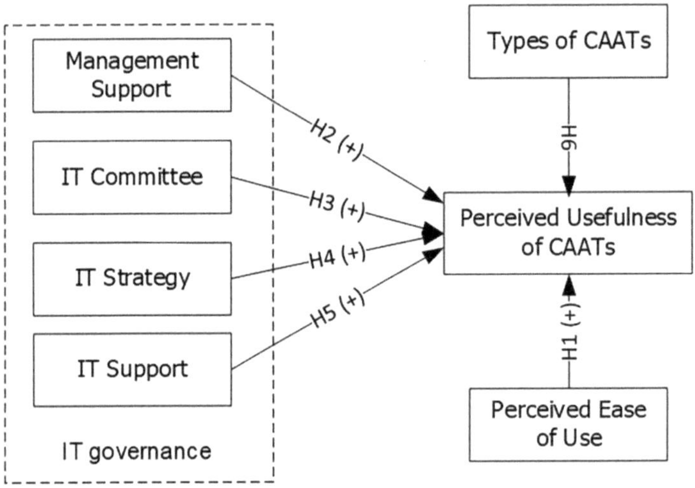

Fig. 2 Research model

Hypothesis 6. CAATs' perceived usefulness of auditors varied between types of CAATs.

Figure 2 presents the proposed research model.

3 Methodology

3.1 Research Process

This study applied the PLS_SEM data analysis technique with the support of Smart-PLS 4.0. This method has several advantages over other approaches [82]. First, it tends to achieve a higher level of statistics under equilibrium conditions than the traditional covariance method does (Reinartz et al. 2009). Second, PLS_SEM is commonly used in discovery studies. Third, when a complex research model such as the proposed research model has many latent variables that interact with each other and includes many measurement models, using PLS_SEM allows simultaneous processing of the whole model at the same time instead of separately estimating the partial relationships in the model [82]. Finally, PLS_SEM makes fewer rigid assumptions than other methods (Compeau, Higgins, & Huff, 1999) because it accepts latent constructs under conditions of non-normality in small to medium sample sizes (Chin 1998). Figure 3 presents the research process.

The analytical procedure was performed according to the method described by [82]. This process consists of two steps. The first stage evaluated the measurement

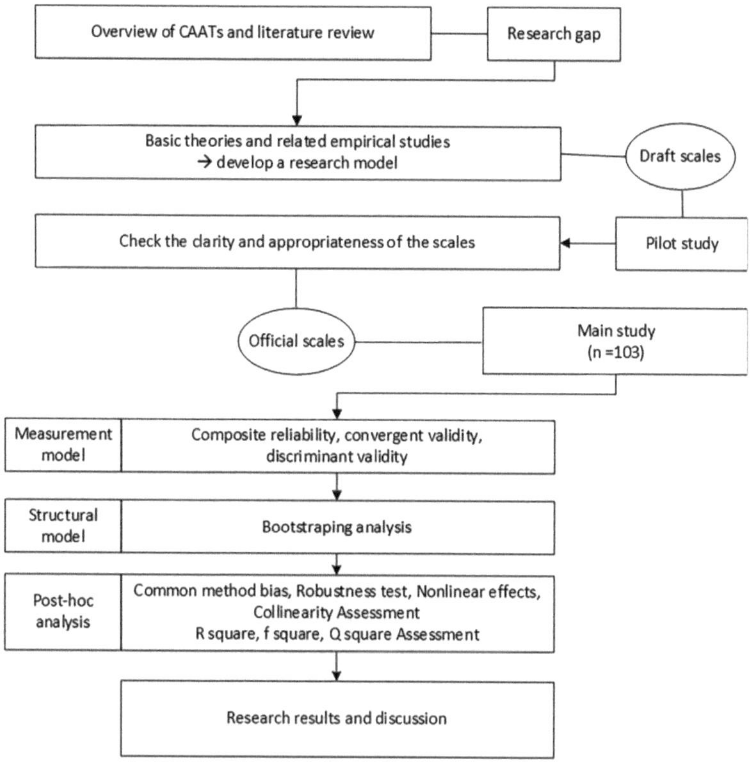

Fig. 3 Research process

models to test the value of the measurement scales using three indicators: internal consistency, reliability, convergent validity, and discriminant validity. The first criterion to be evaluated is internal consistency reliability, which provides an estimate of reliability based on the intercorrelations of the observed indicator variables. Cronbach's alpha (representing the lower bound) and composite reliability (CR) (representing the upper bound) were used to gauge the reliability of the scales. Convergent validity is the degree to which two or more items measuring the same construct agree. It is measured based on outer loadings and average variance extracted (AVE). Discriminant validity is the degree to which an item differentiates between constructs or measures distinct concepts. Fornell and Larcker [83] suggested the Heterotrait-Monotrait (HTMT) ratio criteria for the assessment of discriminant validity.

The second stage examined the structural models used to test the research hypotheses and explanatory power. A bootstrapping technique with a resample size of 5,000 was adopted to evaluate the statistical significance of path estimates [82]. Path coefficients indicate the strength and direction of the relationships between dependent and independent constructs. The path coefficient values are significant and directionally consistent with the hypotheses. The signs (positive or negative) and statistical significance of the values were assessed to support the hypotheses. A statistical level of 0.5 (with a $t - value$ of 1.67) was considered acceptable for an exploratory study.

The post-hoc analyses, such as common method bias, robustness test, nonlinear effects, and collinearity, were conducted in this study. The R^2 value represents the variance in the dependent variable explained by the model. A larger R^2 value indicates better predictive power for the model. The f^2 effect size was tested to gain additional insight into the quality of the suggested path model estimations [82].

3.2 Measurements

All items used to measure the constructs in this study were drawn from the literature. However, they have all been modified to fit the Vietnamese CAATs research context. Specifically, four items of perceived usefulness (PU) and four items of perceived ease of use (PEOU) were adapted from Venkatesh and Davis [44].

Four ITG mechanisms, management support (MAS) (three items), IT strategy (ITS) (two items), IT committee (ITC) (three items), and IT support (ITSU) (three items), were adapted from Li et al. (2015), Ali and Green [75], Ferguson et al. [51], and Ahmi and Kent [15], respectively. Each variable was framed using a seven-point Likert scale, with responses ranging from 1 ("strongly disagree") to 7 ("strongly agree").

3.3 Data Collection

Following the initial development of the survey questionnaire, it was distributed to 15 prospective participants to assess its comprehensibility. The instrument was meticulously refined and finalized based on the provided information. The final version of the questionnaire comprised two primary sections. The first section included demographic questions aimed at gathering data on gender, age, educational attainment, occupational position, and work experience within the current organization. The second section contained items assessing the adoption variables.

The data were collected using an online survey. The survey questionnaire was initially developed in English and then translated into Vietnamese. It was then randomly sent to potential respondents, including external and internal auditors. Participation was both anonymous and voluntary. Questionnaires were returned by 103 of the 200 invited recipients. All invalid and suspicious responses were excluded based on the following criteria. Duplicate email addresses, incomplete responses, and similar responses were excluded. Table 1 presents the demographic information of the respondents and enterprises.

There is a diversity of positions for auditors, including inter, assistant, partner, senior, manager, and internal control, and the age of respondents. Females constituted 56.31%. Almost all auditors have graduated. In terms of experience, 61.17% of the

Table 1 Demographic profiles of respondents and industrial enterprises

	Frequency	Percentage (%)
Positions of respondents		
Intern	15	14.56
Assistant	32	31.07
Senior	25	24.27
Manager	16	15.53
Partner	4	3.88
Internal control	11	10.68
Gender		
Male	45	43.69
Female	58	56.31
Age		
From 20 to 25	42	40.78
From 26 to 30	27	26.21
Over 30	34	33.01
Education		
Graduation	75	72.82
MSC, MBA	14	13.59
Other	14	13.59
Experiences of respondents		
Less than 3 years	63	61.17
3 years to 5 years	27	26.21
Greater than 5 years	13	12.62
Firm size		
Small	30	29.13
Medium	48	46.60
Large	25	24.27
Type of enterprise		
Foreign direct investment (FDI) enterprise	34	33.01
Government enterprises	8	7.77
Private enterprises	61	59.22

respondents had less than three years of experience. In addition, the interviewees worked in businesses of different sizes. Private enterprises account for the highest percentage of the sample. Thus, most respondents had renewable energy experience with CAATs.

Table 2 presents the types of CAATs implemented in the public audit companies and industrial enterprises in the sample. Almost all organizations use spreadsheets for audit activities. Only 37.42% of the respondents operated professional audit software.

Table 2 Types of CAATs

Types of CAATs	Frequency	Percentage (%)
Spreadsheets: Microsoft Excel	101	61.96
Expert software	61	37.42
ACL	5	3.07
IDEA	4	2.45
AS2	10	6.13
SQL	24	14.72
Power BI	16	9.82
Other	2	1.23
In-house software	1	0.61
Total	**163**	**100.00**

Nonresponse bias may be caused by differences in key demographic and research variables, including sample distributions of early and late respondents, the difference in gender, and the positions of respondent groups. To assess this phenomenon, an independent sample $t - test$ was performed using SPSS 24.0, which showed no significant differences between the groups. The test confirmed both the absence of any non-response bias and that differences did not constitute a serious problem in the study (Armstrong and Overton 1977). Moreover, the robustness method was implemented to minimize the risk of common method bias (CMB), for example via reversed questions. Thus, the dataset was qualified for the next analytical process.

4 Results

4.1 Assessment of the Measurement Model

The instrument's psychometric properties, including its validity and reliability, were tested to ensure that the measurements were correct. According to Hair et al. [84], three criteria can be used to examine construct validity and reliability: construct reliability, convergent validity, and discriminant validity. Construct reliability was evaluated using Cronbach's Alpha (CA) and Composite Reliability (CR). As Table 3 shows, all variables had CA indices between 0.837 and 0.945 and CR values between 0.903 and 0.960, thus exceeding the acceptable 0.7 thresholds proposed by Nunnally [85]. These results indicated that the constructs were reliable. Next, the measurement models were assessed for convergent validity with outer loadings and average variance extracted (AVE) as key criteria. The factor loadings of all measurement items were 0.757 or higher [86], and the AVE for each construct was at least 0.598 [87], confirming the high convergence levels of the measurement constructs.

Table 3 Assessment of measurement model

Structures	Mean	Standard deviation	Cronbach's alpha	Composite reliability	Average variance extracted (AVE)
ITC	4.09	0.150	0.926	0.953	0.872
MAS	4.18	0.125	0.918	0.948	0.859
ITSU	4.00	0.129	0.837	0.903	0.757
ITS	4.28	0.139	0.869	0.938	0.884
PEOU	3.73	0.068	0.892	0.929	0.813
PU	3.81	0.072	0.945	0.960	0.859

Table 4 Heterotrait-Monotrait Ratio (HTMT)

	ITC	MAS	ITSU	ITS	PEOU
MAS	0.618				
ITSU	0.577	0.552			
ITS	0.742	0.506	0.578		
PEOU	0.399	0.364	0.605	0.499	
PU	0.680	0.541	0.806	0.672	0.365

Table 5 Fornell-Larker criterion

	ITC	MAS	ITSU	ITS	PEOU	PU
ITC	**0.934**					
MAS	0.571	**0.927**				
ITSU	0.524	0.488	**0.870**			
ITS	0.667	0.451	0.502	**0.940**		
PEOU	0.392	0.357	0.538	0.463	**0.902**	
PU	0.639	0.505	0.721	0.613	0.376	**0.927**

The heterotrait-monotrait ratio (HTMT) suggested by Henseler [88] was used to test discriminant validity. Table 4 presents the HTMT values, which range from 0.01 to 0.801 (below the recommended threshold of 1.00). The Fornell-Larker criterion [83] was examined by considering the square roots of the AVE for each construct, which should exceed the recommended minimum value of 0.5. Accordingly, the square root of the AVE results in Table 5 confirms that each variable is larger than the elements off the diagonal, confirming discriminant validity. Consequently, the conditions for discriminant validity are supported. The above analysis shows that the measurement model achieved higher-than-satisfactory levels of reliability and validity.

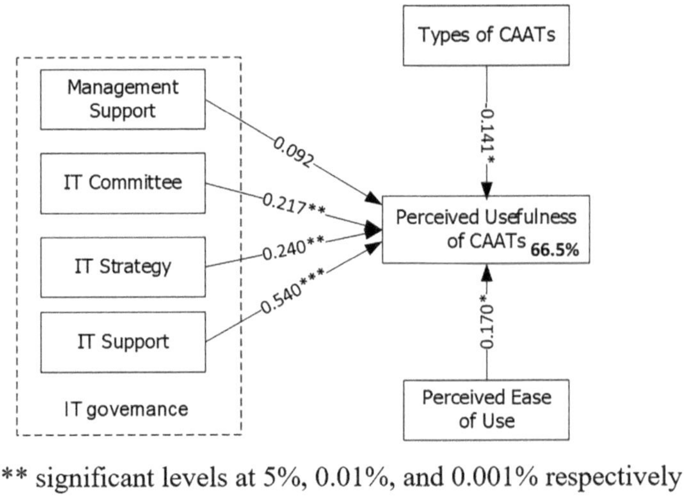

Fig. 4 The results of the structural model

4.2 *Assessment of the Structural Model*

Overall, 5,000 resamples were generated on SmartPLS 4.0 via the bootstrap method to validate the established hypothesis. The path coefficient β and $p - value$ were used to evaluate the relationships between endogenous and exogenous variables based on the recommendations of Hair et al. [84]. Figure 4 shows that CAATs' perceived usefulness is positively and directly affected by five exogenous variables: perceived ease of use ($\beta = 0.170$; $p = 0.043$), IT committee ($\beta = 0.217$; $p = 0.004$), IT strategy ($\beta = 0.240$; $p = 0.005$), IT support ($\beta = 0.540$; $p = 0.000$), and type of CAATs ($\beta = -0.141$; $p = 0.031$). Accordingly, **H1**, **H3**, **H4**, **H5**, and **H6** are supported. Management support (MAS) (H2) did not have a significant impact on CAATs' perceived usefulness. An R^2 of 0.665 implies that CAATs' perceived usefulness is considerably explained (at approximately 66.5%) by its four exogenous variables. Among these factors, IT support is the most significant predictor of CAATs' perceived usefulness.

The f^2 values in Table 6 convey the effects of IT support, IT strategy, and IT committees, which are 0.483, 0.085, and 0.063, respectively. This shows that while IT support has a strong effect on CAATs' perceived usefulness, IT strategy and IT committee have medium effects on CAATs' perceived usefulness. These conclusions are based on Hair et al. [82], who suggest that f^2 values of 0.35, 0.15, and 0.02 reflect a high, medium, and low level of impact on the exogenous latent variables, respectively. As the f^2 values of perceived ease of use and types of CAATs are over 0.02, these results provide more evidence that the variables in question have a small impact on CAATs' perceived usefulness.

Table 6 VIF and *f* square

	VIF	f square
ITS	2.019	0.085
ITSU	1.808	0.483
PEOU	1.638	0.052
MAS	1.655	0.015
ITC	2.235	0.063
Types of CAATs		0.058

We used the variance inflation factor (VIF) to further evaluate the multicollinearity of all constructs. All VIFs were less than the conservative threshold of 5, ranging between 1.638 and 2.235. These values suggest that there is no multicollinearity issue [89]. The results also confirmed that CMV was not a serious problem in our data.

Regarding the robustness of the structural model estimate, the FIMIX-PLS technique was employed to evaluate unobserved heterogeneity, based on a suggestion from Hair et al. [82]. As the size of the third segment reaches 0.217, which is lower than the threshold value of 25%, it is necessary to stop at a two-segment solution. The results show that the AIC, AIC3, AIC4, BIC, and CAIC indices were higher in the first-segment solution than in the second-segment solution, confirming that the unobserved heterogeneity of this data is not a serious problem. In addition, nonlinear relationships were analyzed using the quadratic effects method. No $p - values$ of the connections between the exogenous quadratic variables and their exogenous variables are significant; thus, the relationships between the constructs in the proposed model are linear.

5 Discussion

This empirical study aims to investigate some important antecedents of CAATs' perceived usefulness of auditors by combining the TAM and ITG from the agency theory perspective. Our main finding, which has not been previously reported, is that CAATs' perceived usefulness of auditors is heterogeneous among the three software groups (Microsoft Excel, professional software, and in-house software). This finding is supported by previous studies (e.g., [78–81]). This strongly confirms the results of Kim et al. [14] and Mahzan and Lymer (2009), as the more complex the type of CAATs, the more difficult it is for users to find it. However, this study discovered that in Vietnam, expert software supporting auditing is not more useful than simple software, such as spreadsheets. This may have caused Excel to be more popular than other professional auditing tools.

Three out of the four ITG mechanisms have been shown to affect CAATs' perceived usefulness, confirming the significant role of ITG. IT support, IT strategy, and IT committee have diminishing effects on the perceived usefulness of CAATs by external and internal auditors. As using CAATs is a challenge, auditors require IT support in a timely, convenient, and useful manner. Some scholars have also proven this result (e.g., [15, 23, 76, 77]). Moreover, the literature reviews such as Weill and Ross [32], Wilkin and Chenhall [74], Nfuka and Rusu [50], Ali and Green [75], and Veerankutty et al. [38] show the impact of IT strategy on the perceived usefulness of CAATs. Both public audit companies and industry companies have set up an IT committee to solve IT-related issues [51]. Aligned with Ferguson et al. [51], this study found a similar result. However, we find no evidence to support the relationship between management support and CAATs' perceived usefulness of auditors. This finding is not consistent with empirical arguments [15, 30, 38, 47, 51]. It is interesting to note that in the Vietnamese auditing context, the IT committee is more important than top management support in promoting the adoption and use of CAATs by both external auditors and internal auditors.

Finally, we found a significant positive relationship between CAATs' perceived ease of use and auditors' perceived usefulness. This finding is consistent with the TAM [27, 44, 45] and preliminary empirical literature [47, 48].

Through the PLS analyses, all exogenous variables in the model can predict an additional 65.5% of the variance in CAATs' perceived usefulness. The predictive power of this model was significant. According to Hair et al. (2011) and Henseler et al. (2009), the coefficient of determination R^2 should be greater than 0.20 in exploratory studies. In this study, the R^2 coefficient reached an average (moderate) near the threshold of 0.7, as suggested by the aforementioned authors.

5.1 Applied Implications

This study makes a significant contribution to the CAATs literature. Our findings have theoretical and practical implications. These results contribute to the theoretical literature by providing initial empirical support for combining TAM and agency theory to explain CAATs' perceived usefulness of auditors. Similarly, this study also provides initial empirical support and confirms the relationship between the types of CAATs and the perceived usefulness of CAATs from internal and external auditors.

Public audit companies and industrial enterprises always struggle with CAATs' adoption decisions because of the complexity and effectiveness of these techniques. However, those with a clear ITG comprising IT support, IT strategy, and IT committee, as well as perceived ease of use of CAATs, may find it easier to adopt CAATs. From a practical perspective, this is an important finding for audit companies, industrial enterprises, and CAAT suppliers. Audit companies and industrial enterprises can use their time and resources more effectively if they focus on what they need to understand and modify the perceived usefulness of their audit staff, considering IT support, IT strategy, IT committee, perceived ease of use, and types of CAATs.

CAAT suppliers and implementation consultants should design products to enhance users' perceived ease of use. Support services and training should be considered to increase customer acceptance. Since professional CAATs have not yet been appreciated for their usefulness, software vendors must pay more attention to these strategies.

6 Conclusion

As the adoption of new technologies in auditing has caused obvious changes in all organizations, there is not much evidence of the relationship between ITG and auditor acceptance of CAATs. The present study, therefore, represents a special attempt to contribute to the CAATs literature in an emerging business environment, using the case of Vietnam as a case study. The results of the application of an integrated TAM and agency theory through ITG explored important determinants of CAATs' perceived usefulness. We observe that IT support, IT strategy, IT committee, perceived ease of use, and types of CAATs have significant influences on the perceived usefulness of auditors. Based on the present findings, we also highlight several useful practical implications that can help public audit companies, industrial enterprises, and CAAT suppliers improve their ability to adopt CAATs in independent and internal audit activities.

Our study has some limitations. First, the particularities of our data and the cross-sectional nature of this study may have caused a lack of confidence in causality exploration, preventing us from building a maturity model to evaluate the relationships among structures. Next, the research was conducted in a Vietnamese context, in which businesses focus on adopting simple CAATs rather than expert CAATs. Therefore, our findings may not be generalizable to other contexts. Future research should consider building on the present study by testing the impact of these factors on the adoption of CAATs in individual cases or identifying other factors that may influence CAATs' adoption to achieve further insights and understanding.

Appendix

See Table 7.

Table 7 Appendix A. Measurement scales

Construct/Item	Description
PU: Perceived usefulness [44]	
PU1	Using CAATs improves my performance in my job
PU2	Using CAATs in my job increases my productivity
PU3	Using CAATs enhances my effectiveness in my job
PU4	I find CAATs to be useful in my job
PEOU: perceived ease of use [44]	
PEOU1	My interaction with CAATs is clear and understandable
PEOU2	Interacting with CAATs does not require a lot of mental effort
PEOU3	I find CAATs to be easy to use
PEOU4	I find it easy to get CAATs to do what I want it to do
MAS: management support (Li et al. 2015)	
MAS1	Senior managements believe the use of audit technology is a good idea
MAS2	Senior managements are interested in audit technology usage during the audit task
MAS3	Senior managements support the use of audit technology in audit task
ITS: IT strategy [75]	
ITS1	IT strategy provides strategic direction and the alignment of IT and organisation activities
ITS2	IT strategy provides direction for sourcing and use of IT resources, skills and infrastructure to meet the strategic objective
ITC: IT committee [51]	
ITC1	IT steering committee provides strategic direction for IT related audit engagement that is in line with the strategic directions of the organisation
ITC2	IT steering committee provides a mechanism for coordinating IT related audit engagement
ITC3	IT steering committee provides leadership in managing IT related activities
ITSU: IT support [15]	
ITSU1	Auditors assisted by strong IT support from IT staff
ITSU2	Auditors aided through the availability of internal IT audit expertise
ITSU3	Auditors aided by external technical support

References

1. Özcan, E.Ç., Akkaya, B.: The effect of industry 4.0 on accounting in terms of business management. In: Agile Business Leadership Methods for Industry 4.0. Emerald Publishing Limited (2020)
2. Gulin, D., Hladika, M., Valenta, I.: Digitalization and the challenges for the accounting profession. In: Proceedings of the ENTRENOVA-ENTerprise REsearch InNOVAtion Conference (2019)
3. Frey, C.B., Osborne, M.A.: The future of employment: how susceptible are jobs to computerisation? Technol. Forecast. Soc. Chang. **114**, 254–280 (2017)
4. Braun, R.L., Davis, H.E.: Computer-assisted audit tools and techniques: analysis and perspectives. Manag. Audit. J. **18**(9), 725–731 (2003)
5. Alles, M.G.: Drivers of the use and facilitators and obstacles of the evolution of big data by the audit profession. Account. Horiz. **29**(2), 439–449 (2015)
6. Cao, M., Chychyla, R., Stewart, T.: Big data analytics in financial statement audits. Account. Horiz. **29**(2), 423–429 (2015)
7. Janvrin, D., Bierstaker, J., Lowe, D.J.: An examination of audit information technology use and perceived importance. Account. Horiz. **22**(1), 1–21 (2008)
8. Lowe, D.J., Bierstaker, J.L., Janvrin, D.J., Jenkins, J.G.: Information technology in an audit context: have the Big 4 lost their advantage? J. Inf. Syst. **32**(1), 87–107 (2018)
9. Tiberius, V., Hirth, S.: Impacts of digitization on auditing: a Delphi study for Germany. J. Int. Account. Audit. Tax. **37**, 100288 (2019)
10. Manita, R., Elommal, N., Baudier, P., Hikkerova, L.: The digital transformation of external audit and its impact on corporate governance. Technol. Forecast. Soc. Chang. **150**, 119751 (2020)
11. Salijeni, G., Samsonova-Taddei, A., Turley, S.: Big Data and changes in audit technology: contemplating a research agenda. Account. Bus. Res. **49**(1), 95–119 (2019)
12. Debreceny, R., Lee, S.L., Neo, W., Toh, J.S.: Employing generalized audit software in the financial services sector: challenges and opportunities. Manag. Audit. J. **20**(6), 605–618 (2005)
13. Curtis, M.B., Payne, E.A.: An examination of contextual factors and individual characteristics affecting technology implementation decisions in auditing. Int. J. Account. Inf. Syst. **9**(2), 104–121 (2008)
14. Kim, H.J., Mannino, M., Nieschwietz, R.J.: Information technology acceptance in the internal audit profession: impact of technology features and complexity. Int. J. Account. Inf. Syst. **10**(4), 214–228 (2009)
15. Ahmi, A., Kent, S.: The utilisation of generalized audit software (GAS) by external auditors. Manag. Audit. J. **28**(2), 88–113 (2013)
16. Pedrosa, I., Costa, C.J., Aparicio, M.: Determinants adoption of computer-assisted auditing tools (CAATs). Cogn. Technol. Work **22**, 565–583 (2020)
17. Rosli, K., Yeow, P.H., Siew, E.G.: Factors influencing audit technology acceptance by audit firms: a new I-TOE adoption framework. J. Account. Auditing **2012**, 1 (2012)
18. Rosli, K., Siew, E.G., Yeow, P.H.: Technological, organisational, and environmental aspects of audit technology acceptance. Int. J. Bus. Manag. **11**(5), 140–145 (2016)
19. Li, H., Dai, J., Gershberg, T., Vasarhelyi, M.A.: Understanding usage and value of audit analytics for internal auditors: an organizational approach. Int. J. Account. Inf. Syst. **28**, 59–76 (2018)
20. Siew, E.G., Rosli, K., Yeow, P.H.: Organizational and environmental influences in the adoption of computer-assisted audit tools and techniques (CAATTs) by audit firms in Malaysia. Int. J. Account. Inf. Syst. **36**, 100445 (2020)
21. Ferri, L., Spanò, R., Ginesti, G., Theodosopoulos, G.: Ascertaining auditors' intentions to use blockchain technology: evidence from the Big 4 accountancy firms in Italy. Meditari Account. Res. **29**(5), 1063–1087 (2021)
22. Bierstaker, J., Janvrin, D., Lowe, D.J.: What factors influence auditors' use of computer-assisted audit techniques? Adv. Account. **30**(1), 67–74 (2014)

23. Smidt, L., Ahmi, A., Steenkamp, L., Van der Nest, D.P., Lubbe, D.: A maturity-level assessment of generalised audit software: internal audit functions in Australia. Aust. Account. Rev. **29**(3), 516–531 (2019)
24. Protiviti. A Global Look at IT Audit Best Practices: Assessing the International Leaders in an Annual ISACA/Protiviti Survey (2016)
25. Afsay, A., Tahriri, A., Rezaee, Z.: A meta-analysis of factors affecting acceptance of information technology in auditing. Int. J. Account. Inf. Syst. **49**, 100608 (2023)
26. Wassie, F.A., Péter, L.L.: Audit technology as a catalyst for improving non-financial performance in ethiopian audit firms. J. Open Innov.: Technol. Market, and Complex. 100556 (2025)
27. Davis, F.D.: Perceived usefulness, perceived ease of use, and user acceptance of information technology. MIS Quart. 319–340 (1989)
28. Tripathi, S., Mishra, V.: Determinants of cloud computing adoption: a comparative study. Pacific Asia J. Assoc. Inf. Syst. **11**(3), 3 (2019)
29. Janvrin, D., Lowe, D.J., Bierstaker, J.: Auditor acceptance of computer-assisted audit techniques. Technical Report 4, Iowa State University, Arizona State University and Villanova University (2008)
30. Mahzan, N., Lymer, A.: Examining the adoption of computer-assisted audit tools and techniques: cases of generalized audit software use by internal auditors. Manag. Audit. J. **29**(4), 327–349 (2014)
31. Zainol, S.S., Samsuri, A.S., Tunku Arifin, T.R., Hussin, S., Othman, M.S., Jie, S.J.: Determinants of computer assisted audit techniques (Caats) adoption. A study in small and medium practices in Malaysia. European J. Bus. Soc. Sci. **6**(2):135–150 (2017)
32. Weill, P.: Don't just lead, govern: how top-performing firms govern IT. MIS Q. Exec. **3**(1), 1–17 (2004)
33. Bhattacharjya, J., Chang, V.: Evolving ITG practices for aligning IT with business-a case study in an Australian Institution of Higher Education. J. Inf. Sci. Technol. **4**(1) (2007)
34. Mukhtar, R., Ali, N.A.: Quality governance of human aspects of quality initiatives in the public service sector. Current Issues of Bus. Law **6**(1) (2011)
35. Weill, P., Ross, J.W.: ITG: How Top Performers Manage IT Decision Rights for Superior Results. Harvard Business Press (2004)
36. Nawi, H.S.A., Rahman, A.A., Ibrahim, O.: Government's ICT project failure factors: a revisit. In: 2011 International Conference on Research and Innovation in Information Systems, pp. 1–6. IEEE (2011)
37. Amid, A., Moalagh, M., Ravasan, A.Z.: Identification and classification of ERP critical failure factors in Iranian industries. Inf. Syst. **37**(3), 227–237 (2012)
38. Veerankutty, F., Ramayah, T., Ali, N.A.: Information technology governance on audit technology performance among Malaysian public sector auditors. Soc. Sci. **7**(8), 124 (2018)
39. Black, S., Gerard, G.J.: Accelerating the future of audit technologies: introducing the special issue and emphasizing future research directions. Int. J. Account. Inf. Syst. 100740 (2025)
40. Jakšić, D.: Implementation of computer assisted audit techniques in application controls testing. Manag. Inf. Syst. **4**(1), 9–12 (2009)
41. Sayana, S.A., Cisa, C.: Using CAATs to support IS audit. Inf. Syst. Control J. **1**, 21–23 (2003)
42. Dowling, C., Leech, S.: Audit support systems and decision aids: current practice and opportunities for future research. Int. J. Account. Inf. Syst. **8**(2), 92–116 (2007)
43. Bradley, J.: Chapter 2: if we build it they will come? The technology acceptance model. In: Dwivedi, Y.K., Wade, M.R., Schneberger, S.L. (eds.) Information Systems Theory: Explaining and Predicting Our Digital Society, vol. 1. Springer, USA (2012)
44. Venkatesh, V., Davis, F.D.: A theoretical extension of the technology acceptance model: four longitudinal field studies. Manage. Sci. **46**(2), 186–204 (2000)
45. Venkatesh, V., Bala, H.: Technology acceptance model 3 and a research agenda on interventions. Decis. Sci. **39**(2), 273–315 (2008)
46. Lee, Y., Kozar, K.A., Larsen, K.R.T.: The technology acceptance model: past, present, and future. Commun. Assoc. Inf. Syst. **12**(1), 50 (2003)

47. Huang, S.M., Hung, Y.C., Tsao, H.H.: Examining the determinants of computer-assisted audit techniques acceptance from internal auditors' viewpoints. Int. J. Serv. Stand. **4**(4), 377–392 (2008)
48. Curtis, M.B., Payne, E.A.: Modeling voluntary CAAT utilization decisions in auditing. Manag. Audit. J. **29**(4), 304–326 (2014)
49. Love, P., Reinhard, J., Schwab, A., Spafford, G.: Global technology audit guide (GTAG) 15 information security governance. Technical report, The Institute of Internal Auditors, Altamonte Springs, FL (2010)
50. Nfuka, E.N., Rusu, L.: The effect of critical success factors on ITG performance. Ind. Manag. Data Syst. **111**(9), 1418–1448 (2011)
51. Ferguson, C., Green, P., Vaswani, R., Wu, G.: Determinants of effective information technology governance. Int. J. Audit. **17**(1), 75–99 (2013)
52. Premkumar, G., Roberts, M.: Adoption of new information technologies in rural small businesses. Omega **27**(4), 467–484 (1999)
53. Thong, J.Y.: An integrated model of information systems adoption in small businesses. J. Manag. Inf. Syst. **15**(4), 187–214 (1999)
54. Lee, S., Kim, K.J.: Factors affecting the implementation success of Internet-based information systems. Comput. Hum. Behav. **23**(4), 1853–1880 (2007)
55. Racherla, P., Hu, C.: eCRM system adoption by hospitality organisations: a technology-organisation-environment (TOE) framework. J. Hosp. Leis. Mark. **17**(1–2), 30–58 (2008)
56. Wang, Y.M., Wang, Y.S., Yang, Y.F.: Understanding the determinants of RFID adoption in the manufacturing industry. Technol. Forecast. Soc. Chang. **77**(5), 803–815 (2010)
57. Ramdani, B., Chevers, D., Williams, D.A.: SMEs' adoption of enterprise applications: a technology-organisation-environment model. J. Small Bus. Enterp. Dev. **20**(4), 735–753 (2013)
58. Borgman, H.P., Bahli, B., Heier, H., Schewski, F.: Cloudrise: exploring cloud computing adoption and governance with the TOE framework. In: 2013 46th Hawaii International Conference on System Sciences, pp. 4425–4435. IEEE (2013)
59. Lin, H.F.: Understanding the determinants of electronic supply chain management system adoption: using the technology–organisation–environment framework. Technol. Forecast. Soc. Chang. **86**, 80–92 (2014)
60. Leung, D., Lo, A., Fong, L.H.N., Law, R.: Applying the technology-organisation-environment framework to explore ICT initial and continued adoption: an exploratory study of an independent hotel in Hong Kong. Tour. Recreat. Res. **40**(3), 391–406 (2015)
61. Yeh, C.H., Lee, G.G., Pai, J.C.: Using a technology-organisation-environment framework to investigate the factors influencing e-business information technology capabilities. Inf. Dev. **31**(5), 435–450 (2015)
62. Haneem, F., Kama, N., Bakar, N.A.A.: Critical influential determinants of IT innovation adoption at organisational level in local government context. IET Softw. **13**(4), 233–240 (2019)
63. AlBar, A.M., Hoque, M.R.: Factors affecting cloud ERP adoption in Saudi Arabia: an empirical study. Inf. Dev. **35**(1), 150–164 (2019)
64. Alsharari, N.M., Al-Shboul, M., Alteneiji, S.: Implementation of cloud ERP in the SME: evidence from UAE. J. Small Bus. Enterp. Dev. **27**(2), 299–327 (2020)
65. Daoud, L., Marei, A., Al-Jabaly, S., Aldaas, A.: Moderating the role of top management commitment in usage of computer-assisted auditing techniques. Accounting **7**(2), 457–468 (2021)
66. Flora, P.E., Raj, S.: Navigating Technology's top 10 Risks: Internal Audit's role. Technical report, The Institute of Internal Auditors Research Foundation, Altamonte Springs, FL (2015)
67. Institute, I.T.G.: Information security Governance: Guidance for information security managers. Technical report, ITG Institute, Rolling Meadows, IL (2008)
68. Kayworth, T., Whitten, D.: Effective information security requires a balance of social and technology factors. MIS Q. Exec. **9**(3), 163–175 (2010)
69. Steinbart, P.J., Raschke, R.L., Gal, G., Dilla, W.N.: The influence of a good relationship between the internal audit and information security functions on information security outcomes. Acc. Organ. Soc. **71**, 15–29 (2018)

70. Low, C., Chen, Y., Wu, M.: Understanding the determinants of cloud computing adoption. Ind. Manag. Data Syst. **111**(7), 1006–1023 (2011)
71. Bergeron, F., Raymond, L., Rivard, S.: Ideal patterns of strategic alignment and business performance. Inf. Manag. **41**(8), 1003–1020 (2004)
72. Das, S.R., Zahra, S.A., Warkentin, M.E.: Integrating the content and process of strategic MIS planning with competitive strategy. Decis. Sci. **22**(5), 953–984 (1991)
73. Chan, Y.E., Huff, S.L., Barclay, D.W., Copeland, D.G.: Business strategic orientation, information systems strategic orientation, and strategic alignment. Inf. Syst. Res. **8**(2), 125–150 (1997)
74. Wilkin, C.L., Chenhall, R.H.: A review of ITG: a taxonomy to inform accounting information systems. J. Inf. Syst. **24**(2), 107–146 (2010)
75. Ali, S., Green, P.: Effective information technology (IT) governance mechanisms: an IT outsourcing perspective. Inf. Syst. Front. **14**, 179–193 (2012)
76. Blanton, J.E., Watson, H.J., Moody, J.: Toward a better understanding of information technology organization: a comparative case study. MIS Quarterly, pp. 531–555 (1992)
77. Law, C.C., Chen, C.C., Wu, B.J.: Managing the full ERP life-cycle: considerations of maintenance and support requirements and ITG practice as integral elements of the formula for successful ERP adoption. Comput. Ind. **61**(3), 297–308 (2010)
78. Kholeif, A.O., Abdel-Kader, M., Sherer, M.: ERP customization failure: institutionalized accounting practices, power relations, and market forces. J. Account. Organis. Change **3**(3), 250–269 (2007)
79. Elbardan, H., Kholeif, A.: Enterprise Resource Planning, Corporate Governance, and Internal Auditing. Springer (2017)
80. Ali, M., Miller, L.: ERP system implementation in large enterprises—a systematic literature review. J. Enterp. Inf. Manag. **30**(4), 666–692 (2017)
81. Hasan, N., Miah, S.J., Bao, Y., Hoque, M.R.: Factors affecting post-implementation success of enterprise resource planning systems: a perspective of business process performance. Enterprise Inf. Syst. **13**(9), 1217–1244 (2019)
82. Hair Jr, J.F., Sarstedt, M., Ringle, C.M., Gudergan, S.P.: Advanced issues in partial least squares structural equation modeling. Sage Publications (2017)
83. Fornell, C., Larcker, D.F.: Evaluating structural equation models with unobservable variables and measurement error. J. Mark. Res. **18**(1), 39–50 (1981)
84. Hair Jr, J.F., Hult, G.T.M., Ringle, C., Sarstedt, M.: A primer on partial least squares structural equation modeling (PLS-SEM). Sage Publications (2016)
85. Nunnally, J.C.: An overview of psychological measurement. Clinical diagnosis of mental disorders: A handbook, pp. 97–146 (1978)
86. Gholami, R., Sulaiman, A.B., Ramayah, T., Molla, A.: Senior managers' perception on green information systems (IS) adoption and environmental performance: results from a field survey. Inf. Manag. **50**(7), 431–438 (2013)
87. Loeser, F., Recker, J., Brocke, J.V., Molla, A., Zarnekow, R.: How IT executives create organisational benefits by translating environmental strategies into Green IS initiatives. Inf. Syst. J. **27**(4), 503–553 (2017)
88. Henseler, J., Ringle, C.M., Sarstedt, M.: A new criterion for assessing discriminant validity in variance-based structural equation modeling. J. Acad. Mark. Sci. **43**(1), 115–135 (2015)
89. Rogerson, P.A.: Monitoring point patterns for the development of space–time clusters. J. R. Stat. Soc. A. Stat. Soc. **164**(1), 87–96 (2001)

Micro-Level Determinants of Household Financial Portfolio Choices in Rural Vietnam

Truong Anh Tuan, Nguyen Quoc Nhat, Nguyen Thi Ngoc Hai, and Tran Van Thanh

Abstract This study analyzes the micro-level determinants of household financial portfolio choices in rural Vietnam. It uses data collected from 1,115 households in the provinces of Ha Tinh and Dak Lak, Vietnam. The multivariate probit model is employed to investigate the determinants affecting the choices of five financial services: credit, savings, lending, investment, and insurance. The regression results show that the household head's education, risk attitude, number of impaired members, participation in socio-political organizations, self-employment, and shock losses are important determinants of financial portfolio choices. We also find that financial services are not entirely independent but interconnected to a certain degree. The findings offer important insights into the driving factors influencing household decisions regarding the use of financial instruments. They establish the foundation for developing more impactful policies to improve household finance.

Keywords Financial services · Household financial portfolios · Multivariate probit · Vietnam

---

T. Anh Tuan (✉) · T. Van Thanh
University of Labour and Social Affairs – Ho Chi Minh City Campus, Ho Chi Minh City, Vietnam
e-mail: tuanta@ldxh.edu.vn

T. Van Thanh
e-mail: thanhtv@ldxh.edu.vn

N. Quoc Nhat · N. T. Ngoc Hai
School of Finance and Accounting, Industrial University of Ho Chi Minh City, Ho Chi Minh City, Vietnam
e-mail: nguyenquocnhat@iuh.edu.vn

N. T. Ngoc Hai
e-mail: nguyenthingochai@iuh.edu.vn

H. T. T. Nguyen et al. (eds.), *Data Science in Finance and Accounting*, Studies in Big Data 181, https://doi.org/10.1007/978-3-032-06179-9_15

1 Introduction

Literature has acknowledged that rural households often turn to credit, savings, investments, and insurance as financial portfolio choices to cope with economic shocks [1–3]. These financial tools function as both ex-ante and ex-post strategies for smoothing income fluctuations. In the long run, during periods of economic abundance, households have the opportunity to prepare for income shocks ex ante by accumulating precautionary savings, investing in durable assets, or acquiring insurance coverage. In times of financial constraints, on the other hand, households may consider various ex-post coping measures such as borrowing (credit), dissaving, liquidating assetsc, or activating insurance policies.

Research on the factors influencing household decisions regarding financial tools is essential for informing policies that enhance the effectiveness of household financial strategies. In actuality, a substantial body of literature has examined the determinants of household financial tool use. Regarding credit, [4], conducting research in Malawi, demonstrated that household size and educational level are important factors influencing access to formal credit, while exposure to shocks, landholding size, and access to extension services are essential factors influencing access to informal credit. [5] proved that socioeconomic characteristics such as gender, marital status, household size, education, and rice varieties significantly affected the accessibility of credit for smallholder rice producers in the West Region of Cameroon. In terms of savings, [6] explored the microeconomic factors influencing household savings in Nigeria, discovering household size, occupation, location, age, marital status, living conditions, land ownership, and infrastructure availability as key determinants. [7] applied a mixed method to empirically examine the factors influencing household savings behavior in the Metekele region of Beneshangu Guzu state, Ethiopia. The results indicated that many factors, such as age, educational level, household size, number of livestock units, income, irrigation use, production season, involvement in non-farm activities, access to credit, training, and education, had a significant influence on savings decisions. In regard to insurance, many studies included in systematic reviews by [8, 9] have shown that demographic and socioeconomic factors influence insurance uptake.

It is evident from the literature on factors affecting household financial tools that the vast majority of studies have concentrated on credit, savings, and insurance. Investments in durable goods are often seen as one of the strategies for dealing with income shocks [10]. However, there is a surprising lack of research specifically focusing on the determinants of them. In addition, informal lending to relatives and friends has become a prevalent practice in certain areas, particularly in rural Vietnam [11–13]. Household lending to other individuals is considered one of the most prominent forms of informal savings, in parallel with formal savings, which involve depositing money into bank accounts [14, 15]. While the literature acknowledges household lending as a form of informal saving, there is currently a dearth of research that examines the variables affecting this financial instrument.

Findings from the literature have also shown the existence of interconnections between pairs of financial instruments in household portfolios. The study conducted by [16] showed a positive relationship between micro life insurance and formal savings, as well as between formal credit and formal savings. However, [17] found that household savings decrease the probability of access to microcredit. [18] revealed a significant positive correlation between each of the two financial tools within a household portfolio, including private insurance, life insurance, investments, credit, and savings. We, therefore, argue that households do not use a single financial service but instead adopt a simultaneous portfolio of different services in their financial strategies, in which each service may function as either a complement (positive correlation) or a substitute (negative correlation) for others to a certain extent [19]. This implies a potential correlation between the adoption of one service and the adoption of other services.

This research aims to contribute to the current discourse regarding the micro-level determinants affecting the financial portfolio choices of rural households. Our approach stands out from previous research in three key respects. First, we investigate investments and lending, two important household financial instruments in rural areas that have received little attention in earlier research. Second, instead of concentrating only on one specific financial instrument like prior research [5, 20–22] we examine five simultaneously, including credit, lending, savings, investments, and insurance. Our technique offers a more comprehensive picture of the factors impacting household financial services. Additionally, it enhances understanding of the effects of a certain factor on multiple financial services concurrently. Third, previous studies methodologically employed the univariate logit/probit model to explore the determinants of financial tools separately [5, 7, 22, 23]. This approach, however, may result in biased and inefficient estimates because it fails to take into account the interrelated nature of financial instruments [24, 25]. This research fills this gap by adopting a multivariate probit model that accounts for these interrelationships. The multivariate probit model has the advantage of examining the determinants of financial services in multiple equations, allowing correlations between the error terms.

The rest of this paper is structured as follows. Section 2 provides the theoretical and empirical models. Section 3 includes data and variable definitions. Section 4 provides the results. The conclusion and policy implications are presented in Sect. 5.

2 Theoretical and Empirical Models

2.1 Theoretical Model

The theoretical framework for this research is based on the random utility model [26]. This model assumes that a household will opt for the alternative (financial service) that maximizes its utility. A household's choice to adopt a financial instrument hinges on comparing the indirect utility of adoption with that of non-adoption. Let U_{i1} stand for the utility of choosing to use a financial service of household i, and U_{i0} for the utility of choosing not to use this service. The utility functions can be expressed as follows:

$$U_{i1} = V_{i1} + \varepsilon_{i1} = X'_{i1}\beta + \varepsilon_{i1}$$
$$U_{i0} = V_{i0} + \varepsilon_{i0} = X'_{i0}\beta + \varepsilon_{i0}$$

where, V_i is the observable or deterministic component of the estimated utility, whereas ε_i represents the stochastic and unobservable component. X'_i represents a $1 \times k$ vector of explanatory variables affecting the decision to use the financial service, β denotes a $k \times 1$ vector of unknown parameters. Household i will decide to access a financial service if its net utility is positive $(U_{i1} - U_{i0} > 0)$; otherwise, it will choose not to access it. The net utility function can be defined in the following way:

$$AF_i^* = U_{i1} - U_{i0} = X'_{i1}\beta + \varepsilon_{i1} - X'_{i0}\beta - \varepsilon_{i0} = (X'_{i1}\beta - X'_{i0}\beta) + (\varepsilon_{i1} - \varepsilon_{i0}).$$

Let $g(X'_i\beta) = X'_{i1}\beta - X'_{i0}\beta$ and $\mu_i = \varepsilon_{i1} - \varepsilon_{i0}$, then:

$$AF_i^* = g(X'_i\beta) + \mu_i \tag{1}$$

where, AF_i^* represents the difference in utility between the alternative of accessing a financial service and the alternative of not having access; $g(X'_i\beta)$ captures the observable heterogeneity; and μ captures the unobservable observable heterogeneity.

Equation (1) implies that the decision made by household i is guided by the difference between the access alternative and the non-access alternative. The household will have access to the financial service, if $AF_i^* > 0$. Given that the decision to access or not is binary, the general model for estimating the probability of household access based on the observable variables may be expressed as

$$P(AF_i = 1) = P(AF_i^* > 0) = g(X'_i\beta_i)$$

where, $AF_i = 1$ if $AF_i^* > 0$ (the financial service is taken), and $AF_i = 0$ if $AF_i^* \leq 0$ (the financial service is not taken).

2.2 Empirical Model

$$\begin{cases} AF_{im}^* = X'_{im}\beta_m + \mu_{im}, m = 1, \ldots, 5 \\ AF_{im} = 1 \text{ if } AF_{im}^* > 0, \text{ and 0 otherwise} \end{cases} \tag{2}$$

where, $m = 1, ..., 5$ indexes five financial services: credit, lending, savings, investments, and insurance. AF^*_{im} is a latent variable for the observed AF_{im} (financial access status). $AF_{im} = 1$ if the financial service m is taken, and $AF_{im} = 0$ otherwise. X'_{im} is a $1 \times k$ vector of observable micro-level household characteristics assumed to affect the choice of financial service m. The $k \times 1$ vector of corresponding coefficients to be estimated is denoted by β_m. The stochastic error terms μ_{im} are responsible for capturing unobservable characteristics.

In the MVP model, the error terms μ_{im} follow a multivariate normal distribution with mean zero and a variance-covariance matrix Δ, specified as follows:

$$\Delta = \begin{pmatrix} 1 & \rho_{12} & \rho_{13} & \rho_{14} & \rho_{15} \\ \rho_{21} & 1 & \rho_{23} & \rho_{24} & \rho_{25} \\ \rho_{31} & \rho_{32} & 1 & \rho_{34} & \rho_{35} \\ \rho_{41} & \rho_{42} & \rho_{43} & 1 & \rho_{45} \\ \rho_{51} & \rho_{52} & \rho_{53} & \rho_{54} & 1 \end{pmatrix}$$

where, ρ is the pairwise correlation coefficient of the error terms between any pair of equations due to unobservable characteristics [27]. If the off-diagonal correlation coefficients ($\rho_{jk} = \rho_{kj}$) in the variance-covariance matrix become non-zero, it justifies the use of a multivariate probit model rather than using separate logit/probit models for each financial service. A statistically significant positive correlation coefficient may imply a complementary relationship between the two financial services, whereas a significant negative coefficient may indicate the potential for substitution between them.

The coefficients in the system of equations (2) are estimated by simulated maximum-likelihood (SML) five-equation probit models, applying the Geweke-Hajivassiliou-Keane smooth recursive simulator for determining the maximum likelihood [25]. One limitation of the SML estimator is that it is biased when the number of simulation draws is low [28]. The simulation bias is reduced by increasing the number of simulation draws to a minimum equivalent to the square root of the sample size [27]. We follow this rule in our estimation procedure to improve the accuracy of calculated coefficients. Prior to conducting the MVP model, we calculate the variance inflation factor (VIF) and tolerance (TOL) values to assess the existence of multicollinearity among covariates. The standard errors in our model are clustered at the village level to account for possible heteroscedasticity and intra-cluster error correlation [29].

3 Data and Variable Definitions

The dataset used in this study originates from the research project "*Poverty Dynamics and Sustainable Development: A Long-term Panel Project in Thailand and Vietnam*", commonly referred to as the Thailand-Vietnam Socioeconomic Panel (TVSEP), which is financed by the German Research Foundation (for further information, visit https://www.tvsep.de). The primary aim of the TVSEP is to provide a longitudinal dataset for studying socioeconomic development and vulnerability to poverty dynamics in the rural areas of Thailand and Vietnam, two developing countries in Southeast Asia [30]. The TVSEP in Vietnam has been gathering data from approximately 2,200 households in three provinces: Ha Tinh, Thua Thien Hue, and Dak Lak, all of which are located in the central region of the country (Fig. 1 illustrates the study sites). The three provinces were selected based on their high reliance on agriculture, below-average per capita income, and inadequate infrastructure [31, 32]. According to [30], these three provinces may be regarded as representatives of rural Vietnamese areas. The three-phase stratified random data-collecting process adheres to the guidelines established by the United Nations Department of Economic and Social Affairs [33].

From 2007 to 2022, the TVSEP conducted eight waves of surveys in Vietnam; waves before 2022 covered all three provinces, but wave 2022 only covered Ha Tinh and Dak Lak, excluding Thua Thien Hue. Two questionnaires were used to collect TVSEP data: one was designed for households and the other for villages. The household questionnaire contains the necessary information for this study, such as household demographic characteristics, borrowing, savings, lending, asset investments, and insurance.

This study uses data from the 2022 survey wave, which originally includes 1,227 rural households. After excluding households with missing information, our final sample consists of 1,115 households.

Our regression model distinguishes five categories of dummy dependent variables, which indicate whether or not households engaged in credit, savings, lending, investments, and insurance within the past year. The credit category includes all the loans owed in the past year. Savings refers to the actions taken by households to reserve a portion of their income in bank accounts, at home, or in other locations. The lending category relates to households providing loans of cash or goods (such as rice and fertilizer). Investments encompass households' decisions to invest in durable assets. For the insurance category, we exclude government health insurance since it is nearly universal in Vietnam [34]. We also exclude compulsory insurance, such as social insurance or vehicle insurance. Therefore, insurance encompasses life insurance, property insurance, private health insurance, disability insurance, and accident insurance that households had in the previous year. Regarding covariates, we include an array of micro-level factors that are assumed to affect household financial tool choices. The definitions of all variables used in our analysis are presented in Table 1. All monetary values are measured in 2005 purchasing power parity dollars (PPP USD).

Table 1 Variable definitions and descriptive statistics

Variables	Definitions and measurement	Mean	Std. Dev.
Dependent variables			
Credit	The household had loans in the previous year (yes = 1, no = 0)	0.412	0.492
Savings	The household saved parts of their income in the previous year (yes = 1, no = 0)	0.661	0.474
Lending	The household informally lent out cash or goods in the previous year (yes = 1, no = 0)	0.239	0.426
Investments	The household invested in durable goods in the previous year (yes = 1, no = 0)	0.386	0.487
Insurance	The household had at least one insurance, including life insurance, property insurance, private health insurance, disability health insurance, and accident insurance, in the previous year (yes = 1, no = 0)	0.242	0.429
Covariates			
Head_gender	Household head's gender (male = 1; female = 0)	0.778	0.415
Head_marital_status	Household head's marital status (married = 1, other = 0)	0.790	0.407
Head_ethnicity	Household head's ethnicity (Kinh = 1, other = 0)	0.849	0.358
Head_age	Household head's age (in years)	59.613	11.704
Head_highest_edu	Educational level of household head (0: below primary school, 1: primary school, 2: secondary school, 3: high school, 4: vocational school, 5: university)	1.980	1.048
Risk	The general willingness of the household head to take risks (measured on a scale of 0 to 10)	6.490	2.494
Hh_size	Household size	3.282	1.675
ChildShare	The proportion of individuals aged under 6 in the household	0.038	0.093
ElderlyShare	The proportion of individuals aged over 60 in the household	0.319	0.388
No_impair	Number of household members with impairments	0.948	0.906
SocialGroup	Number of household members participating in socio-political organizations	1.343	1.061
Self_employ	Household engaged in self-employment activities (yes = 1, no = 0)	0.260	0.439
Livestock	Household kept livestock (yes = 1, no = 0)	0.402	0.490
Loss_shocks	Income and asset losses due to shocks in the previous year (PPP USD)	744.949	3,963.059
Remittance	Total remittances received by the household in the previous year (PPP USD)	344.365	1,247.964
Total_asset	Value of household assets (PPP USD)	6,938.097	16,875.115
Total_area	Total land area of the household (in hectares)	0.854	1.220
No_smartphone	Number of smartphones that the household owned	1.552	1.282

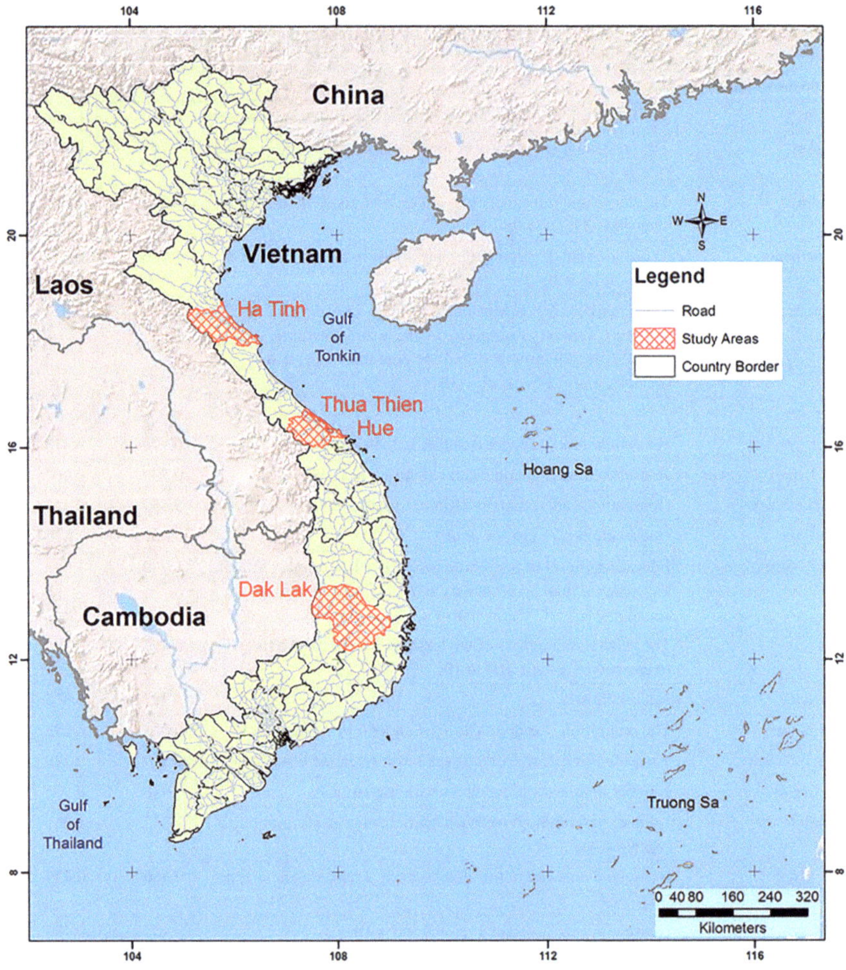

Fig. 1 Map of the surveyed provinces [33]

4 Empirical Results and Discussion

4.1 Results of Descriptive Analysis

Table 1 presents the summary statistics of the variables used in our analysis. Initially, looking at the dependent variables, approximately 66% of households had savings, representing the highest percentage. The proportion of households with access to credit stands at nearly 41%, making it the second-highest percentage. The third position corresponds to the proportion of households making durable asset investments, with 38.6%. The proportions of households having insurance and lending are 24.2%

Table 2 Number of households across a number of financial services used

No. of financial services used	No. of households	Percentage	Cumulative percentage
0	94	8.43	8.43
1	320	28.70	37.13
2	384	34.44	71.57
3	215	19.28	90.85
4	81	7.26	98.12
5	21	1.88	100.00
Total	1,115	100.00	

and 23.9%, respectively. Table 2 illustrates the number of households distributed across the financial services used. The majority of households (91.57%) utilize at least one financial service. It is noteworthy that around 63% of households concurrently adopt two or more financial services. This indicates that they rarely use a single financial instrument exclusively; instead, they use a combination of multiple instruments simultaneously.

Regarding covariates, Table 1 indicates that 77.8% of household heads are male, 79% are married, and 84.9% belong to the Kinh majority. Household heads' average age is 59.6 years. The education level of household heads is notably low, with an average score of 1.88 on a scale from 0 to 5. This result aligns with the finding of [35], who discovered that the educational attainment of household heads in rural Vietnam is poor, averaging 6.85 years of schooling. The mean risk preference of household heads in the sample is 6.49 on an eleven-point Likert scale ranging from 0 to 10, suggesting a modest inclination toward risk-seeking behavior relative to the risk-neutral midpoint. The average number of members in a household is 3.3. The proportion of children and individuals aged 60 or older is recorded at 3.8 and 31.9%, respectively. In households, the mean numbers of members with impairments and members participating in socio-political organizations are 0.9 and 1.3, respectively. Each household possesses an average of 1.6 smartphones and 0.85 hectares of land. In our sample, 26% of households engage in self-employment activities, and 40.2% kept livestock. On average, the asset value, remittance amount, and loss due to shocks are documented at 6938.1, 344.4, and 745 PPP USD, respectively.

4.2 *Determinants of Household Financial Portfolio Choices*

The results of the multicollinearity test are presented in Table 3. All variables show VIF values below 10, and no TOL coefficient is below 0.1, confirming the absence of multicollinearity and validating the assumption of independence among covariates. Therefore, the estimated coefficients in our regression model are not biased

Table 3 Multicollinearity test among covariates

Variables	Variance inflation factor (VIF)	Tolerance (TOL)
ElderlyShare	2.38	0.42
Head_marital_status	2.26	0.44
Head_age	2.19	0.46
Head_gender	2.17	0.46
Hh_size	2.05	0.49
Total_asset (ln)	1.71	0.58
No_smartphone	1.70	0.59
Head_ethnicity	1.44	0.69
ChildShare	1.34	0.75
SocialGroup	1.26	0.79
Head_highest_edu	1.18	0.85
Total_area	1.17	0.85
Self_employ	1.16	0.86
Remittance (ln)	1.14	0.88
No_impair	1.13	0.88
Livestock	1.10	0.91
Loss_shocks (ln)	1.06	0.94
Risk	1.06	0.95
Mean VIF	1.53	

due to multicollinearity. The findings from the multivariate probit model, presented in Tables 4 and 5, provide some valuable insights. The Wald test for the five equations of the multivariate probit model produces a highly significant result (Wald χ^2, (90) $=$ 1,764.05; Prob $> \chi^2 = 0.000$), rejecting the null hypothesis that all regression coefficients across the five equations are equal to zero. In addition, the likelihood ratio test of $\rho_{21} = \rho_{31} = \rho_{41} = \rho_{51} = \rho_{32} = \rho_{42} = \rho_{52} = \rho_{43} = \rho_{53} = \rho_{54} = 0$ rejects the null hypothesis that the error terms across equations are not correlated ($\chi^2(10) = 87.1702$, Prob $> \chi^2 = 0.0000$). These statistical results justify our decision to apply a multivariate probit model for examining determinants of household financial portfolio choices.

Factors affecting household financial portfolio choices are presented in Table 4. The marital status of the head of a household negatively influences the possibility of obtaining credit, meaning that single-headed households have an increased chance of accessing credit. This stands in contrast to the result of [36], who conducted a study in Ethiopia and claimed that single-headed households received less microcredit compared to their married counterparts, as lenders perceived married-headed households as more stable, responsible, and capable of repaying loans. The explanations of this finding may be grounded in the context of rural Vietnam. That is, there exists a wide variety of preferential microcredit programs, both formal and semi-formal, that are aimed at alleviating poverty in rural Vietnam [37, 38]. In Vietnam, single-headed households are regarded as more vulnerable to poverty [39], making them more likely to be eligible for obtaining credit from these programs than married-headed households.

Age is one factor that influences households' motivation to obtain credit. Head age and the proportion of members over 60 are negatively correlated with the possibility of the household having access to credit at the 1 and 5% levels of significance, respectively. According to these findings, households with a greater share of younger members are better able to have access to credit. A plausible rationale for this is that younger individuals are more engaged with production and business activities, resulting in an increased need for loans. One additional potential explanation is that the younger adults indicate a greater level of active participation in socio-political organizations [40], which raises their chances of being eligible for preferential credit and having access to information regarding credit sources. Similar findings regarding the correlation between the age of the household head and the probability of having access to credit have been found in previous literature [40–43].

Education shows an adverse relationship with the probability of acquiring credit, lending out cash or goods, and investing in durable goods. The negative association between education and the chance of obtaining credit means that the higher the head's education, the lower the probability of accessing credit. It is possible that heads with a higher level of education are more productive, which enables them to raise capital [41] and reduce their dependence on credit sources. Likewise, the negative association between head education and lending indicates that higher-educated heads are less likely to lend their cash and goods. It is probable that highly educated individuals are more aware of the informal nature of lending activities and the associated risk of borrower default. Moreover, the inverse correlation between education and investment in durable assets could suggest that educated people are better at identifying more productive channels for their financial resources rather than just investing in non-productive assets. This finding aligns closely with [44], who discovered that an increase in the education level of the household head correlates with a decrease in spending on non-productive assets.

Risk preferences have a positive correlation with lending, implying that those who are more risk-seeking tend to engage more in lending activities. This may be due to the perception that informal lending is risky, leading more risk-averse people to participate less. Our finding indicates that households led by more risk-seeking heads have a tendency to invest more in durable assets. This result is consistent with that of [45]. The insurance coefficient's negative sign denotes that those who are less risk-averse are more likely to purchase insurance. This appears contrary to the conventional theory of insurance demand [46]. This could potentially be attributed to the fact that many people lack knowledge about insurance products [47]. Previous literature has provided mixed findings regarding the association between risk preferences and insurance. For example, [47] revealed that less risk-averse household heads are more inclined to acquire insurance; in contrast, [48] proved the reverse relationship, while [49] identified no significant connection.

Table 4 Determinants of household financial portfolio choices from the multivariate probit model

	Credit	Savings	Lending	Investments	Insurance
Head_gender	0.080 (0.580)	0.187 (0.170)	0.177 (0.265)	−0.206 (0.160)	−0.214 (0.249)
Head_marital_status	−0.395 (0.004)***	−0.107 (0.408)	−0.184 (0.205)	−0.159 (0.280)	0.133 (0.480)
Head_ethnicity	−0.047 (0.728)	0.212 (0.211)	−0.104 (0.440)	0.177 (0.198)	0.037 (0.834)
Head_age	−0.022 (0.000)***	0.001 (0.917)	0.007 (0.271)	−0.008 (0.163)	0.005 (0.401)
Head_highest_edu	−0.111 (0.006)***	0.033 (0.461)	−0.117 (0.008)***	−0.092 (0.049)**	0.051 (0.293)
Risk	−0.020 (0.264)	−0.015 (0.402)	0.055 (0.002)***	0.030 (0.067)*	0.047 (0.049)**
Hh_size	0.095 (0.001)***	−0.053 (0.160)	−0.051 (0.158)	−0.007 (0.833)	0.020 (0.660)
ChildShare	−0.425 (0.366)	−1.093 (0.014)**	0.093 (0.863)	−0.004 (0.994)	−0.657 (0.274)
ElderlyShare	−0.306 (0.043)**	0.248 (0.108)	−0.021 (0.913)	0.198 (0.177)	−0.212 (0.345)
No_impair	0.182 (0.000)***	−0.118 (0.010)***	−0.229 (0.000)***	−0.008 (0.867)	0.095 (0.159)
SocialGroup	−0.008 (0.853)	0.078 (0.082)*	0.175 (0.000)***	0.095 (0.022)**	−0.043 (0.433)
Self_employ	−0.033 (0.732)	0.212 (0.033)**	0.229 (0.040)**	−0.048 (0.635)	−0.130 (0.286)
Livestock	0.211 (0.014)**	0.000 (0.999)	−0.052 (0.568)	0.135 (0.175)	−0.094 (0.379)
Loss_shocks (ln)	0.048 (0.000)***	0.048 (0.001)***	0.045 (0.001)***	0.003 (0.867)	0.033 (0.044)**
Remittance (ln)	0.019 (0.189)	0.024 (0.121)	0.021 (0.159)	0.037 (0.013)**	0.023 (0.279)
Total_asset (ln)	−0.016 (0.585)	0.163 (0.000)***	0.108 (0.005)***	0.348 (0.000)***	0.096 (0.026)**
Total_area	0.053 (0.153)	0.089 (0.032)**	−0.010 (0.748)	−0.052 (0.364)	−0.015 (0.741)
No_smartphone	−0.027 (0.534)	0.044 (0.323)	0.066 (0.136)	0.132 (0.003)***	−0.011 (0.824)
_cons	1.239 (0.003)***	−1.229 (0.003)***	−2.141 (0.000)***	−2.915 (0.000)***	−2.796 (0.000)***
Observations	1.115	1.115	1.115	1.115	1.115

Wald $\chi^2(90) = 1,764.05$
Prob > $\chi^2 = 0.000$
Number of draws = 100
Log pseudolikelihood = $-2,814.1005$
Note Robust standard errors are clustered at the village level and reported in parentheses; significance levels: *<10%, **<5%, ***<1%

The size of a household is positively associated with access to credit at a 1% significance level. This may imply that household members serve as a crucial resource for both farm and non-farm productions, facilitating profit generation that ensures credit payback. This finding aligns with the results of [50] and [51]. The proportion of children in the household has an adverse connection with household savings, which potentially suggests that a high child share results in high costs associated with child-rearing, leading to a low disposable income available for savings. The number of household members with impairments correlates positively with access to credit, while negatively with the possibility of savings and lending. The impairments of household members often result in a significant portion of household expenditures [52]; therefore, credit access can help smooth their spending. Expenditures for members with impairments also reduce the amount of disposable income available for savings and lending.

Our results suggest that the probability of savings, lending, and investments is positively correlated with the number of household members who participate in socio-political organizations. The explanation may be that membership in socio-political organizations provides household members with enhanced skills, social connections, and employment opportunities. These benefits enable an increase in income, which in turn increases savings, lending, and investments. The results also indicate that the chances of household savings and lending are positively correlated with participation in self-employment. It means that self-employment has the potential to enhance the income of households [53], thereby increasing their likelihood of savings and lending. Furthermore, possessing livestock positively relates to access to credit, suggesting that households use credit as an instrument to acquire the cash necessary for livestock production [53].

The loss from shocks positively influences the likelihood of obtaining credit, showing that households regard credit as an ex-post strategy to smooth their consumption in response to shocks. This finding is consistent with [4], who found that experiencing shocks raises the likelihood of households in Malawi accessing both formal and informal credit. The positive coefficients of shock loss for savings, lending, and insurance show that experiencing shocks in the previous year increases the probabilities of saving, lending, and purchasing insurance. This suggests that households perceive these financial tools as an ex-ante coping strategy against shocks. These findings roughly align with those of [54], who reported a positive correlation between shock experience and savings in Thailand, and [55], who concluded that experiencing a flood or rain shock positively influences the decision to purchase insurance in Cambodia.

The positive coefficient of remittances for investments suggests that rural households allocate a portion of their remittances toward investing in durable assets. This conforms with the findings of [56] in India and [57] in Sri Lanka. The likelihood of savings, lending, investments, and insurance acquisition is positively correlated with the value of household assets. These findings align with expectations, as the accumulation of assets increases wealth, thereby increasing the likelihood of savings, lending, investments, and the purchase of insurance. Our findings show that the total land area positively influences household savings. Interestingly, the likelihood of investments is positively associated with the number of smartphones owned by the household.

4.3 Correlations Among Financial Services

As displayed in Table 5, the pairwise correlation coefficients across error terms of the MVP model suggest that financial services are interrelated. We observe the existence of negative coefficients of credit and savings, as well as of credit and investments. This means that there exists, to a certain extent, a substitution relationship between credit and savings, as well as credit and investments. These substitution relationships may be justified by the fact that if households increase involvement in savings and investments, it indicates that they have more disposable income, which consequently reduces their dependence on credit money sources. Conversely, if households decrease their participation in savings and investments, it means there is a shortage of disposable income, leading them to rely more on credit money sources. Positive coefficients exist between savings and lending, savings and investments, and lending and investments. Such relationships indicate a level of complementarity between these financial services. The explanation for these complementary relationships may

Table 5 Correlation coefficients (ρ) between error terms in multivariate probit regressions

	Credit (1)	Savings (2)	Lending (3)	Investments (4)	Insurance (5)
Credit (1)	1				
Savings (2)	−0.308(0.052)***	1			
Lending (3)	0.063 (0.048)	0.184(0.066)***	1		
Investments (4)	−0.118(0.047)**	0.127 (0.055)**	0.234 (0.054)***	1	
Insurance (5)	−0.048 (0.069)	−0.037 (0.070)	−0.128(0.072)*	−0.244(0.068)***	1

Likelihood ratio test of $\rho_{21} = \rho_{31} = \rho_{41} = \rho_{51} = \rho_{32} = \rho_{42} = \rho_{52} = \rho_{43} = \rho_{53} = \rho_{54} = 0$: $\chi^2(10) = 87.1702$, Prob > $\chi^2 = 0.0000$

Note Robust standard errors are clustered at the village level and reported in parentheses; significance levels: * <10%, ** <5%, *** <%

be that savings, lending, and investments are equally important to households. Therefore, when disposable income is abundant, they increase their participation in all three financial services and vice versa. The correlations between insurance and lending, as well as between insurance and investments, are negative, showing that lending and investments serve as substitutes for insurance to some extent.

5 Conclusion and Policy Implications

The determinants of household financial choices have been the subject of extensive research worldwide. However, most studies only focus on one specific financial service. This study differs from previous research in that it simultaneously examines multiple types of financial instruments in the financial portfolios of rural households. Specifically, this study explores a variety of micro-level factors influencing household financial portfolio choices in rural Vietnam. Using the TVSEP dataset, we apply the MVP model to account for the interdependencies across financial instruments. The findings of the MVP regression demonstrate that the household head's educational attainment, willingness to take risks, the number of members with impairments, participation in socio-political organizations, self-employment, and shock losses are key determinants of household financial portfolio choices. These findings are related to earlier research conducted in different countries. Our results also provide evidence of the mutual relationships between the financial uptakes of rural households. To some degree, there is a substitution relationship within pairs of financial instruments, such as credit and savings, credit and investments, insurance and lending, and insurance and investments. However, there exists a certain complementarity within the following pairs of financial service choices: savings and lending, savings and investments, and lending and investments.

The results described in this study carry important implications for policymaking. First, our findings suggest that rural households view credit as a tool to stabilize their consumption when they experience shock-related losses or have family members with impairments. Credit is also a beneficial source of capital for households that raise livestock. It is therefore crucial to implement policies that enhance rural households' access to credit programs. Second, the absence of a correlation between the number of individuals with impairments and the adoption of insurance indicates that households fail to use insurance as a coping mechanism. Therefore, policies aimed at lowering insurance premium rates may improve households' ability to reach insurance coverage. Third, our research shows the interconnectedness of household financial services, indicating the need for joint analysis. Policymakers should take these complementarities and substitutions into account when making policies to improve household finance.

References

1. Ansah, I.G.K., Gardebroek, C., Ihle, R.: Using assets as resilience capacities for stabilizing food demand of vulnerable households. Int. J. Disaster Risk Reduct. **82**, 103352 (2022)
2. Benami, E., Carter, M.R.: Can digital technologies reshape rural microfinance? implications for savings, credit, and insurance. Appl. Econ. Perspect. Policy **43**(4), 1196–1220 (2021)
3. Kouandou, A., Pérolde, I.: Zeh: Insuring consumption against shocks: The role of informal savings in Nigeria. Food Policy **129**, 102737 (2024)
4. Wezi Salima, Laston Petro Manja, Levison S. Chiwaula, and Gowokani Chijere Chirwa. The impact of credit access on household food security in malawi. *Journal of Agriculture and Food Research*, 11:100490, March 2023
5. Ndaghu Ndonkeu Nathanel, Udeme Henrietta Ukpe, Djomo Choumbou Raoul Fani, and Esuh Nnoko Divine. *Determinants of Credit Accessibility and Performance of Smallholder Rice Farmers: A Case of the West Region of Cameroon*, pages 57–67. Springer International Publishing, 2023
6. Nwosu, E.O., Anumudu, C.N., Nnamchi, C.E.: Microeconomic determinants of household savings in nigeria. J. Int. Dev. **32**(2), 150–167 (September 2019)
7. Mohd Abass Bhat, Geleta Demera Gomero, and Shagufta Tariq Khan. Antecedents of savings behaviour among rural households: A holistic approach. *FIIB Business Review*, 13(1):56–71, July 2024
8. Bhatia, R., Bhat, A.K., Tikoria, J.: Life insurance purchase behaviour: A systematic review and directions for future research. Int. J. Consum. Stud. **45**(6), 1149–1175 (March 2021)
9. Lambregts, T.R., Schut, F.T.: Displaced, disliked and misunderstood: A systematic review of the reasons for low uptake of long-term care insurance and life annuities. The Journal of the Economics of Ageing **17**, 100236 (October 2020)
10. Ngoma, H., Finn, A., Kabisa, M.: Climate shocks, vulnerability, resilience and livelihoods in rural zambia. Climate Dev. **16**(6), 490–501 (August 2023)
11. Dale W Adams. *Taking a Fresh Look at Informal Finance*, pages 5–23. Routledge, March 2019
12. Evidence from rural vietnam: Le Phuong Xuan Dang, Viet-Ngu Hoang, Son Hong Nghiem, and Clevo Wilson. Social networks with organisational resource, generalised trust and informal loans. Economic Analysis and Policy **77**, 388–402 (March 2023)
13. Migheli, M.: Land-use rights and informal credit in rural vietnam. Italian Economic Journal **10**(1), 409–434 (April 2023)
14. Jann Goedecke, Isabelle Guérin, Bert D'Espallier, and Govindan Venkatasubramanian. Why do financial inclusion policies fail in mobilizing savings from the poor? lessons from rural south india. *Development Policy Review*, 36(S1), November 2017
15. Wang, F.: An empirical equilibrium model of formal and informal credit markets in developing countries. Rev. Econ. Dyn. **46**, 224–243 (October 2022)
16. Giesbert, L., Steiner, S., Bendig, M.: Participation in micro life insurance and the use of other financial services in ghana. Journal of Risk and Insurance **78**(1), 7–35 (January 2011)

17. Pham Tien Thanh, Katsuhiro Saito, and Pham Bao Duong. Impact of microcredit on rural household welfare and economic growth in vietnam. *Journal of Policy Modeling*, 41(1):120–139, January 2019
18. Lisa Farrell, Tim R.L. Fry, and Leonora Risse. The significance of financial self-efficacy in explaining women's personal finance behaviour. *Journal of Economic Psychology*, 54:85–99, June 2016
19. Truong Anh Tuan, Pham Khanh Nam, and Le Thanh Loan. The impact of health insurance on households' financial choices: Evidence from vietnam. *Research in Economics*, 76(3):264–276, September 2022
20. Dick Chune Midamba, Accram Jjengo, and Kevin Okoth Ouko. Assessing the determinants of saving behaviour: evidence from rural farming households in central uganda. *Discover Sustainability*, 5(1), June 2024
21. Staal, K.: Household savings and negative interest rates. Int. Adv. Econ. Res. **29**(1–2), 1–13 (April 2023)
22. Nadia van Huyssteen and Sharon Rudansky-Kloppers. Factors influencing consumers' purchase decisions regarding personal motor vehicle insurance in south africa. *Cogent Business & Management*, 11(1), December 2023
23. Samuel Wai Johnson: Determinants of access to credit: Evidence from a fragile conflict-affected setting. The European Journal of Development Research **33**(6), 1475–1492 (September 2020)
24. Adejoke Yewande Bakare, Ayodeji Sunday Ogunleye, and Ayodeji Damilola Kehinde. Impacts of microcredit access on climate change adaptation strategies adoption and rice yield in kwara state, nigeria. *World Development Sustainability*, 2:100047, June 2023
25. W. H. Greene. *Econometric Analysis*. Prentice Hall, Upper Saddle River, 7th ed. edition, 2012
26. D. McFadden. Conditional logit analysis of qualitative choice behavior. In P. Zarembka, editor, *Frontiers in Econometrics*, pages 105–142. Academic press, 1974
27. Cappellari, L., Jenkins, S.P.: Multivariate probit regression using simulated maximum likelihood. The Stata Journal: Promoting communications on statistics and Stata **3**(3), 278–294 (September 2003)
28. Hajivassiliou, V., McFadden, D., Ruud, P.: Simulation of multivariate normal rectangle probabilities and their derivatives theoretical and computational results. Journal of Econometrics **72**(1–2), 85–134 (May1996)
29. A. Colin Cameron and Douglas L. Miller. A practitioner's guide to cluster-robust inference. *Journal of Human Resources*, 50(2):317–372, 2015
30. Bernd Hardeweg, Stephan Klasen, and Hermann Waibel. *Establishing a Database for Vulnerability Assessment*, pages 50–79. Palgrave Macmillan UK, 2013
31. Trung Thanh Nguyen, Thanh-Tung Nguyen, Van-Hanh Le, Shunsuke Managi, and Ulrike Grote. Reported weather shocks and rural household welfare: Evidence from panel data in northeast thailand and central vietnam. *Weather and Climate Extremes*, 30:100286, December 2020
32. Trung Thanh Nguyen, Viet Tuan Tran, Thanh-Tung Nguyen, and Ulrike Grote. Farming efficiency, cropland rental market and income effect: evidence from panel data for rural central vietnam. *European Review of Agricultural Economics*, September 2020
33. Trung Thanh Nguyen, Manh Hung Do, and Dil Rahut. Shock, risk attitude and rice farming: Evidence from panel data for thailand. *Environmental Challenges*, 6:100430, January 2022
34. Nguyen Thao Thi Nguyen, Long Hoang Nguyen, Tham Thi Nguyen, Linh Gia Vu, Thuc Minh Thi Vu, Minh Ngoc Le Vu, Giang Thu Vu, Carl A. Latkin, Cyrus S. H. Ho, and Roger C. M. Ho. Preference and willingness to pay for reproductive health services among adults in urban-rural transition settings of a developing country: evidence from a cross-sectional study in a rural district of hanoi, vietnam. *BMC Health Services Research*, 23(1), November 2023
35. Ninh, L.K.: Economic role of education in agriculture: evidence from rural vietnam. Journal of Economics and Development **23**(1), 47–58 (2020)
36. Amanuel Berhanu, Abayineh Amare, Birki Gurmessa, Yadeta Bekele, and Tamiru Chalchisa. Does microcredit use helps farmers win battle against food insecurity: evidence from jimma zone of southwest ethiopia. *Agriculture & Food Security*, 10(1), November 2021

37. Duong, A., Antriyandarti, E.: The impact of the vietnam bank for social policies preferential credit on household welfare in vietnam: a panel data analysis. Journal of Economics and Development **24**(1), 18–32 (April 2021)
38. Duong, H.A., Nghiem, H.S.: The impact of microfinance on poverty reduction in vietnam: A dynamic panel-data analysis. Poverty & Public Policy **14**(4), 405–422 (December 2022)
39. Duc Hong Vo, Chi Minh Ho, and Anh The Vo. The economic circumstances of widows in vietnam. *PLOS ONE*, 18(5):e0285595, May 2023
40. Luan, D.X., Bauer, S., Kuhl, R.: Income impacts of credit on accessed households in rural vietnam: Do various credit sources perform differently? Agris on-line Papers in Economics and Informatics, VII **I**(1), 57–67 (March 2016)
41. Tri Haryanto, Wahyu Wisnu Wardana, Iqram Ramadhan Jamil, Annisaa Rizky Dwi Brintanti, and Kabiru Hannafi Ibrahim. Impact of credit access on farm performance: Does source of credit matter? *Heliyon*, 9(9):e19720, September 2023
42. Aregash Esayas Lelisho and Mesfin Esayas Lelisho. Impact of credit use on crop productivity and gross income of smallholder farmers: a propensity score matching approach. *Discover Agriculture*, 2(1), July 2024
43. Chung Phan, Stefano Filomeni, and Seng Kiong Kok. The impact of technology on access to credit: A review of loan approval and terms in rural vietnam and thailand. *Research in International Business and Finance*, 72:102504, October 2024
44. Rosemary Emegu Isoto and David Simon Kraybill: Gender disaggregated effects of microcredit on capital accumulation in rural households: Evidence from uganda. Agric. Financ Rev. **79**(2), 204–216 (April 2019)
45. He, Z., Ye, J., Shi, X.: Housing wealth and household consumption in urban china. Urban Studies **57**(8), 1714–1732 (June2019)
46. Harris Schlesinger. *The Theory of Insurance Demand*, pages 167–184. Springer New York, 2013
47. Takahashi, K., Noritomo, Y., Ikegami, M., Jensen, N.D.: Understanding pastoralists' dynamic insurance uptake decisions: Evidence from four-year panel data in ethiopia. Food Policy **95**, 101910 (August 2020)
48. Kwame Adjei-Mantey and Charles Yuji Horioka: Determinants of health insurance enrollment and health expenditure in ghana: an empirical analysis. Rev. Econ. Household **21**(4), 1269–1288 (August 2022)
49. Cristian Mihai Dragos, Simona Laura Dragos, Codruta Mare, Gabriela Mihaela Muresan, and Alexandra-Anca Purcel. Does risk assessment and specific knowledge impact crop insurance underwriting? evidence from romanian farmers. *Economic Analysis and Policy*, 79:343–358, September 2023
50. Bonsa Hinkosa. Determinants of adoption of female headed household in microfinance program: a case study on oromia credit and saving share company (ocssco) in guto gida district, ethiopia. *Journal of Innovation and Entrepreneurship*, 12(1), November 2023
51. Koomson, I., Asongu, S.A., Acheampong, A.O.: Financial inclusion and food insecurity: Examining linkages and potential pathways. J. Consum. Aff. **57**(1), 418–444 (December 2022)
52. Felician Andrew Kitole, Robert Michael Lihawa, and Eliaza Mkuna. Comparative analysis on communicable and non-communicable diseases on catastrophic spending and impoverishment in tanzania. *Global Social Welfare*, 11(2):123–134, August 2022
53. Pankaj C. Patel and Marcus T. Wolfe. Will self-employment set you free from poverty? evidence from two national surveys from economically developed countries. *Journal of Poverty*, pages 1–40, April 2024
54. Menglan Wang and Manh Hung Do: Reported shocks, households' resilience and local food commercialization in thailand. Journal of Economics and Development **25**(2), 153–170 (January 2023)
55. Falco, C., Rotondi, V., Kong, D., Spelta, V.: Investment, insurance and weather shocks: Evidence from cambodia. Ecol. Econ. **188**, 107115 (October 2021)
56. Anu Abraham. *Role of Household Socio-economic Status in Determining the Impact of Remittances on Human Capital Investment*, pages 231–251. Routledge India, May 2024

57. Jayaweera, R., Verma, R.: Are remittances a solution to housing issues? a case study from sri lanka. Social Sciences & Humanities Open **7**(1), 100392 (2023)

The Impact of Information and Communication Technology on Economic Growth and Productivity Paradox in Southeast Asia, Analyzed with Bayesian Methods

Anh Hong Thi Nguyen

Abstract This study aims to examine the impact of Information and Communication Technology (ICT) on economic growth. Specifically, the author employs Principal Component Analysis (PCA) to calculate the ICT index based on four indicators in the ICT sector for Southeast Asian countries during the period from 1990 to 2022, using Bayesian methods. The findings indicate that a productivity paradox has emerged in these countries during the study period, showing that ICT has a negative effect on economic growth. However, an interesting discovery is that implementing reasonable public expenditure policies in the ICT sector can lead to a positive impact on economic growth, with a probability of effectiveness reaching 99%. Furthermore, the research analyzes the probability of ICT's impact on each country and finds that each nation, due to its unique socio-economic characteristics, experiences different effects on economic growth. This suggests that specific policy implications should be developed for each country.

Keywords ICT · Economic growth · Productivity paradox · Bayesian methods · PCA

1 Introduction

In economic growth theories, the importance of technology has been mentioned and analyzed in various growth models, each with different roles. Growth theories such as the Solow-Swan Model, Endogenous Growth Models, Schumpeterian Growth Theory, and Innovation-driven Growth Models all affirm that technology contributes to economic growth. Technology plays a key role in enhancing productivity, driving long-term growth, and explaining the differences in development rates between countries.

A. H. T. Nguyen (✉)
School of Finance and Accounting, Industrial University of Ho Chi Minh City, Ho Chi Minh City, Vietnam
e-mail: nguyenthihonganh@iuh.edu.vn

H. T. T. Nguyen et al. (eds.), *Data Science in Finance and Accounting*, Studies in Big Data 181, https://doi.org/10.1007/978-3-032-06179-9_16

A prominent feature of global development in the 21st century is the widening gap between countries due to the digital divide. Developing countries have fewer opportunities for economic growth and improving living standards compared to developed economies, largely due to the lack of information and communication technology (ICT), particularly limited internet connectivity. Thus, strengthening ICT is seen as a crucial factor in narrowing the gap between developed and developing countries. One study [1] discussed the challenges faced by the service industry in the early 1990s and the necessity of restructuring as it confronted pressures related to efficiency and productivity. He argued that the service sector needed to improve productivity to meet the challenges of global competition and the rapid advancement of technology. In particular, he emphasized the importance of adopting new technology, reorganizing business structures, and enhancing the skills of the workforce in the service industry to ensure competitiveness and sustainable growth.

However, empirical studies on this topic have yielded inconsistent findings. One work [2] studied 42 developing countries and 24 high-income developed countries from 1985 to 1999 and found that the impact of ICT on economic growth was statistically significant and positive in developed, high-income countries, whereas it was not statistically significant in developing countries. A research [3] using the GMM method to analyze dynamic panel data from OPEC member countries during 1990-2007, found a significant impact of ICT investment on economic growth in OPEC member countries. In this study [4] on the impact of ICT patents on economic growth in 43 economies, including 26 advanced and 17 emerging economies, during 1998-2016, using the two-step system GMM method, found that ICT patents had a stronger impact on economic growth in advanced economies but had a negative and significant effect on growth in emerging economies.

According to World Bank data, global figures show that the number of Internet users worldwide increased from 0% in 1990 to 63% in 2022, and similarly, the number of fixed and mobile subscriptions per 100 people rose from 0 in 1990 to 102 in 2022. In some Southeast Asian countries, such as Vietnam, in 1990, the percentage of the population using the internet was almost nonexistent, and by 1996, it was 0.00013%. By 2022, 78.59% of Vietnam's population was using the Internet. The number of fixed and mobile subscriptions per 100 people in Vietnam was 0 in 1990 but increased to 140 per 100 people in 2022. Since 1990, Vietnam has transformed from a country with very weak ICT infrastructure to one with a robust information and communication technology system, playing a crucial role in its development.

Singapore, a developed nation in Southeast Asia and the world, had an internet usage rate of just 0.16% in 1991, but by 2022, it had reached 95.95%. Clearly, ICT enhances access to information, creates new approaches, and innovates production processes in the economies of these countries. However, the improvements brought by ICT have had contradictory effects on economic growth in different nations. Empirical studies have shown that ICT does not always have a positive impact on economic growth, a concept that economist Robert Solow introduced in the late 1980s as the Productivity Paradox: "You can see the computer age everywhere but in the productivity statistics" [5]. The Productivity Paradox stems from the contradiction between the explosion of information technology and the lack of corresponding increases in labour productivity or growth, as expected by managers who believe that information and communication technology (ICT) should stimulate productivity gains and economic growth. For these reasons, the author conducts a study on the impact of ICT development on GDP growth in Southeast Asian countries from 1990 to 2022, using the Bayesian method to examine whether the Productivity Paradox appears in Southeast Asian countries (excluding Myanmar and East Timor due to insufficient data).

The rest of the paper is organized as follows. Section 2 provides an overview of the theoretical background and existing literature on this topic. Section 3 outlines the model, data, and methodology. Section 4 presents the results and provides a discussion. Finally, Sect. 5 concludes the paper.

2 Theoretical Underpinnings and Some Stylized Fact

2.1 Theoretical Underpinnings

ICT has become a crucial factor in driving economic growth, particularly through modern economic growth theories. According to [6], in the exogenous growth model, the long-term growth of the economy largely depends on the exogenous factor of technological progress. However, this model does not delve into the mechanisms through which ICT impacts productivity. Later, the endogenous growth theories of [7, 8] added to this by arguing that technology can be developed from within the economy through investments in knowledge and research and development. ICT plays a key role in enhancing knowledge, improving labor productivity, and thus raising total factor productivity.

Another important contribution to this field is Innovation Theory [9], which emphasizes that innovation is the primary driver of economic growth and development. Schumpeter argued that old technologies are replaced by newer and more efficient ones. ICT is a prime example of this substitution, as ICT innovation helps increase operational efficiency, expand markets, and create incentives for businesses and the entire economy.

Furthermore, Technology Spillover Theory [10] extended Romer's theory by demonstrating that technology can spread from one country to another through channels such as foreign direct investment (FDI) and international trade. ICT helps promote this process by increasing connectivity and enabling firms in developing countries to access knowledge and technology from developed countries. Aitken and Harrison [11] also showed that FDI enables developing countries to benefit from advanced technology, thereby improving productivity in the economy.

However, Solow's Productivity Paradox (1987) presents a contradiction, as Solow observed that despite increased investment in ICT, overall productivity did not improve markedly. This paradox suggests that ICT adoption requires appropriate infrastructure and human resources. It highlights the importance of complementary factors such as human capital and infrastructure to fully harness the benefits of ICT.

The theories above suggest that ICT plays a vital and multidimensional role in economic growth. Not only does it help increase productivity through knowledge accumulation and production optimization, but ICT also fosters innovation following Schumpeter's perspective, creating new development opportunities and enhancing the economy's competitiveness.

2.2 Some Stylized Fact

The rapid development of ICT globally over the past decades has truly impressed and drawn attention not only from policymakers and business managers but also from researchers interested in how ICT affects economic growth in different countries or continents. Many studies have confirmed the significant positive impact of ICT on economic growth, while on the other hand, there exist cases where ICT negatively affects economic growth.

Jorgenson and Stiroh [12] discussed the role of information technology in driving economic growth in the United States during the 1990s. The authors pointed out that the development and widespread adoption of information technology became a key factor in promoting economic growth, improving productivity and efficiency across various industries. Additionally, information technology not only contributed directly to labour and capital productivity but also enhanced the efficient use of other factors in the economy, especially in the context of the technological boom at the end of the 20th century. Dewan and Kraemer [13] based on data from over 36 countries from 1985 to 1993, observed a significant difference between developed and developing countries. Specifically, in developed economies, there is a strong and significant positive relationship between ICT and economic growth, but no such evidence was found for developing countries. The authors attributed this disparity to the low levels of ICT investment in developing countries, or in other words, while high-income countries are producers of digital technologies, middle-income countries are merely users. Bahrini and Qaffas (2019) studied the impact of ICT on economic growth in developing countries in the Middle East and North Africa (MENA) region and Sub-Saharan Africa (SSA) from 2007 to 2016 using the GMM model. Their findings showed that, except for fixed-line telephones, ICT technologies such as mobile phones, internet usage, and broadband adoption were the main drivers of economic growth in developing countries in the MENA and SSA regions during this period. Aghaei and Rezagholizadeh [14] using dynamic and static panel data approaches within the framework of a growth model and applying them to the economies of Organisation of Islamic Cooperation (OIC) countries over the period 1990-2014, found that for every 1% increase in ICT investment, economic growth in OIC countries increased by 0.52%.

Using the GMM model, [15] studied the impact of ICT, specifically the development of mobile phones, on economic growth in African countries from 1988 to

2007. The results confirmed that ICT, including mobile phone development, significantly contributed to economic growth in African countries. Hammed and Ademosu [16] utilized the ARDL method to study the impact of ICT innovation, FDI, and economic growth in BRICS countries from 1990 to 2021. The results indicate that ICT consistently and significantly contributes to the economic growth rate of BRICS countries, and that ICT infrastructure, combined with foreign-oriented investment, can play a crucial role in boosting the growth process of BRICS economies. Additionally, implementing a policy to attract human resources from developing countries to transform various resources aids in economic growth.

Dumor et al. [17] assessed the impact of ICT on trade and economic growth in 65 countries of the Belt and Road Initiative (BRI) in East Africa, including 27 sub-Saharan African countries, during the period from 2000 to 2019. The empirical results indicate that better access to ICT enhances bilateral export and economic growth among participating countries. The development of ICT and infrastructure has provided numerous opportunities for the East African Community (EAC) to promote intra-regional trade. Thus, BRI countries need to invest more heavily in ICT infrastructure to foster a sustainable and continuous economic development model in the region. Appiah-Otoo and Song [18], by using panel data from 123 countries, including 45 high-income countries, 58 middle-income countries, and 20 low-income countries from 2002 to 2017, studied the impact of ICT on economic growth, comparing rich and poor countries. This study's findings show that ICT increases economic growth in both groups of countries, but poor countries tend to gain more benefits from the ICT revolution. Inani and Tripathi [19], in studying The Nexus of ICT, Electricity Consumption, and Economic Growth in India: An ARDL Approach, found that ICT promotes economic growth in the short term but has no impact in the long term. Therefore, India should focus on expanding ICT to stimulate short-term economic growth. The results of this study are similar to those for Nigeria, showing a significant impact of ICT investment on economic growth. Therefore, if Nigeria wants to boost economic growth, the country needs to implement additional specific policies to facilitate ICT investment [20].

However, some empirical studies on the relationship between ICT and economic growth in certain countries have shown either negative or no impact. Ishida [21] studied the impact of ICT development on economic growth and energy consumption in Japan from 1980 to 2010. The results indicated that ICT did not spur economic growth as suggested by previous studies on the same topic. Pohjola [22] found no statistically significant correlation between ICT investment and economic growth in 43 countries from 1985 to 1999. This is attributed to the weak accessibility and availability of communication technologies and technology-based products in many developing countries.

Additionally, some studies have shown that different ICT indicators have varying effects on economic growth, or that ICT only positively affects economic growth in developed or industrialized economies, but not in developing countries. For instance, [23] examined the dynamic relationship between ICT, the Human Development Index, and economic growth in Asian economies from 1990 to 2016. The findings indicated that mobile phone usage boosted economic growth, while internet usage had no impact on growth. Niebel [24] analyzed the impact of ICT on economic growth in 59 countries, including developing, emerging, and developed nations, from 1995 to 2010. The study found that ICT investments in developing and emerging economies did not yield as much benefit as in developed countries.

In another study, [25] evaluated the overall impact of ICT on the economy of 20 countries, including developing countries such as China, India, Indonesia, Malaysia, and the Philippines. The developed or newly industrialized economies included Australia, Austria, Canada, Denmark, Finland, France, Ireland, Italy, Japan, South Korea, Singapore, Spain, Sweden, the United Kingdom, and the United States. The results revealed that ICT contributed to economic growth in many developed countries and newly industrialized economies, but not in developing countries. Albiman and Sulong [26] studied the long-term impact of ICT on economic growth in the Sub-Saharan region over 27 years (from 1990 to 2014). The study's results indicated that ICT proxies such as mobile phones and the internet had a direct positive effect on economic growth in this region. However, when considering the nonlinear impact, the mass penetration of ICT appeared to slow down economic growth in the Sahara region.

3 Data and Methodology

3.1 Overview of Methodology

This study employs a comprehensive Bayesian analytical framework to examine the impact of Information and Communication Technology (ICT) on economic growth in Southeast Asian countries during the period 1990–2022. The methodology comprises the following steps:

Step 1: Construction of the ICT Index: A composite ICT index was developed using Principal Component Analysis (PCA) based on four ICT indicators: fixed telephone subscriptions, fixed broadband subscriptions, mobile cellular subscriptions, and internet usage. The PCA ensures data dimensionality reduction while retaining maximum variance, confirmed through the Kaiser-Meyer-Olkin (KMO) test for factor adequacy.

Step 2: Model Specification: The study specifies a panel regression model where the GDP growth rate (GRO) is the dependent variable, while ICT and macroeconomic variables such as education (EDU), foreign direct investment (FDI), government consumption expenditure (GCE), and gross fixed capital formation (INV) are independent variables.

Step 3: Bayesian Estimation: To address sample size limitations and to obtain probabilistic insights, the study utilizes Bayesian estimation with the Metropolis-Hastings algorithm via Gibbs sampling (10,000 iterations, 3 chains). Five prior specifications are tested to select the most appropriate distribution using Bayesian model comparison criteria such as DIC, log marginal likelihood, and posterior probability.

Step 4: Model Diagnostics and Validation: Posterior convergence is assessed using trace and autocorrelation plots, Monte Carlo Standard Errors (MCSE), and the Gelman-Rubin statistic (R_c). The selected model is further validated via sensitivity analysis.

Step 5: Country-level Probability Analysis: A final step involves computing posterior probabilities for the impact of ICT and other variables at the country level, highlighting heterogeneous effects across Southeast Asia.

A visual representation of this methodology is presented in Fig. 1.

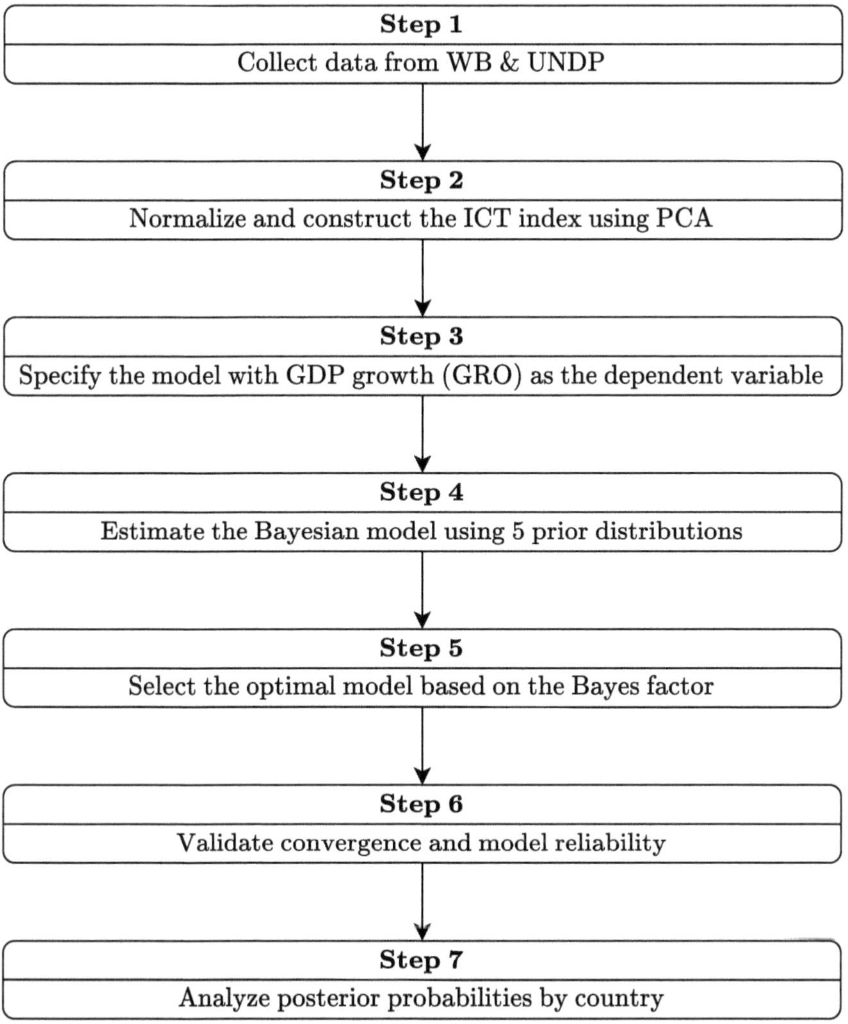

Fig. 1 Flowchart of the Bayesian research methodology

3.2 Model Specification

To capture the effects of ICT and macroeconomic variable on economic growth, this study estimates an econometric model of the following form from current research:

$$\mathrm{GRO}_{it} = \alpha 0_i + \beta_{1i}\mathrm{ICT}_{it} + \beta_{2i}\mathrm{EDU}_{it} + \beta_{3i}\mathrm{FDI}_{it} + \beta_{4i}\mathrm{GCE}_{it} + \beta_{5i}\mathrm{INV}_{it} + \varepsilon_{it}$$

where GRO is GDP growth rate, ICT is ICT development index, ε is an error term, α, β and γ are parameters to be estimated, EDU is education index, FDI is percentage of foreign direct investment net inflows to the GDP, GCE is government consumption as % of GDP, INV is gross fixed capital formation as % of GDP. Table 1 presents descriptive statistics for the main variables.

Table 1 shows that ICT had the lowest value in Singapore in 2002 and the highest value in Cambodia in 2019. EDU has an average value of 0.58, with the lowest value observed in the Philippines in 2002 and the highest in Singapore in 2004. FDI has an average value of 5.46, with the lowest recorded in Indonesia in 2000 and the highest in Singapore in 2021. GCE reached its lowest value in Cambodia in 2022 and its highest in Laos in 2029. INV has an average value of 6.68, with the lowest value observed in Thailand in 1998 and the highest in Brunei Darussalam in 1996. Regarding GRO, the average value is 4.85; Singapore achieved the highest value in 2010, while Cambodia had the lowest in 1994.

The study employs Bayesian methods to address the small sample size and to analyze the probabilistic impact of independent variables on the dependent variable. We used the Metropolis-Hastings algorithm through Gibbs sampling, with an MCMC chain size of 10,000, to conduct Bayesian analysis. Since prior studies were primarily conducted using frequentist methods, prior information was not available. In such cases, [27] suggest defining standard Gaussian distributions for various priors and conducting Bayesian factor analysis to select the best-informed prior simulations. To implement this process, the authors first constructed five simulations with the priors for the model as follows:

Next, the author proceeds to regress the simulations using the five prior specifications presented in Table 2. The author will then conduct Bayesian factor analysis

Table 1 Description of the data sample

Variable	Mean	Standard deviation	Min	Max
GRO	4.855165	4.304086	−34.80864	14.51975
ICT	−3.54e−10	1.690835	−1.713493	5.267768
EDU	0.5797244	0.1908379	0.181	0.937
FDI	5.464802	6.01327	−2.75744	31.62074
GCE	6.020184	10.78875	−48.06519	85.35944
INV	6.682307	13.29478	−44.02647	86.21863

Table 2 Prior distributions for the likelihood function GRO $\sim N(\mu, \sigma)$

Simulation 1	$\beta_i \sim N(0, 1)$
	$\sigma^2 \sim \text{Invgamma}(0.01, 0.01)$
Simulation 2	$\beta_i \sim N(0, 10)$
	$\sigma^2 \sim \text{Invgamma}(0.01, 0.01)$
Simulation 3	$\beta_i \sim N(0, 100)$
	$\sigma^2 \sim \text{Invgamma}(0.01, 0.01)$
Simulation 4	$\beta_i \sim N(0, 1000)$
	$\sigma^2 \sim \text{Invgamma}(0.01, 0.01)$
Simulation 5	$\beta_i \sim N(0, 10000)$
	$\sigma^2 \sim \text{Invgamma}(0.01, 0.01)$

and posterior Bayesian analysis to select the most appropriate non-informative prior distribution. In the following step, the author will analyze convergence to ensure the model's stationarity. Sensitivity analysis will also be conducted to ensure the robustness of the Bayesian regression model.

3.3 Data Descriptions

The author uses a research sample consisting of data over 33 years from 1990 to 2022 sourced from the World Bank (WB) and UNDP for Southeast Asian countries (excluding Myanmar and East Timor due to insufficient data), including Vietnam, Laos, Thailand, Brunei, Singapore, Malaysia, the Philippines, Indonesia, and Cambodia. The variables used in this study include economic growth (GRO), measured by the annual GDP growth rate. The main independent variable in the study is the information and communication technology (ICT) development index. The ICT index is calculated based on key indicators including the number of fixed telephone subscriptions (FTS; per 100 people), fixed broadband subscriptions (FBS; per 100 people), mobile cellular subscriptions (MCS; per 100 people), and the percentage of individuals using the Internet (IUI; % of the population). To create a composite index from these indicators, data normalization is performed to ensure that each indicator contributes equally to the final result. Subsequently, Principal Component Analysis (PCA) is applied to compute the correlation matrix of the indicators and to identify the principal components–combinations of these indicators aimed at retaining most of the original information. The principal components with the highest eigenvalues (typically those that explain over 70-80% of the variance) are selected to construct the ICT index. Finally, the ICT index is calculated by taking the weighted sum of the indicators based on the selected principal components, resulting in an index that reflects the overall level of ICT development in a country.

The author used PCA to create the ICT variable from 4 variables (fixed telephone subscriptions, fixed broadband subscriptions, mobile cellular subscriptions,

Table 3 Kaiser-Meyer-Olkin (KMO) test

Variable	KMO	Overall
FTS	0.7487	0.7261
FBS	0.8158	
MCS	0.7088	
IUI	0.6682	

individuals using the internet) and the Kaiser-Meyer-Olkin (KMO) test results in Table 3 indicate that the data is suitable for factor analysis:

The results of the Kaiser-Meyer-Olkin (KMO) test indicate the adequacy of the data for principal component analysis (PCA) based on four variables: FTS (fixed telephone subscriptions), FBS (fixed broadband subscriptions), MCS (mobile cellular subscriptions), and IUI (percentage of individuals using the Internet). Specifically, the KMO value for the FTS variable is 0.7487, indicating good adequacy, allowing this variable to be effectively used in PCA. The FBS variable has a KMO value of 0.8158, which is very high, indicating very good adequacy and that this variable can contribute significantly to the analysis. The MCS variable, with a KMO value of 0.7088, also shows good adequacy, suggesting that it can be effectively analyzed. However, the KMO value for the IUI variable is 0.6682, indicating average adequacy, close to the minimum threshold of 0.7. Nevertheless, the overall KMO value is 0.7261, indicating that the data can be effectively used for PCA and will yield reliable results when conducting the analysis. Additionally, there are other independent variables described in detail in Table 4.

4 Results and Analysis

We conducted regression on the simulations using the five priors outlined in Table 2. Subsequently, Bayesian factor analysis and posterior Bayesian analysis were performed to select the most appropriate prior distribution. The results of the posterior Bayesian analysis are Table 5.

The priority order for Bayesian factor analysis is the simulation with the highest average Log Bayes Factor (Log BF), highest Log Maximum Likelihood (Log ML), and the smallest average Deviance Information Criterion (DIC). Therefore, the results of the Bayesian factor analysis in Table 5 indicate that Simulation 1 is the most advantageous. Additionally, Simulation 1 has a superior posterior probability compared to the other simulations, making it the most suitable prior. The regression results from Simulation 1 was selected.

Table 4 Variables description and data sources

Variable	Definitions	Measure	Source
GRO	Economic growth	Real GDP growth (annual %)	WDI
ICT	ICT index	Authors Estimates through PCA	WDI
FTS	Fixed telephone subscriptions (per 100 people)		WDI
FBS	Fixed broadband subscriptions (per 100 people)		WDI
MCS	Mobile cellular subscriptions (per 100 people)		WDI
IUI	Individuals using the Internet (% of population)		WDI
EDU	Education Index	The average of the literacy index and the education enrollment index.	UNDP
FDI	Foreign direct investment	Percentage of foreign direct investment net inflows to the GDP	WDI
GCE	Government consumption expenditure	General government final consumption expenditure as % of GDP	WDI
INV	Investment	Gross fixed capital formation as % of GDP	WDI

Table 5 Summary results of the posterior Bayesian analysis

	Chains	Avg DIC	Avg log (ML)	P(M)	P(M|y)
Simulation 1	3	1345.7098	−694.4048	0.2000	0.8906
Simulation 2	3	1342.0179	−696.5030	0.2000	0.1093
Simulation 3	3	1342.1008	−703.2792	0.2000	0.0001
Simulation 4	3	1342.2013	−711.1841	0.2000	0.0000
Simulation 5	3	1342.2100	−719.2621	0.2000	0.0000

To ensure that Bayesian inference based on Markov Chain Monte Carlo (MCMC) sampling is valid, the author assessed the convergence of MCMC parameter estimates through visual diagnostic plots. The next step involves analyzing convergence through diagnostic checks to ensure the model's stationarity. The author will also conduct sensitivity analysis to ensure the robustness of the Bayesian regression model.

Figure 2 shows that the model is appropriate. The trace plots and autocorrelation plots are at low levels; the shapes of the plots are consistent with a normal distribution. Additionally, the plots indicate alignment with the simulated distribution density

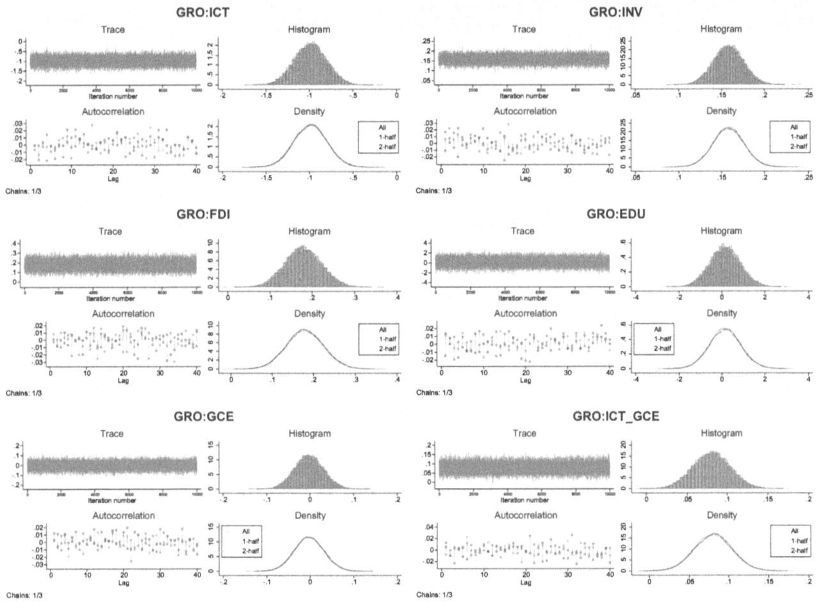

Fig. 2 The results of the *bayesgraph* diagnostics

Table 6 Regression results for Simulation 1

GRO	Mean	Std. dev.	MCSE	Median	95% cred. interval	
ICT	−0.98611	0.19024	0.00110	−0.98523	−1.36117	−0.61256
EDU	0.15252	0.73070	0.00424	0.14858	−1.28552	1.58393
GCE	−0.00201	0.03392	0.00019	−0.00197	−0.06853	0.06508
INV	0.15823	0.01777	0.00010	0.15805	0.12366	0.19330
FDI	0.17927	0.04409	0.00025	0.17921	0.09333	0.26629
ICT_GCE	0.08039	0.02357	0.00013	0.08041	0.03415	0.12642
Avg acceptance rate		1				
Avg efficiency:		0.9781				
Max Gelman-Rubin R_c		1				

and reflect that all lags are within effective bounds, confirming the robustness of the Bayesian statistics.

The results in Table 6 show that the average acceptance rate of the models is consistently 1. The minimum efficiency of the models is 0.9781, so it can be concluded that the models meet the required standards. Moreover, the Monte Carlo Standard Error (MCSE) is very small, and according to [28], as the MCSE approaches zero, the Markov Chain Monte Carlo (MCMC) chain becomes more stable. Additionally, the

Table 7 The results of the *bayestest* interval

	Mean	Dev MCSE	MCSE
prob (GRO:ICT) < 0	1.00000	0.00000	0.00000
prob (GRO:EDU) > 0	0.58263	0.49313	0.00286
prob (GRO:GCE) < 0	0.52416	0.49942	0.00288
prob (GRO:INV) > 0	1.00000	0.00000	0.00000
prob (GRO:FDI) > 0	1.00000	0.00000	0.00000
prob (GRO:ICT_GCE) > 0	0.99970	0.01732	0.00010

optimal R_c values for all parameters are 1, confirming that the MCMC chain meets the convergence requirement because, according to [29, 30], if the R_c diagnostic value for any model parameter is greater than 1.2, it is considered non-convergent.

The regression results in Table 6 indicate that ICT and GCE have a negative impact on growth, but ICT is statistically more significant due to a tighter confidence interval, while GCE has a relatively large standard deviation and a wide confidence interval, indicating greater volatility and difficulty in determining its precise impact. In contrast, EDU, INV, FDI, ICT_GCE have positive effects with varying degrees of influence; however, INV, FDI, ICT_GCE play a more important role in driving growth due to their larger regression coefficients, positive and stable confidence intervals, and small Monte Carlo Standard Error, whereas EDU has a wide confidence interval, suggesting that the impact of openness is unclear. To examine the probability of the independent variables affecting the dependent variable, specifically how much, the authors calculated the impact probabilities, which are presented in Table 7.

The analysis results in Table 7 show that there is a 100% probability that ICT has a negative impact on economic growth in Southeast Asian countries during the period 1990-2022. This result is consistent with previous studies such as [21, 22, 24]. These findings indicate that the productivity paradox has emerged in this region. This could be because, in Southeast Asian countries, the investment and application of ICT in the production and service sectors do not automatically yield immediate results due to the unsynchronized application of technology or the lack of suitable technical infrastructure and human resources, leading to wasted resources. The technical infrastructure, workforce skills, and the restructuring process of enterprises are not ready to effectively absorb new technologies. For these reasons, ICT does not immediately increase economic productivity but instead creates cost burdens for businesses and governments, thus slowing down economic growth. Additionally, ICT may widen the gap between social classes and economic sectors. Regions or groups that are unable to access ICT may be left behind, leading to inequality and hindering overall development. Furthermore, the introduction of digitalization and automation processes reduces the demand for labour, causing unemployment and short-term economic instability.

Thus, while ICT offers great potential, if not implemented and managed effectively, it still leads to high costs and reduced productivity, resulting in negative impacts on the economy. In addition, GCE (government consumption expenditure) also negatively impacts economic growth, accounting for 52%, indicating that, on average, public spending in Southeast Asian countries is inefficient. This could be due to inappropriate allocation or focus on sectors that do not generate economic value, leading to wasted resources and negative impacts on economic growth.

Interestingly, while ICT and GCE individually have negative effects on economic growth when not properly managed and implemented, the interaction variable between ICT and GCE creates a 100% positive effect on economic growth because the government can play a supporting, encouraging, and regulatory role in the ICT implementation process. Proper public spending can help ICT reach its full potential, thereby promoting effective and sustainable economic growth.

Furthermore, both INV (investment) and FDI (foreign direct investment) have a 100% positive impact on economic growth. In practice, INV and FDI play crucial roles in driving economic growth in Southeast Asian countries, particularly by enhancing infrastructure, improving labour skills, transferring technology, and creating jobs, which help boost productivity and increase exports.

The study's results also show that EDU (education) only has a 58% positive effect on economic growth, suggesting that although education plays an important role in driving economic growth, the quality and effectiveness of the educational system in the region have not yet reached their optimal levels. This may be due to the education systems in the region not fully meeting labor market demands, with the workforce's skills after training still lacking and weak, thus not contributing to key economic sectors.

Alternatively, there may be disparities in education quality among Southeast Asian countries. These nations are at different stages of development, leading to uneven resource allocation within their educational systems. For example, Singapore and Malaysia have more developed educational systems, while countries like Laos, Cambodia, and Indonesia face challenges regarding infrastructure, teacher quality, and financial resources.

Therefore, to gain deeper and more specific insights into the extent of the impact of ICT and other factors in the research model on economic growth in each country, the author continues with a Bayestest interval analysis, with the results shown in Table 8.

The research results in Table 8 indicate that the probability of the impact of variables on economic growth varies by country. For ICT, there is a negative impact on economic growth in all countries except Malaysia, where it is positive. This finding suggests that investing in ICT does not always yield positive outcomes. According to the Productivity Paradox, investing in information technology requires time for implementation, workforce training, and process optimization, leading to negative short-term impacts. Moreover, many IT projects do not meet requirements, resulting in wasted resources. Investing in technology without a clear plan can increase costs without a corresponding increase in productivity. Additionally, Southeast Asian

Table 8 Results of probability impact analysis in each country

GRO	Prob (ICT)	Prob (EDU)	Prob (GCE)	Prob (INV)	Prob (FDI)	Prob (ICT_GCE)
Singapore	−0.8702	+0.7294	+0.6852	+0.9996	+0.9932	−0.6133
Brunei	−0.91006	+0.5806	+0.9403	+0.9213	+0.7481	−0.8574
Cambodia	−0.8257	+0.5960	−0.0967	+0.9722	+0.9749	+0.9032
Indonesia	−0.7207	−0.5957	+0.6791	+1.0000	+0.9952	−0.8862
Laos	−0.9977	+0.7936	+0.6457	−0.6302	+1.0000	+0.6159
Malaysia	+0.6135	+0.7804	+0.7469	+1.0000	+0.9999	−0.7393
Philippines	−0.8198	+0.7751	−0.7701	+1.0000	+0.9421	−0.6888
Thailand	−0.6508	+0.6050	+0.7209	+1.0000	+0.9896	−0.8616
Vietnam	−0.7323	+0.7504	+0.9679	+0.9140	+0.9920	+0.5093

countries may lack appropriate policies to support the integration of technology into the economy and may face shortcomings in managing and executing IT projects.

In contrast to the other countries, the impact of ICT is positive for economic growth in Malaysia. Malaysia may have developed and implemented effective IT support policies, allowing them to reap the benefits of IT investment, leading to positive economic growth. Furthermore, Malaysia is an attractive destination for foreign technology companies. Multinational corporations such as Intel, IBM, and Microsoft are all present in Malaysia. The influx of FDI into the technology sector has brought not only capital investment but also technology transfer, knowledge, and skills. Additionally, the Malaysian government has established robust support policies for the ICT sector through programs and initiatives, such as the Digital Economy Program launched in 1996, which has been expanded and further developed in 2021 with a roadmap for digital economic transformation to make Malaysia a country with a highly developed digital economy by 2030.

The analysis in Table 8 also shows that EDU positively impacts economic growth, except in Indonesia. In Singapore, the probability of the impact of EDU on economic growth is 73%, which is understandable given Singapore's high-quality and developed education system that provides skilled and knowledgeable human resources for the economy. Other countries with an EDU impact probability on economic growth above 70% include Laos, Malaysia, the Philippines, and Vietnam. In reality, the governments of Laos, Malaysia, the Philippines, and Vietnam have considered education a core element for sustainable development and poverty reduction. Singapore, Brunei, Thailand, Vietnam, Malaysia, and the Philippines are among the countries that invest heavily in developing a high-quality education system. Education is regarded as the key to developing a quality workforce, contributing to economic growth. Meanwhile, Laos and Cambodia, while not as developed as the aforementioned countries, have focused on basic education and vocational training, improving workforce quality and boosting economic productivity. On the other hand, Indonesia has a negative impact probability on economic growth of 59%, which may be due to its heavy reliance on the mining and agriculture industries. These sectors typically do not require high labour skills, making education less crucial for driving economic growth. Furthermore, despite increased investment in education, the investment level remains insufficient to improve the quality of education across all regions, particularly in remote areas of Indonesia. This results in many workers being inadequately trained and unable to contribute effectively to economic growth in Indonesia.

The analysis results in Table 8 further show that government expenditure (GCE) positively impacts economic growth in several countries, including Singapore, Brunei, Indonesia, Laos, Malaysia, Thailand, and Vietnam, while negatively affecting countries like Cambodia and the Philippines. This indicates that countries such as Brunei and Vietnam have made reasonable and effective government investments to promote economic growth, with positive impact probabilities on economic growth in these two countries being 94% and 96%, respectively. In contrast, the Philippines and Cambodia struggle to optimize public expenditure, with the Philippines showing the lowest impact probability on growth at −77%, reflecting inefficient management or waste in the state budget, leading to decreased effectiveness of public investment. Although Cambodia's impact is not as low as the Philippines, its GCE still has a negative impact on economic growth at 9.6%, possibly due to ineffective budget usage. Thus, ensuring that government spending is reasonable and transparent can foster economic growth in these countries.

The analysis results in Table 8 also indicate that investment generally yields positive results for economic growth, except in Laos. While investment usually has a positive effect on economic growth, in Laos, this may be due to inefficiencies in investment management, a focus on traditional industries, and institutional issues that can negatively impact domestic investment's contribution to economic growth. Additionally, it is undeniable that FDI consistently contributes positively to economic growth in countries, with a high probability of impact nearing 100%. An interesting finding from the research is that 6 out of 9 countries utilizing government expenditure for ICT have not done so effectively, including Singapore and Malaysia. This suggests that public spending on ICT projects may be inefficient and that there is a lack of capacity for technology implementation. Particularly in Southeast Asian countries, which are developing (excluding Singapore), the basic needs for developing physical infrastructure, healthcare, and education may be more critical than investing in ICT. If governments focus too heavily on ICT without meeting these basic needs, the combination of ICT and public spending may not yield significant benefits and could even reduce economic growth.

5 Conclusion and Policy Implications

In this study, the author focuses on analyzing the impact of information and communication technology (ICT) on economic growth in Southeast Asian countries. The author utilizes Principal Component Analysis (PCA) to calculate ICT based on

four indicators: fixed telephone subscriptions, fixed broadband subscriptions, mobile cellular subscriptions, and individuals using the Internet.

The research findings reveal an interesting phenomenon of the productivity paradox in Southeast Asian countries, where ICT has a negative impact on economic growth. However, when government expenditure interacts with ICT, it positively influences economic growth, indicating that effective government spending on ICT projects can lead to numerous benefits. Therefore, governments in Southeast Asian countries need to make reasonable investments in technology to establish a foundation for long-term economic development, fostering innovation and enhancing the efficiency and competitiveness of their economies in an increasingly digital context. Additionally, the quality of education is another factor contributing to economic growth. This is particularly evident in countries that have made strong investments in developing a high-quality education system, which is key to cultivating a skilled workforce that contributes to economic growth. In Indonesia, despite the government recognizing the importance of education, the implementation of educational policies faces challenges, lacking coherence and effectiveness, which hinders the positive impact of education on the country's economic growth. To promote economic growth in Southeast Asian countries, governments should enhance ICT projects across various sectors and support the integration of digital technology into production and business processes. Creating a transparent investment environment with stable policies is crucial, alongside developing strategies to attract foreign direct investment (FDI) in high-tech and sustainable industries. Additionally, it is essential to implement policies aimed at improving the quality of education, particularly in vocational and higher education, to cultivate a skilled workforce. Finally, optimizing the use of ICT in economic growth necessitates increased investment in ICT infrastructure and the effective, strategic leveraging of technology.

Regarding some policy implications for each country based on the research results, the author proposes the following recommendations. For Singapore, it is essential to focus on optimizing the combination of government spending and investment in ICT to enhance productivity and innovation. For Brunei, the government should increase spending on education and ICT to develop a high-quality workforce while creating an investment environment and developing high-tech industries to attract FDI. In Cambodia, foreign investment should be leveraged to develop ICT infrastructure and optimize public spending on ICT projects effectively. For Indonesia, there should be a focus on educational reform to improve training quality while optimizing investment in ICT to enhance economic efficiency and reduce dependence on traditional industries. In Laos, there should be further promotion of ICT projects and reforms, along with the development of educational programs to train high-quality human resources. Additionally, the government needs to strengthen support for promoting investment in technology and improve project management. For Malaysia, the government should continue to promote investment in ICT and education to sustain sustainable development. In the Philippines, effective improvement of public spending and investment in education is needed to enhance quality and efficiency. For Thailand, it is crucial to coordinate public spending on ICT effectively. For Vietnam, there should be a push to enhance investment in ICT and develop vocational education programs to improve skills while further leveraging opportunities from FDI to develop the technology sector and improve ICT infrastructure.

References

1. Roach, S.S.: Services under siege-the restructuring imperative. Harv. Bus. Rev. **69**(5), 82–91 (1991)
2. Mankiw, N.G., Romer, D., Weil, D.N.: A contribution to the empirics of economic growth. Q. J. Econ. **107**(2), 407–437 (1992)
3. Nasab, E.H., Aghaei, M.: The effect of ict on economic growth: further evidence. Int. Bull. Bus. Admin. **5**(2), 46–56 (2009)
4. Nguyen, C.P., Doytch, N.: The impact of ict patents on economic growth: an international evidence. Telecommun. Policy **46**(5), 102291 (2022)
5. Solow, R.M.: We'd better watch out. New York Rev. Books **12**, 36 (1987)
6. Solow, R.M.: A contribution to the theory of economic growth. Q. J. Econ. **70**(1), 65–94 (1956)
7. Romer, P.M.: Increasing returns and long-run growth. J. Polit. Econ. **94**(5), 1002–1037 (1986)
8. Lucas, R.E.: On the mechanics of economic development. J. Monet. Econ. **22**(1), 3–42 (1988)
9. Schumpeter, J.A.: The Theory of Economic Development: An Inquiry into Profits, Capital. Interest, and the Business Cycle. Harvard University Press, Credit (1934)
10. Grossman, G.M., Helpman, E.: Innovation and Growth in the Global Economy. MIT Press (1993)
11. Aitken, B.J., Harrison, A.E.: Do domestic firms benefit from direct foreign investment? Evidence from venezuela. Am. Econ. Rev. **89**(3), 605–618 (1999)
12. Jorgenson, D.W., Stiroh, K.J.: Information technology and growth. Am. Econ. Rev. **89**(2), 109–115 (1999)
13. Dewan, S., Kraemer, K.L.: Information technology and productivity: evidence from country-level data. Manage. Sci. **46**(4), 548–562 (2000)
14. Aghaei, M., Rezagholizadeh, M.: The impact of information and communication technology (ict) on economic growth in the oic countries. Econ. Environ. Stud. **17**(2 (42)), 257–278 (2017)
15. Andrianaivo, M., Kpodar, K.: Ict, financial inclusion, and growth: evidence from african countries. Unpublished (2011)
16. Hammed, Y.S., Ademosu, S.T.: Ict innovation, fdi and economic growth: Evidence from brics. J. Econ. Behav. Stud. **15**(2 (J)), 20–32 (2023)
17. Dumor, K., Shurong, Z., Dumor, H.K., Ampaw, E.M., Amouzou, E.K., Okae-Adjei, S., Boadi, E.K.: Evaluating the effect of ict on trade and economic growth from the perspective of Eastern African belt and road countries. Inf. Technol. Dev. **30**(3), 452–471 (2024)
18. Appiah-Otoo, I., Song, N.: The impact of ict on economic growth-comparing rich and poor countries. Telecommun. Policy **45**(2), 102082 (2021)
19. Inani, S.K., Tripathi, M.: The nexus of ict, electricity consumption and economic growth in India: an ardl approach. Int. J. Indian Culture Bus. Manag. **14**(4), 457–479 (2017)
20. Oladipo, O., Murphy, A., Anannsingh-Jamieson, F., Amaghionyeodiwe, L.: Information and communication technology, and economic growth in Nigeria: a time series analysis. Int. J. Technol. Policy Manage. **16**(2), 149–162 (2016)
21. Ishida, H.: The effect of ict development on economic growth and energy consumption in Japan. Telematics Inform. **32**(1), 79–88 (2015)
22. Pohjola, M.: The new economy: facts, impacts and policies. Inf. Econ. Policy **14**(2), 133–144 (2002)
23. Zhang, J., Danish, evidence from Asian economies: the dynamic linkage between information and communication technology, human development index, and economic growth. Environ. Sci. Pollut. Res. **26**, 26982–26990 (2019)

24. Niebel, T.: Ict and economic growth–comparing developing, emerging and developed countries. World Dev. **104**, 197–211 (2018)
25. Lee, S.-Y.T., Gholami, R., Tong, T.Y.: Time series analysis in the assessment of ict impact at the aggregate level–lessons and implications for the new economy. Inf. Manag. **42**(7), 1009–1022 (2005)
26. Albiman, M.M., Sulong, Z.: The role of ict use to the economic growth in sub saharan African region (SSA). J. Sci. Technol. Policy Manag. **7**(3), 306–329 (2016)
27. Block, J.H., Jaskiewicz, P., Miller, D.: Ownership versus management effects on performance in family and founder companies: a bayesian reconciliation. J. Fam. Bus. Strat. **2**(4), 232–245 (2011)
28. Flegal, J.M., Haran, M., Jones, G.L.: Markov chain monte carlo: can we trust the third significant figure? Stat. Sci. 250–260 (2008)
29. Gelman, A., Rubin, D.B.: Inference from iterative simulation using multiple sequences. Stat. Sci. **7**(4), 457–472 (1992)
30. Brooks, S.P., Gelman, A.: General methods for monitoring convergence of iterative simulations. J. Comput. Graph. Stat. **7**(4), 434–455 (1998)

Big Data Analytics for Financial Decisions of Companies: A Systematic Literature Review

Thi Thu Hoan Nguyen and Bao Ngoc Nguyen

Abstract This systematic literature review aims to explore the application of big data analytics in financial decision-making within companies, with a focus on investment and financing decisions. The review followed specific inclusion criteria to target only articles that are English, accessible, up-to-date, and relevant to the application of BDA, AI, ML, DL, or NN in financial decision-making at the firm level. Based on the PRISMA 2020 guidelines for systematic reviews, the study identified 48 studies out of 6,102 initial records found on Scopus, following study identification, screening, and eligibility checks, ensuring the broad coverage of BDA's applications in investment and financing decisions of companies. Among these, 92.9% of studies highlighted BDA's transformative role in enhancing investment decisions through its capabilities in improved credit risk evaluation, risk management, investment analysis, fraud detection, demand forecasting, inventory management, and working capital management, as well as financing decisions. Numerous studies consistently showed that investment and financing decisions of companies are facilitated by the adoption of BDA along with AI, ML, DL, and NN techniques. Limitations include methodological variability across studies and the lack of detailed datasets, making generalization challenging. Despite this, the findings emphasize BDA's potential to enhance financial decision-making, enabling firms to balance growth and operational stability. Future research should address biases and explore more specific applications of BDA in finance.

Keywords Big data analytics · Artificial intelligence · Machine learning · Deep learning · Neural networks · Financial decision-making · Decision-making

---

T. T. H. Nguyen (✉)
School of Finance and Accounting, Industrial University of Ho Chi Minh City, Ho Chi Minh City, Vietnam
e-mail: nguyenthithuhoan@iuh.edu.vn

B. N. Nguyen
Fulbright University Vietnam, Ho Chi Minh City, Vietnam

H. T. T. Nguyen et al. (eds.), *Data Science in Finance and Accounting*, Studies in Big Data 181, https://doi.org/10.1007/978-3-032-06179-9_17

1 Introduction

The rapid advancements in information technology have led to the availability of vast amounts of financial data, involving financial transactions, corporate-level financial and non-financial information, macroeconomic data, and estimates [1]. This surge of data has fueled the rise of big data analytics (BDA) which enables companies to translate complex datasets into actionable insights regarding markets and customer behaviors [2]. Studies emphasize that companies have increasingly invested in BDA to optimize their decision-making process, recognizing BDA as a critical tool for enhancing business performance [3–8]. These advancements in BDA enable companies to leverage actionable business intelligence for more informed and accurate decision-making [6, 9]. BDA is particularly influential in guiding business decisions of companies across various industries, including but not limited to finance [10], e-commerce [11], manufacturing [12, 13], and energy and utilities [14]. Its impacts are profound in facilitating predictive analytics and the development of sophisticated decision-making models [15].

Therefore, this paper follows the format of a systematic literature review to gain a comprehensive understanding and synthesize findings from various studies about how BDA is applied in corporate making financial decisions. This involves providing a snapshot of existing studies that are relevant to the research topic and presenting either consistent or disputed findings to provide a balanced view of the current state of research. By consolidating and synthesizing existing studies, this systematic literature review provides a comprehensive overview of how BDA is applied in financial decision-making at the firm level. This approach could help companies stay abreast of BDA developments and integrate such BDA practices into their businesses, contributing to the advancement of knowledge in the fields of BDA and finance.

The review is structured as follows: **Part 2** provides the research background, **Part 3** outlines the methodology, **Part 4** reviews the application of BDA in financial decisions within companies, **Part 5** synthesizes key findings to explore consistent and disputed results among included studies, **Part 6** summarizes the review while highlighting limitations and future research directions.

2 Background

Within the corporate context, previous studies have consistently emphasized how BDA has enhanced decision-making [3, 5, 6, 8]. Various studies highlight the significance of advanced technologies such as artificial intelligence (AI), machine learning (ML), deep learning (DL), and neural networks (NN) in the integration with BDA in informing the business decisions of companies across industries. For instance, Manta et al. [10] analyzed the impact of AI in informing strategic business decisions to anticipate evolving shifts in the banking market. Aslam et al. [16] explored the applications of AI and ML in fraud detection for insurance companies to mitigate financial risks. Also, Nibareke and Laassiri [17] explored the application of ML models to inform more precise decisions on flight schedules using BDA of historical flight records, weather records, and airport traffic data. Additionally, Lee, Kim, and Jung [18] explored DL-based stock market prediction models that incorporated environmental-social-governance (ESG) sentiments to interpret stock market trends and risks, thus guiding stakeholders to make more precise decisions. Moreover, Eachempati et al. [19] emphasized the role of NN in enhancing the transparency and interpretability of accounting disclosures, thereby informing more precise decisions of investors.

Despite the growing body of literature explicitly exploring how companies have adopted BDA to enhance their decision-making in their business domains, there is a lack of systematic literature review that works as a one-stop solution for researchers and professionals to leverage their understanding of BDA for financial decision-making of companies across industries. Thus, this study emerges as the intersection of BDA and financial decision-making of companies under the format of the systematic literature review, aiming to systematically identify and evaluate the literature using established inclusion criteria [20].

Thereby, within the scope of this review, BDA incorporates a wide range of techniques and technologies aimed at capturing, storing, transferring, analyzing, and visualizing large volumes of structured, unstructured, and other types of data [21, 22]. Additionally, BDA involves a wide range of techniques including AI, ML, DL, and NN, This is foundational to the advancement of decision-making within companies by leveraging complex huge datasets into more precise decisions. Thus, beyond the BDA, the study examines the state-of-the-art application of AI, ML, DL, and NN techniques which are under the umbrella of BDA for financial decision-making of companies due to its synergistic relationship with BDA [23]. In terms of financial decision-making, given the complex nature of interpretation, which could be defined differently from multiple perspectives, the study adopts the categorization of financial decisions proposed by Doumpos and Zopounidis [1] as illustrated in Fig. 1.

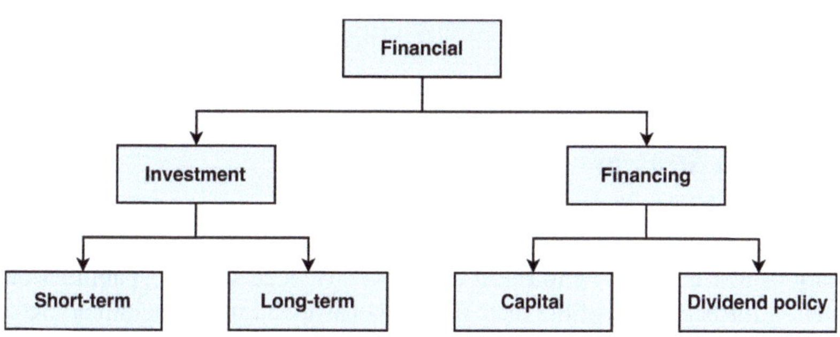

Fig. 1 Adapted from corporate financial decision categories by Doumpos and Zopounidis [1], this figure illustrates two primary types of financial decisions, serving as a foundation for exploring how BDA impacts the financial decisions of corporations

There are two major financial decisions including investment decisions and financing decisions [1, 24]. Within the scope of this review, we consider investment options to involve evaluating investment projects, allocating capital to these projects, and managing current investments using BDA. On top of that, financial decisions involve not only long-term investments that are typically associated with capital budgeting but also short-term investments that refer to working capital management such as managing inventory, finished products, cash, deposits, and receivables [25]. Besides, financial decisions also involve financing options that are associated with seeking and securing funds for both long-term investments and daily operation demands [1]. Notably, according to Doumpos and Zopounidis [1], it can be determined that the interpretation of financial decisions varies across different industries; however, in this review, it will be categorized into two main types of financial decision-making including (1) Investment decisions and (2) Financing decisions.

3 Methodology

The study employs a systematic review approach to examine the application of BDA in enhancing financial decision-making within companies. Adapted from the flow diagram of Moher et al. [26], a summary of the methodological framework is illustrated in Fig. 2, which outlines four key stages including (1) identification through database searching, (2) screening, (3) eligibility check, and (4) included studies for data synthesis.

In addition, the review is carried out in accordance with the Preferred Reporting Items for Systematic Reviews and Meta-Analyses (PRISMA) 2020 guidelines, which provide a standardized protocol for systematically collecting, screening, and synthesizing findings from studies that address the research question [27–29]. Additionally, guidance from the EQUATOR Network has been incorporated to enhance the reporting quality and ensure transparency in the review process [26, 27]. This combined approach ensures that this study is comprehensive with trustworthy and reproducible findings [26, 27].

3.1 Search Strategy

To provide broad coverage of interdisciplinary fields, Scopus was chosen to be the primary database for article retrieval due to its advanced search capabilities and its reputation for being one of the most comprehensive and multidisciplinary research databases available [30, 31]. Second, an advanced search strategy was carried out to identify studies on Scopus based on titles, abstracts, or author-specified keywords. Beyond the key search terms including *"Big Data Analytics"* and *"Financial Decision-making"* to capture the most relevant studies, the search query expanded to other keywords such as *"Artificial Intelligence"*, *"Machine Learning"*, *"Deep*

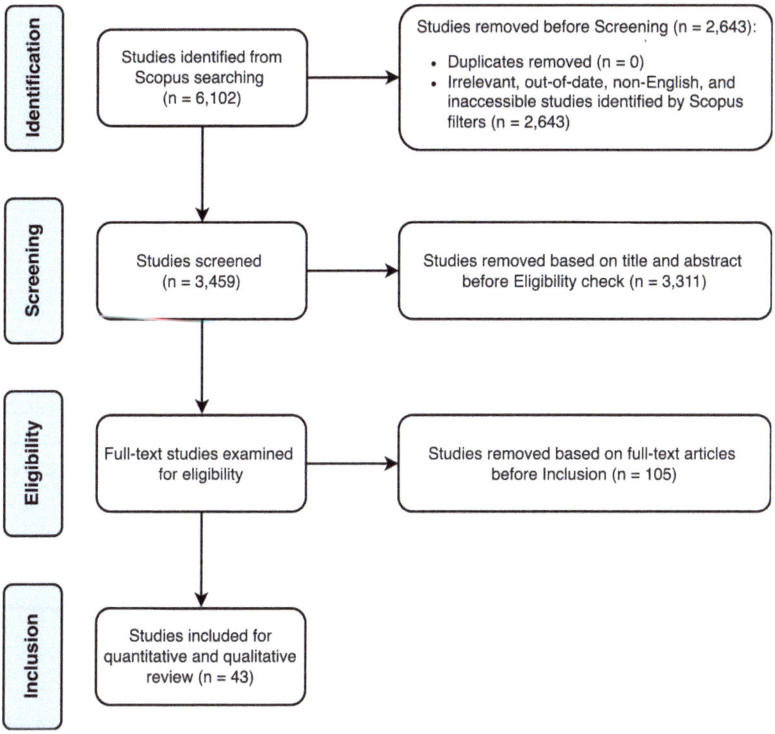

Fig. 2 Flow diagram adapted from Moher et al. [26] and the PRISMA 2020 template of Page et al. [27] for systematic review

Learning", *"Neural Networks"* due to their synergistic relationships with BDA [23]. These keywords were combined with the Boolean operators "AND" and "OR" with terms such as *"Financial Decision-making"* OR *"Corporate Finance"* OR *"Financial Decision"*. To assess the suitability of the search query, several relevant publications were manually selected and checked if they appeared in the search results. As a result, all of the manually selected studies were found in the sample, highlighting that the search query is appropriate.

3.2 *Identification of Studies for Review via Scopus*

The search strategy using keywords and Boolean operators initially identified 6,102 results on Scopus. The search results were refined using various database filters, including refining the search, publication year, document type, language of original document (English only), and open access (refer to Table 1). In detail, the publication year was restricted to between 2015 and 2024 to capture the latest findings about BDA

Table 1 Identification of studies via Scopus

No.	Filters of scopus	Filter criteria
1	Refine search (search within results)	Big data analytics for financial decision-making of company
2	Publication year	From 2015 to 2024
3	Document type	Article only
5	Language of original document	English
6	Open access	Yes

applications in financial decision-making of companies, in line with the increasing acceleration in scientific publications [32, 33]. The document type was restricted to only articles while excluding conference articles, dissertations, editorials, and reviews, following the standard research practices [34, 35]. Other filters such as refined search, language, and access status were defined to ensure the relevance and accessibility for a comprehensive review. This refinement within search results identified 3,459 studies from Scopus, which were then moved to the screening phase.

3.3 Screening Process

The screening phase aims to remove irrelevant articles that do not align with the scope of this review. During the screening of titles and abstracts, several articles were excluded as they did not relate to financial decision-making or delve into BDA, AI, ML DL, or NN practices within a corporate context. This limitation arose due to the broad interpretation of keywords in the search query. For example, the keyword *"Corporate Finance"* was expected to capture finance-related activities within companies; however, some articles refer to it in contexts such as corporate digital responsibility or corporate social responsibility, which are not relevant to this review. As a result, numerous articles were included in the Scopus identification but required exclusion due to their irrelevance with the review topic. In detail, out of 3,459 exported articles, a total of 2,718 articles were excluded since they did not relate to BDA, AI, ML, DL, or NN applications in financial decisions (see Fig. 1). Consequently, the remaining 741 articles were examined to determine whether they were relevant to any financial decisions at the firm level. Ultimately, there were 3,311 articles excluded due to their irrelevance with financial decisions or any BDA applications, leaving 148 articles for eligibility checks.

Table 2 Inclusion and exclusion criteria for final eligibility check

Criteria	
Inclusion criteria	• Studies must explicitly address the application of BDA, AI, ML, DL or NN in the financial decision-making of corporations/firms/companies/enterprises (within the private sector) in any industry
Exclusion criteria	• Studies focus on financial decision-making at the personal/individual level or state-owned level rather than corporations/firms/companies/enterprises •Studies focus on the decision-making processes of corporations, firms, or enterprises, but not specifically any of investment or financing decisions

3.4 Eligibility Check

Following the PRISMA 2020 of Page et al. [27], the third phase required an eligibility check that involved a thorough full-text scanning of 148 articles to examine their appropriateness for inclusion. During this phase, 148 articles were evaluated according to the pre-defined inclusion and exclusion criteria (see Table 2), to ensure the selection of the most relevant studies for this review. Each article was thoroughly examined based on these criteria, resulting in the final set of studies included in the review (43 publications).

Overall, the search strategy resulted in 6,102 initial records. Following the study selection flow diagram adapted from the PRISMA 2020 template of Page et al. [27] (see Fig. 2), 43 publications were selected for review. In line with Stoll et al. [36] to avoid mistakenly excluding relevant studies, researchers independently conducted dual reviews during both the title/abstract and full-text screening stages. This approach ensured a thorough evaluation and increased the number of studies included in the systematic review. The final selected publications are in English, accessible, up-to-date, and relevant to the application of BDA, AI, ML, DL, or NN in financial decision-making at the firm level.

3.5 Data Synthesis

To ensure that based on high-quality data and valid study findings, the data extraction process adhered to the systematic review principles outlined by Li, Higgins, and Deeks [37]. The guideline ensures that the included studies are well-documented, supports the development of tables and figures, mitigates bias risks, and enables data synthesis [37].

Additionally, the research team collaboratively designed a data extraction template to capture necessary information from the included studies, such as title, abstract, publication year, journal, source title, and other key research attributes to present results. These attributes include the dataset used for BDA, industry focus, and the specific application of BDA in investment or financing decisions at the firm level. The template was thoroughly assessed for completeness by randomly selecting three studies from the final set of the included studies with both researchers independently extracting data. A few discrepancies arose, but they were resolved through discussion within the research team. For instance, disagreements primarily over whether to group certain financial decisions under umbrella terms. Once the data extraction template for systematic review was finalized to ensure a coherent dataset, multiple researchers worked independently with one researcher extracting data and another verifying it, in line with the dual-review approach recommended by Stoll et al. [36] and the data extraction process of Li, Higgins and Deeks [37].

4 Results

4.1 Snapshot of Included Studies

The review starts by presenting key statistics about the current landscape of research on BDA for financial decision-making of companies, summarizing a number and percentage of studies published between 2015 and 2024 (see Table 3). Notably, there has been a significant surge in research on this topic since 2022, highlighting its growing significance in this research topic (37.5%).

Additionally, it can be seen that the distribution of the 43 reviewed studies is across various academic journals. Most journals published a single study (2.3% each), while Applied Mathematics and Nonlinear Sciences and IEEE Access had the highest representation, each accounting for 9.3% of the total (see Table 4). This dispersion indicates a wide interdisciplinary interest in the research topic across operations research, computer science, business, and management domains.

Last but not least, look at Table 5 for categorizing 43 studies based on the quartile rankings from Q1, Q2 to Q3 of the journals (based on the Scimago Journal Rank (SJR)) in which they were published. Over half of the studies (51.2%) appeared in Q1 journals, reflecting a strong presence in high-impact outlets. Q2 journals accounted for 16.3% of the publications, while 32.6% were published in Q3 journals. This distribution highlights the overall quality and academic credibility of the sources included in the review.

Table 3 Distribution of studies by year of publication

Year of publication	#	%	Studies
2015	1	2.3%	Khashei and Mirahmadi [38]
2016	2	4.7%	Maté et al. [39], Zhu et al. [40]
2018	1	2.3%	Acharya et al. [41]
2019	1	2.3%	Antonio et al. [42]
2020	6	14.0%	Hu et al. [43], Köchling and Wehner [44], Mariani and Samuel [45], Mcmahon et al. [46], Nibareke and Laassiri [17], Seyedan and Mafakheri [47]
2021	2	4.7%	Ala'raj et al. [48], Xu et al. [49]
2022	16	37.2%	Ahmad et al. [14], Al-Alwan et al. [50], Ali et al. [51], Aslam et al. [16], Beltrán et al. [52], Blohm et al. [53], Gaytan et al. [54], Karn et al. [55], Kyeong et al. [56], Mourmouris and Poufinas [57], Sharma et al. [58], Sujith et al. [59], Xiuguo and Shengyong [60], Zhang et al. [61], Zhu et al. [62]
2023	8	18.6%	Gravili et al. [63], Gu [64], Gu and Zhou [65], Peng and Li [66], Shen et al. [67], Tong and Tian [68], Xu et al. [69]
2024	6	14.0%	Akour et al. [70], Giannakopoulos et al. [71], Khan et al. [71], Khan et al. [72], Wenting et al. [73], Zhang [74]
Grand Total	**43**	**100.0%**	

4.2 *Applications of Big Data Analytics in Investment Decisions of Companies*

Table 6 also indicates the financial decisions across the 43 reviewed studies. Out of the 43 total studies, only four (9.3%) addressed this area with two studies were situated in the finance industry, one focused on the information technology sector, and another fell into a miscellaneous category. This limited distribution suggests that financing decisions have received relatively low attention across industries within the reviewed literature.

Meanwhile, the remaining 39 reviewed studies focus on investment decisions with industry-wise distribution (see Table 6). The finance sector is the most represented with 34.9% of the studies, followed by energy utilities, human resources, and information technology, each representing 9.3%. Other industries, such as agriculture, aviation, manufacturing, and pharmaceuticals, are less represented, each accounting for 2.3–4.7% of the total. This distribution shows that while finance dominates investment decision research, there is a growing interest across a diverse range of industries.

Table 4 Distribution of studies by journal

Journal name	#	%	Studies
Annals of Operations Research	1	2.3%	Sharma et al. [58]
Applied Energy	1	2.3%	Beltrán et al. [52]
Applied Mathematics and Nonlinear Sciences	4	9.3%	Gu and Zhou [64], Xu et al. [69], Zhang [74]
Business Research	1	2.3%	Köchling and Wehner [44]
Computational Intelligence and Neuroscience	1	2.3%	Gaytan et al. [54]
Computers, Materials and Continua	1	2.3%	Shen, Zhang and Song [67]
Cornell Hospitality Quarterly	1	2.3%	Antonio, de Almeida and Nunes [42]
Entrepreneurship: Theory and Practice	1	2.3%	Blohm et al. [53]
Entropy	1	2.3%	Zhu et al. [40]
European Journal of Innovation Management	1	2.3%	Gravili et al. [63]
Future Generation Computer Systems	1	2.3%	Maté et al. [39]
Health Systems	1	2.3%	Mourmouris and Poufinas [57]
IEEE Access	4	9.3%	Ahmad et al. [14], Mcmahon et al. [46], Xiuguo and Shengyong [60]
Information (Switzerland)	2	4.7%	Giannakopoulos et al. [71], Zhu et al. [62]
Intelligent Automation and Soft Computing	1	2.3%	Ali et al. [51]
International Journal of Business Intelligence Research	1	2.3%	Khan et al. [72]
International Journal of Data and Network Science	2	4.7%	Akour et al. [70], Al-Alwan et al. [50]
International Journal of Financial Studies	1	2.3%	Khashei and Mirahmadi [38]
International Journal of Information Management	1	2.3%	Acharya et al. [41]
Journal of Big Data	3	7.0%	Ala'raj, Abbod and Majdalawieh [48], Nibareke and Laassiri [17], Seyedan and Mafakheri [47]
Journal of Business Research	1	2.3%	Mariani and Samuel [45]
Journal of Food Quality	1	2.3%	Sujith et al. [59]
Journal of Intelligent Systems	1	2.3%	Tong and Tian [68]
Journal of Organizational and End User Computing	1	2.3%	Wenting et al. [73]
Malaysian Journal of Computer Science	1	2.3%	Karn et al. [55]
Mathematics	1	2.3%	Zhang, Xiao and Niu [61]
Problems and Perspectives in Management	1	2.3%	Khan, Faisal and Thomas [75]
Research in International Business and Finance	1	2.3%	Aslam et al. [16]
Scientific Reports	2	4.7%	Gu [64], Xu et al. [49]
Sustainability (Switzerland)	2	4.7%	Kyeong, Kim and Shin [56], Peng and Li [66]
Technological and Economic Development of Economy	1	2.3%	Hu et al. [43]
Grand Total	**43**	**100.0%**	

Table 5 Distribution of studies by journal quartile ranking

Journal Quantile	#	%	Studies
1	22	51.2%	Akour et al. [70], Ala'raj et al. [48], Al-Alwan et al. [50], Ali et al. [51], Antonio et al. [42], Beltrán et al. [52], Blohm et al. [53], Gaytan et al. [54], Giannakopoulos et al. [71], Gravili et al. [63], Gu [65], Köchling and Wehner [44], Kyeong et al. [56], Mariani and Samuel [45], Maté et al. [39], Nibareke and Laassiri [17], Peng and Li [66], Seyedan and Mafakheri [47], Sharma et al. [58], Xiuguo and Shengyong [60], Xu et al. [49], Zhu et al. [62]
2	7	16.3%	Aslam et al. [16], Hu et al. [43], Karn et al. [55], Khashei and Mirahmadi [38], Shen et al. [67], Sujith et al. [59], Zhu et al. [40]
3	14	32.6%	Acharya et al. [41], Ahmad et al. [14], Gu and Zhou [65], Khan et al. [72], Khan et al. [75], Mcmahon et al. [46], Mourmouris and Poufinas [57], Tong and Tian [68], Wenting et al. [73], Xiuguo and Shengyong [60], Xu et al. [69], Zhang [74], Zhang et al. [61]
Grand Total	**43**	**100.0%**	

4.2.1 Credit Risk Evaluation and Risk Management

Credit risk evaluation and risk management are critical factors in the investment decisions of banks and financial institutions, as even a slight improvement in forecasting accuracy can significantly reduce credit risks. Even small performance enhancements of credit scoring directly impact the bank investment by leading to a significant increase in profit due to the high volume of loans issued based on credit scores [38].

On top of that, numerous studies by Khashei and Mirahmadi [38], Zhu et al. [40], Meshref [76], Kyeong, Kim, and Shin [56], Xu, Lu and Xie [49], Karn et al. [55], Gu [64], and Shen, Zhang, and Song [67] have explored the use of BDA for enhancing credit scoring models that support the risk evaluation and, consequently, optimizing the investment decisions for banks and other financial institutions. However, the adoption of BDA for credit scoring and risk management varies significantly depending on the techniques, models, and datasets utilized across studies and applications (refer to Table 7). These discrepancies arise from the unique needs of financial institutions, datasets, and their quality, and the complexity of financial decision-making processes.

Khashei and Mirahmadi [38] used Australian credit card data to develop hybrid soft intelligent models that improved credit scoring, supporting binary financial decisions. Similarly, Karn et al. [55] illustrated the power of DL techniques such as Gated Recurrent Units (GRU) and Convolutional NN (CNN) to outperform traditional ML models (including support vendor machines, random forests, multi-layer perceptron

Table 6 Distribution of studies by industry focus

Industry focus	#	%	Studies
Financing decisions			
Finance	2	4.7%	Gaytan et al. [54]
Information technology	1	2.3%	Information technology
Miscellaneous	1	2.3%	Zhu, Yan and Bai [62]
Investment decisions			
Agriculture	1	2.3%	Giannakopoulos et al. [71]
Aviation	1	2.3%	Nibareke and Laassiri [17]
Consumer goods	1	2.3%	Mariani and Samuel [45]
Energy utility	4	9.3%	Ahmad et al. [14], Beltrán et al. [52], Maté et al. [39], Sharma et al. [58]
Fashion	1	2.3%	Acharya et al. [41]
Finance	15	34.9%	Ala'raj et al. [48], Aslam et al. [16], Blohm et al. [53], Gu [64], Hu et al. [43], Karn et al. [55], Khan et al. [72], Khashei and Mirahmadi [38], Kyeong et al. [56], Mourmouris and Poufinas [57], Shen et al. [67], Xiuguo and Shengyong [60], Xu et al. [49], Zhang et al. [61]
Hospitality	1	2.3%	Antonio, de Almeida and Nunes [42]
Human resources	4	9.3%	Gravili et al. [63], Khan et al. [75], Köchling and Wehner [44], Wenting et al. [73]
Information technology	4	9.3%	Gu and Zhou [65], Tong and Tian [68], Xu et al. [69]
Manufacturing	1	2.3%	Peng and Li [66]
Miscellaneous	1	2.3%	Sujith et al. [59]
Pharmaceutical	1	2.3%	Akour et al. [70]
Supply chain	2	4.7%	Ali et al. [51], Seyedan and Mafakheri [47]
Telecommunication	1	2.3%	Al-Alwan et al. [50]
Transportation	1	2.3%	Mcmahon et al. [46]
Grand Total	**43**	**100.0%**	

NN) in predicting overdue payments of credit card customers. These studies suggest the adoption of BDA could result in more precise credit risk management and investment decisions, yielding a substantial profit increase. To support this finding, Kyeong, Kim, and Shin [56] leveraged real-time system log data from KakaoBank in South Korea to showcase how text mining and ML techniques could enhance credit scoring models, introducing a credit evaluation approach for banks to assess the financial health and risk profile of potential investments. In a complementary effort, Shen, Zhang, and Song [67] adopted electricity consumption data from an

Table 7 Investment decisions of companies derived from credit scoring and risk management using BDA (Authors' own analysis)

Studies	Dataset	Industry focus	Decision-making focus	Model(s)/Technique(s)
Khashei and Mirahmadi [38]	Australian credit card data	Finance	Credit scoring	Hybrid soft intelligent classification model; Artificial NN
Karn et al. [55]	Not explicitly mentioned	Finance	Credit risk evaluation	DL-based technique; Gated recurrent unit; Convolutional NN
Zhu et al. [40]	Not explicitly mentioned	Supply chain	Credit risk prediction	Random Subspace and Real AdaBoost integrated ensemble ML; Decision Trees
Gu [64]	Not explicitly mentioned	Finance	Risk prediction	Back Propagation NN; Particle Swarm Optimization; Convolutional NN
Kyeong et al. [56]	System log data of KakaoBank, capturing customer online behaviors in real-time.	Finance	Credit scoring	Text mining; ML
Shen et al. [67]	Electricity consumption data recorded in an Internet of Things (IoT) cloud center.	Energy utility	Credit scoring and risk assessment	Ranking Extreme Gradient Boosting Net (RankXGB)
Xu et al. [49]	Renrendai dataset including observations and features such as credit score, borrower certification level and asset ownership	Finance	Credit scoring and default risk prediction	Gradient Boosting Model; NN; Extreme Gradient Boosting Tree; Random Forest
Meshref [76]	Bank marketing dataset from Kaggle	Finance	Loan approval prediction and decisions	Bagging and Boosting ensemble ML

IoT cloud center to generate credit scores through a two-stage DL task architecture and, consequently, enables swift and precise credit evaluations for enterprises at a fast pace to make informed decisions. These findings further demonstrate the diverse applications of BDA in investment decision-making.

Zhu et al. [40] further advanced this conversation by adopting an integrated ensemble ML, which is Random Subspace and Real AdaBoost (RS-RAB), for optimizing credit risk prediction in the Chinese supply chain sector. This model indicates how BDA could enhance risk evaluation, a critical component in making informed credit lending decisions. Similarly, Gu [64] adopted Back Propagation NN and Particle Swarm Optimization to improve financial risk prediction, addressing imbalances in financial management systems.

Additionally, Xu, Lu, and Xie [49] adopted NN and ML techniques such as Gradient Boosting Models, NN, and Random Forests to forecast loan defaults using the Renrendai dataset in China. Their work emphasizes how BDA helps identify potential risks associated with borrowers, thereby refining the investment decisions of banks and peer-to-peer companies. Meanwhile, Meshref [76] highlighted the predictive capabilities of ML algorithms such as Bagging and Boosting with Kaggle's bank marketing dataset that enhance loan approval predictions and facilitate informed investment decisions.

4.2.2 Investment Analysis and Evaluation

Beyond the credit risk evaluation and risk management, within the finance industry, BDA plays a pivotal role in enhancing investment decisions by assessing investment opportunities [53, 77], identifying information asymmetry [78], predicting investment performance [61, 71] and trading trends [75], referring to Table 8.

Setty, Elovici, and Schwartz [77] demonstrated how ML models, including Random Forest, Naive Bayes, DT, Gradient Boosting, and K-Nearest Neighbors could improve early-stage investment decisions for venture capital funds and private equity firms. By collecting, processing, and analyzing vast amounts of complex financial and operational data of different startups, BDA is shown to identify promising investment opportunities and mitigate risks associated with early-stage investments. This BDA approach informs investment decisions of VC funds and PE firms, as also supported by Blohm et al. [53], who found that the ML algorithm such as Extreme Gradient Boosted Trees could outperform experienced investors by possibly avoiding cognitive biases in investment decision-making. Besides, Alshehadeh et al. [78] indicated how the application of BDA using data from financial statements from companies listed on the Amman Stock Exchange could mitigate information asymmetry, thereby improving the quality of investment decisions.

Moreover, in the agricultural sector, Giannakopoulos et al. [71] highlighted the combination of BDA with AI models such as Fuzzy Cognitive Mapping and Artificial NN to forecast prices of farming, chemicals, and machinery, aiding investment decisions in agricultural firms. Meanwhile, Zhang, Xiao, and Niu [61] employed ML techniques including Data Envelopment Analysis, Support Vector Regression, and Back Propagation NN to predict company performance within banks and financial institutions. Their study demonstrates how BDA could assess the efficiency and effectiveness of potential investment targets. Moreover, Khan et al. [72] also highlighted the use of ML techniques such as Support Vector Regression, Recurrent NN, and Long Short-Term Memory for predicting currency trading trends, thus enhancing investment decisions in foreign exchange markets. Their findings indicate the significance of predictive modeling in navigating the complexities of currency trading, further amplifying how BDA could refine currency investment decisions.

Table 8 Investment decisions of companies using BDA (Authors' own analysis)

Studies	Dataset	Industry focus	Investment decision	Model(s)/Technique(s)
Blohm et al. [53]	Data obtained from Twitter, LinkedIn, Google Trends, and national trade registers	Information & communications technology	Venture profile analysis	Extreme Gradient Boosted Trees
Zhang et al. [61]	Regional carbon dioxide emissions data	Environment	Carbon-emission performance prediction	Data Envelopment Analysis; Slack-Based Measure; Linear Regression; Support Vector Regression; Back Propagation NN; Decision Trees; Random Forests; Gradient-boosting Decision Trees; Adaboost and Bagging
Alsheh-adeh et al. [78]	Financial statements of industrial companies on Amman Stock Exchange	Finance	Enhancing investment decision via mitigation of information asymmetry	BDA
Setty et al. [77]	Dataset of startups, funding sources, investment rounds, and performance	Finance	Investment return prediction	Random Forest; Naive Bayes; Decision Trees; Gradient Boosting; K-Nearest Neighbors
Giannak-opoulos et al. [71]	Website analytics from established agriculture firms and agro-Economic indexes such as agriculture employment rate, chemical product price index, farm product price index, and machinery equipment price index	Agriculture	Enhanced digital marketing investment decisions.	Fuzzy cognitive mapping; Hybrid modeling; Artificial NN
Khan et al. [75]	Real-time exchange data obtained from Nasdaq Data Link Application Programming Interfaces	Finance	Currency price prediction and enhanced trading decisions	Support Vector Regression; Recurrent NN; Long Short-Term Memory

4.2.3 Fraud Detection

In addition, like credit risk evaluation and risk management, fraud detection is known as a critical factor for sound investment decisions since fraud may result in continual portfolio losses [58]. On top of that, refer to Table 9, it can be seen that BDA plays a pivotal role in enhancing the accuracy of fraud detection in the audit, identifying and detecting fraud, as demonstrated by studies such as Hu et al. [43], Xiuguo and Shengyong [60], Aslam et al. [16], and Raval et al. [79].

BDA plays a critical role in detecting fraud, as exemplified by the studies conducted by Hu et al. [43], Xiuguo and Shengyong [60], Aslam et al. [16], and Raval et al. [79]. Hu et al. [43] introduced an AI-enabled auditing technique, which is a Fusion Multiple Rule-Based Decision-Making model, to improve audit quality and financial assessments, thereby aiding investment decisions. Similarly, employing Long Short-Term Memory and Gated Recurrent Unit models, Xiuguo and Shengyong [60] exemplified the automated financial statement fraud detection process of listed companies in China, ensuring the integrity of financial data.

Meanwhile, Aslam et al. [16] demonstrated the adoption of ML algorithms such as Logistic Regression, Support Vector Machines, and Naive Bayes in effectively detecting different forms of fraud, including securities and commodities fraud, bank

Table 9 Investment decisions of companies enhanced by fraud detection using BDA (Authors' own analysis)

Studies	Dataset	Industry focus	Financial decision-making focus	Model(s)/Technique(s)
Hu et al. [43]	Not explicitly mentioned	Finance	Auditing decision	Fusion multiple rule-based decision making
Xiuguo and Shengyong [60]	Financial and non-financial indices derived from annual reports of Chinese listed companies	Finance	Financial fraud detection	Long Short-Term Memory; Gated Recurrent Unit based DL model
Aslam et al. [16]	Car insurance dataset	Automobile	Securities and commodities fraud, bank fraud, insurance fraud, and other fraud detection	Logistic regression, Support Vector Machine, Naïve Bayes
Raval et al. [79]	Open-access credit card fraud datasets	Finance	Credit card fraud detection	Long short-term memory integrated with explainable artificial intelligence

fraud, insurance fraud, and other frauds in the auto insurance sector. Last but not least in Table 9, Raval et al. [79] concentrated on integrating open-access datasets into Long Short-Term Memory models with Explainable Artificial Intelligence to address credit card fraud patterns, and enhance the fraud detection capabilities, reinforcing the importance of transparency in financial operations.

4.2.4 Demand Forecasting and Inventory Management

Investment decisions at the firm level are influenced not only by improved credit risk evaluation, risk management, investment analysis, and fraud detection but also by enhanced demand forecasting and inventory management, both of which are critical for optimizing working capital management–a key component of investment decision-making [1]. On top of that, BDA has been recognized for its ability to provide demand prediction, trend analysis, and customer behavior, and consequently, manage working capital and enhance investment decisions as studied by Maté et al. [39], McMahon et al. [46], Nibareke and Laassiri [17], Mariani and Wamba [45], Seyedan and Mafakheri [47], Ahmad et al. [14], Ali et al. [51], and Beltrán et al. [52]. These studies not only demonstrate how companies leverage BDA in driving financial decision-making within companies but also reveal common patterns in the integration of data and techniques for more efficient financial planning, resource allocation, and investment strategies (see Table 10).

Starting with the utility energy sector, Maté et al. [39], Ahmad et al. [14], and Beltrán et al. [52] highlighted the use of ML, DL, and AI models for predicting electricity demand and price to optimize the resource allocation and investment decisions. Maté et al. [39] indicated the use of ML models such as Decision Trees, Artificial NN, and Support Vector Machines to analyze structured and unstructured data, yielding the forecasting of electricity consumption for energy companies to better allocate resources. Similarly, Ahmad et al. [14] applied a mix of ML, DL, and AI models for forecasting electricity demands. Besides, Beltrán et al. [52] found the use of BDA in forecasting electricity prices for utility energy companies to make data-driven decisions that optimize financial outcomes.

In the hospitality industry, Antonio, de Almeida, and Nunes [42] indicated the application of BDA using data mined from travel websites and gradient-boosting algorithms to forecast booking cancellations at high accuracy and enhance revenue management. The demand forecasting capability of BDA is confirmed via the findings of Al-Alwan et al. [50] in their study of telecommunication companies and Mariani and Wamba [45] in their studies of consumer goods companies. These studies build on the predictive capabilities established in the utility energy sector, highlighting the ability of BDA to predict demand and customer behavior while translating data into informed investment decisions.

The role of BDA in forecasting demand extends into the transportation sector, where Mcmahon et al. [46] and Nibareke and Laassiri [17] applied BDA techniques in railway management systems and flight delay prediction. For instance, Nibareke and Laassiri [17] found how the adoption of Linear Regression, Naive Bayes, and

Table 10 Investment decisions of companies derived from demand forecasting and inventory management using BDA (Authors' own analysis)

Studies	Dataset	Industry focus	Decision-making focus	Model(s)/Technique(s)
Maté et al. [39]	Structured (US Energy Information Administration), semi-structured (meteorology textual data) and Twitter unstructured data	Utility energy	Electricity consumption predictions	Decision Trees; Artificial NN; SVM
Acharya et al. [41]	Internal operation, customers, and external market trends	Fashion	Inventory management	BDA
Antonio et al. [42]	Popular travel websites (including weather, holidays, events, social reputation and online prices/ inventory)	Hospitality	Bookings cancellation forecasting	Data mining; Gradient Boosting-Based algorithm
Mcmahon et al. [46]	Railway management systems	Transportation	Railway asset management	BDA techniques with five key values (Volume, Velocity, Variety and Veracity)
Nibareke and Laassiri [17]	Flight datasets in the Internet and social media (YouTube, Tumble, Twitter, and Facebook)	Aviation	Flight delays predictions	Linear Regression; Naive Bayes; Decision Trees
Mariani and Wamba [45]	Social media consumer reviews	Consumer goods	Demand forecasting	BDA
Seyedan and Mafakheri [47]	Supply chain datasets including stores, customers, products, sales, orders, shipping and delivery	Supply chain	Demand and sales forecasting	Time-series forecasting; Clustering; KNN; ANN; SVM
Ahmad et al. [14]	Historical load data	Utility energy	Electricity demand forecast	ML; DL; AI; and Hybrid forecasting models
Ali et al. [51]	Supply chain collaboration dataset from Kaggle (category, customer country, order country, and feedback)	Supply chain	Investment decision optimization	K-Nearest Neighbor; SVM
Beltrán et al. [52]	Electricity price (Spanish, French, Iberian Peninsula), natural gas price (Iberian), European emission allowances, Brent oil price and EUR/USD currency rate	Utility energy	Electricity price forecasting	Time series; Gradient Boosting Decision Trees
Al-Alwan et al. [50]	Responses from managers at senior and middle levels of telecommunication companies	Telecommunication	Demand predictions	Fusion-based ML

Decision Trees supports more efficient working capital management, much like the booking forecasts in the hospitality sector.

In the fashion sector, BDA enables companies to translate vast data from different sources such as internal operations, customers, and external market insights into actionable insights that optimize pricing strategies while enhancing inventory management [41]. To support this finding, Seyedan and Mafakheri [47] emphasized the significance of predictive BDA techniques such as time-series forecasting, clustering, and other ML models such as K-Nearest Neighbor, Support Vector Machine and Artificial NN in directly influencing inventory management and improving demand forecasting in the supply chain sector. These techniques in turn impact companies' investment decisions to ensure the efficiency of capital allocation [47]. Ali et al. [51] indicated the application of ML models such as K-Nearest Neighbor and Support Vector Machine to optimize investment decisions in supply chain collaboration.

4.2.5 Human Resources Management

In addition, BDA has increasing influence within the human resources industry, thus improving companies' working capital management through talent acquisition, employee engagement and retention, learning and development, workforce planning, and performance planning [73]. Working capital management is considered one of the key factors of investment decisions at the firm level as defined by Doumpos and Zopounidis [1]. Refer to Table 11, numerous studies by Köchling and Wehner [44], Gravili et al. [63], Wenting et al. [73], Khan, Faisal, and Thomas [72] reveal that algorithmic decision-making becomes more prevalent in talent acquisition and development.

Table 11 The application of BDA in human resources management (Authors' own analysis)

Studies	Dataset	Industry focus	Investment decision	Model(s)/Technique(s)
Köchling and Wehner [44]	Not explicitly mentioned	Human resources management	Talent acquisition	BDA
Gravili et al. [63]	Not explicitly mentioned	Human resources management	Talent acquisition	Data mining; BDA
Wenting et al. [73]	Not explicitly mentioned	Human resources management	Talent acquisition; Performance management; Employee engagement and retention; Learning and development; Workforce planning	AI-based models
Khan, Faisal, and Thomas [72]	Not explicitly mentioned	Human resources management	Talent acquisition	AI-based models

All the included studies [44, 63, 72, 73] with the focus on human resources management primarily employ qualitative or mixed methods that test and confirm their hypotheses through questionnaires, interviews or systematic literature reviews. As a result, their findings tend to focus on broader themes related to BDA or AI, rather than providing detailed information on specific necessary datasets for analytics and techniques used in human resources management (see Table 11).

Notably, Gravili et al. [63] highlight that BDA enhances human resources decision-making by optimizing training, productivity, and employee well-being. Similarly, Köchling and Wehner [44], Wenting et al. [73], and Khan, Faisal, and Thomas [72] found that AI-driven data analysis significantly improves talent acquisition by identifying the most suitable candidates and mitigating biases during the recruitment process. They shared the same findings about the effects of AI-enabled systems on performance management by enabling objective and unbiased evaluations. Wenting et al. [73] further highlight the broader applications of BDA and AI systems, noting their significant contributions to workforce planning through the forecasting of employee needs and optimization of workforce allocation. Moreover, these technologies are found to facilitate the learning and development of employees by identifying skill gaps and tailoring training programs accordingly [73]. Wenting et al. [73] highlighted the facilitation of BDA and AI on enhanced employee engagement and retention by tracking engagement levels and helping develop effective retention strategies. On the other hand, Köchling and Wehner [44] noted a drawback of algorithmic decision-making in recruitment: that it could cause unfair treatment of certain groups and implicit discrimination due to preset algorithms.

Although these studies do not directly address financial decision-making, the practices of talent acquisition, employee engagement and retention, learning and development, workforce planning, and performance planning describe the working capital management at the firm level, which is the key factor of company investment decisions [1]. In other words, given that these domains are integral to working capital management, the included articles emphasize BDA's significant role in shaping investment decisions.

4.3 Applications of Big Data Analytics in Financing Decisions of Companies

Beyond investment decisions, recent studies have shown that BDA has significant impacts on financing decisions, with a focus on capital structure. For instance, Zhang [74] explores how Decision Tree algorithms, integrated with BDA, could enhance financing decisions in enterprise cloud accounting, effectively managing debt and equity financing structures to support both sustainable growth and operational stability. Zhu, Yan, and Bai [62] addressed the prediction of enterprise free cash flow using an improved NN algorithm, demonstrating how accurate cash flow forecasting can enhance financing decisions. Last but not least, Gaytan et al. [54] made a comparison of AI models in predicting capital structure, emphasizing the potential of artificial NN models to optimize the balance between debt and equity financing, thus ensuring that firms could meet their long-term growth objectives while maintaining sufficient liquidity for day-to-day operations. Together, these studies underscore the transformative impact of BDA in making informed financing decisions that balance long-term investment needs with the demands of daily business operations.

5 Discussion

5.1 Key Findings and Implications for Financial Decision-Making

The systematic literature review of BDA for financial decision-making reveals critical findings of how BDA is transforming the way of business decision-making across industries. Notably, BDA plays a pivotal role in enhancing investment decisions, with approximately 92.9% of the included articles focusing on this area. BDA in combination with AI, ML, DL, and NN techniques has been consistently proven to be significantly influential in financial decision-making, in which it supports investment evaluation, credit risk evaluation, fraud detection, demand forecasting, inventory management, and trading optimization strategies. For example, studies by Setty, Elovici, and Schwartz [77] and Blohm et al. [53] demonstrated how ML models could assist VC and PE firms in translating complex datasets into insights about startups, thereby informing more precise investment decisions while mitigating risks at the early stage of investment. This signifies a shift from traditional investment towards more data-driven approaches, enhancing the significance of BDA in financial decision-making. Furthermore, BDA's role extends to financing decisions, particularly in optimizing capital structure. Recent research by Gaytan et al. [54], Zhu, Yan, and Bai [62], and Zhang [74] highlights how BDA can enhance debt and equity financing decisions, supporting the sustainable growth of companies. This development aligns with the trend of using data-driven strategies to balance long-term investment needs with operational stability.

5.2 Strengths and Limitations of Reviewed Studies

The reviewed studies exhibit several strengths. Firstly, they provide a comprehensive overview of BDA applications across various domains, investment evaluation, credit risk evaluation, fraud detection, working capital management, demand forecast, trading optimization strategies, and consequently, enhancing investment and financing decisions of companies. In other words, this broad coverage offers a holistic view of BDA's impact on financial decision-making. On the other hand, it is hard to deny limitations regarding those reviewed articles. One notable limitation is the variability in methodologies across the reviewed studies, which can affect the comparability of results. Different studies employ BDA in combination with various techniques including AI, ML, DL, and NN techniques which makes it challenging to generalize findings across all applications.

6 Conclusion

Drawing upon 43 reviewed articles between 2015 and 2024, the review highlights the profound applications of BDA in the financial decision-making of companies across industries, particularly in the investment and financing domains. The review also reveals that BDA, integrated with AI, ML, DL, and NN techniques, significantly enhances investment evaluations, credit risk evaluation, working capital management, and trading optimization. Notably, the majority of studies focus on BDA for investment decisions, emphasizing the direction of research on BDA applications. Besides, there are significant implications for practice derived from the findings of this review, suggesting companies integrate BDA into their decision-making processes to leverage accuracy while mitigating risks and biases during the process. Advanced analytics such as AI, ML, DL, and NN models could translate complex datasets into comprehensive insights, thus informing investment opportunities, credit risks and fraud detection, demand forecast, working capital management, and financing optimization. For example, implementing sophisticated credit scoring models and fraud detection systems can provide tangible benefits and improve financial performance. Overall, the integration of BDA into financial decision-making holds significant promise for improving investment and financing decisions, with implications for practice emphasizing the need for continued investment in BDA technologies and data-driven approaches.

6.1 Limitations and Further Research Directions

The systematic literature review on BDA for financial decision-making reveals several limitations that necessitate attention. One key limitation lies in the variability in methodologies across the reviewed studies that may affect the consistency and comparability of results. It also poses a challenge to the development of a unified framework for BDA applications. Additionally, due to the strictness of eligibility check along with inclusion and exclusion criteria, the review is constrained by its focus on a specific sample of studies on financial decisions of companies that may not capture the breadth of BDA applications across different industries, potentially overlooking some areas that BDA could have significant impacts. Therefore, further research should aim at addressing these limitations by expanding the scope of research to capture a wider range of BDA applications across different industries. Longitudinal studies could be beneficial in examining the long-term effects of BDA on financial decision-making and performance. Moreover, investigating the advancements and pitfalls regarding the integration of emerging technologies with BDA could offer valuable insights for decision-makers to adopt for their companies. Addressing these areas will contribute to a more nuanced and actionable understanding of how BDA can be leveraged to enhance financial decision-making in diverse contexts.

References

1. Doumpos, M.: Constantin Zopounidis. Multicriteria analysis in finance. Springer (2014)
2. Saltz, J.S.: The need for new processes, methodologies and tools to support big data teams and improve big data project effectiveness. In: 2015 IEEE International Conference on Big Data (Big Data), pp. 2066–2071. IEEE (2015)
3. Phillips-Wren, G., Hoskisson, A.: An analytical journey towards big data. J. Decis. Syst. **24**(1), 87–102 (2015)
4. Akhtar, P., Khan, Z., Tarba, S., Jayawickrama, U.: The Internet of Things, dynamic data and information processing capabilities, and operational agility. Technol. Forecast. Soc. Chang. **136**, 307–316 (2018)
5. Fosso Wamba, S., Akter, S., Trinchera, L., De Bourmont, M.: Turning information quality into firm performance in the big data economy. Manag. Decis. **57**(8) (2018)
6. Zytek, A., Liu, D., Vaithianathan, R., Veeramachaneni, K.: Sibyl: understanding and addressing the usability challenges of machine learning in high-stakes decision making. IEEE Trans. Vis Comput. Graph **28**(1) (2022)
7. Choi, H.-Y., Park, J.: Do data-driven CSR initiatives improve CSR performance? The importance of big data analytics capability. Technol. Forecast. Soc. Change **182**, 121802 (2022)
8. Sivarajah, U., Kamal, M.M., Irani, Z., Weerakkody, V.: Critical analysis of Big Data challenges and analytical methods. J. Bus. Res. **70**(1), 263–286 (2017)
9. Spanaki, K., Gürgüç, Z., Adams, R., Mulligan, C.: Data supply chain (DSC): research synthesis and future directions. Int. J. Prod. Res. **56**(13), 4447–4466 (2017)
10. Manta, A.G., Bădôrcea, R.M., Doran, N.M., Badareu, G., Gherţescu, C., Popescu, J.: Industry 4.0 transformation: analysing the impact of artificial intelligence on the banking sector through bibliometric trends. Electronics **13**(9), 1693 (2024)
11. Chen, D., Esperança, J.P., Wang, S.: The impact of artificial intelligence on firm performance: an application of the resource-based view to e-commerce firms. Front. Psychol. **13** (2022)
12. Li, C., Chen, Y., Shang, Y.: A review of industrial big data for decision making in intelligent manufacturing. Eng. Sci. Technol. Int. J. **29** (2021)
13. Jia, Y., Wang, Z.: Application of artificial intelligence based on the fuzzy control algorithm in enterprise innovation. Heliyon **10**(6), e28116 (2024)
14. Ahmad, N., Ghadi, Y., Adnan, M., Ali, M.: Load forecasting techniques for power system: Research challenges and survey. IEEE Access (2022)
15. Hasan, M.M., Popp, J., Oláh, J.: Current landscape and influence of big data on finance. J. Big Data **7**(1), 1–17 (2020)
16. Aslam, F., Hunjra, A.I., Ftiti, Z., Louhichi, W., Shams, T.: Insurance fraud detection: evidence from artificial intelligence and machine learning. Res. Int. Bus. Financ. **62**, 101744 (2022)
17. Nibareke, T., Laassiri, J.: Using Big Data-machine learning models for diabetes prediction and flight delays analytics. J. Big Data **7**(1) (2020)
18. Lee, H., Kim, J.H., Jung, H.S.: Deep-learning-based stock market prediction incorporating ESG sentiment and technical indicators. Sci. Rep. **14**(1), 10262 (2024)

19. Eachempati, P., Srivastava, P.R., Kumar, A., Tan, K.H., Gupta, S.: Validating the impact of accounting disclosures on stock market: a deep neural network approach. Technol. Forecast. Soc. Chang. **167**, 120903 (2021)
20. Salem, K., Randles, R., Sapre, B., Finnegan, A.: Experiences of ethnic minority personnel in the armed forces: a systematic review. J. Milit. Veteran Family Health **9**(1) (2022)
21. LaValle, S., Lesser, E., Shockley, R., Hopkins, M.S., Kruschwitz, N.: Big data, analytics and the path from insights to value. MIT Sloan Manag. Rev. **52**(2), 21–31 (2011)
22. Cali, U., Kuzlu, M., Pipattanasomporn, M., Kempf, J., Bai, L., Cali, U., Kuzlu, M., Pipattanasomporn, M., Kempf, J., Bai, L.: Foundations of big data, machine learning, and artificial intelligence and explainable artificial intelligence. Digitalization of Power Markets and Systems Using Energy Informatics, pp. 115–137 (2021)
23. Li, J., Ye, Z., Zhang, C.: Study on the interaction between big data and artificial intelligence. Syst. Res. Behav. Sci. **39**(3), 641–648 (2022)
24. Brealey, R.A., Myers, S.C., Allen, F.: Principles of Corporate Finance, 13th edn. McGraw-Hill Education, New York (2020)
25. Kalyebara, B., Islam, S.M.N.: Corporate Governance, Capital Markets, and Capital Budgeting. Contributions to Management Science, Springer, Berlin, Heidelberg (2014)
26. Moher, D., Schulz, K.F., Simera, I., Altman, D.G.: Guidance for developers of health research reporting guidelines. PLoS Med. **7**(2), e1000217 (2010)
27. Page, M.J., McKenzie, J.E., Bossuyt, P.M., Boutron, I., Hoffmann, T.C., Mulrow, C.D., Shamseer, L., Tetzlaff, J.M., Akl, E.A., Brennan, S.E., Chou, R., Glanville, J., Grimshaw, J.M., Hróbjartsson, A., Lalu, M.M., Li, T., Loder, E.W., Mayo-Wilson, E., McDonald, S., McGuinness, L.A.: The PRISMA 2020 statement: an Updated Guideline for Reporting Systematic Reviews. British Med. J. **372**(71) (2021)
28. Higgins, J.P.T., Thomas, J., Chandler, J., Cumpston, M., Li, T., Page, M.J., Welch, V.A. (eds.), Cochrane Handbook for Systematic Reviews of Interventions, 2nd edn. Wiley, Chichester, UK (2019)
29. Ardabili, S., Abdolalizadeh, L., Mako, C., Torok, B., Mosavi, A.: Systematic review of deep learning and machine learning for building energy. Front. Energy Res. **10** (2022)
30. Falagas, M.E., Pitsouni, E.I., Malietzis, G.A., Pappas, G.: Comparison of PubMed, Scopus, Web of Science, and Google Scholar: strengths and weaknesses. FASEB J. **22**(2), 338–342 (2008)
31. Carrera-Rivera, A., Ochoa, W., Larrinaga, F., Lasa, G.: How-to conduct a systematic literature review: a quick guide for computer science research. MethodsX **9**(1), 101895 (2022)
32. Pautasso, M.: Ten simple rules for writing a literature review. PLoS Computational Biol. **9**(7) (2013)
33. Batistič, S., der Laken, P.: History, evolution and future of big data and analytics: a bibliometric analysis of its relationship to performance in organizations. Br. J. Manag. **30**(2), 229–251 (2019)
34. Redine, A., Deshpande, S., Jebarajarkirthy, C., Surachartkumtonkun, J.: Impulse buying: a systematic literature review and future research directions. Int. J. Consum. Stud. **47**(1) (2022)
35. Hassan, S.M., Rahman, Z., Paul, J.: Consumer ethics: a review and research agenda. Psychol. Market. **39**(1) (2021)
36. Stoll, C.R.T., Izadi, S., Fowler, S., Green, P., Suls, J., Colditz, G.A.: The value of a second reviewer for study selection in systematic reviews. Res. Synth. Methods **10**(4), 539–545 (2019)
37. Li, T., Higgins, J.P.T., Deeks, J.J.: Collecting data. In: Higgins, J.P.T., Thomas, J., Chandler, J., Cumpston, M., Li, T., Page, M.J., Welch, V.A. (eds.), Cochrane Handbook for Systematic Reviews of Interventions. Wiley (2019)
38. Khashei, M., Mirahmadi, A.: A soft intelligent risk evaluation model for credit scoring classification. Int. J. Financ. Stud. **3**(3), 411–422 (2015)
39. Maté, A., Peral, J., Ferrández, A., Gil, D., Trujillo, J.: A hybrid integrated architecture for energy consumption prediction. Futur. Gener. Comput. Syst. **98**, 131–147 (2016)
40. Zhu, Y., Xie, C., Wang, G.-J., Yan, X.-G.: Predicting China's SME credit risk in supply chain finance based on machine learning methods. Entropy **18**(5), 195 (2016)

41. Acharya, A., Singh, S.K., Pereira, V., Singh, P.: Big data, knowledge co-creation and decision making in the fashion industry. Int. J. Inf. Manage. **42**, 90–101 (2018)
42. Antonio, N., de Almeida, A., Nunes, L.: Big data in hotel revenue management: exploring cancellation drivers to gain insights into booking cancellation behavior. Cornell Hospit. Quart. **60**(4), 193896551985146 (2019)
43. Hu, K.-H., Chen, F.-H., Hsu, M.-F., Tzeng, G.-H.: Identifying key factors for adopting artificial intelligence-enabled auditing techniques by joint utilization of fuzzy-rough set theory and MRDM technique. Technol. Econ. Dev. Econ. **27**(2), 459–492 (2020)
44. Köchling, A., Wehner, M.C.: Discriminated by an algorithm: a systematic review of discrimination and fairness by algorithmic decision-making in the context of HR recruitment and HR development. Bus. Res. **13**(3), 795–848 (2020)
45. Mariani, M.M., Wamba, S.F.: Exploring how consumer goods companies innovate in the digital age: the role of big data analytics companies. J. Bus. Res. **121**, 338–352 (2020)
46. Mcmahon, P., Zhang, T., Dwight, R.: Requirements for big data adoption for railway asset management. IEEE Access **8**, 15543–15564 (2020)
47. Seyedan, M., Mafakheri, F.: Predictive big data analytics for supply chain demand forecasting: methods, applications, and research opportunities. J. Big Data **7**(1), 1–22 (2020)
48. Ala'raj, M., Abbod, M.F., Majdalawieh, M.: Modelling customers credit card behaviour using bidirectional LSTM neural networks. J. Big Data **8**(1) (2021)
49. Xu, J., Lu, Z., Xie, Y.: Loan default prediction of Chinese P2P market: a machine learning methodology. Sci. Rep. 11(1) (2021)
50. Al-Alwan, M.A.M., Al-Nawafah, S.S., Al-Shorman, H.M., Khrisat, F.A., Faisal Alathamneh, F., Al-Hawary, S.I.S.: The effect of big data on decision quality: Evidence from the telecommunication industry. Int. J. Data Netw. Sci. **6**(3), 693–702 (2022)
51. Ali, N., Ghazal, T.M., Ahmed, A., Abbas, S., Khan, M.A., Alzoubi, H.M., Farooq, U., Ahmad, M., Adnan Khan, M.: Fusion-based supply chain collaboration using machine learning techniques. Intell. Autom. Soft Comput. **31**(3), 1671–1687 (2022)
52. Beltrán, S., Castro, A., Irizar, I., Naveran, G., Yeregui, I.: Framework for collaborative intelligence in forecasting day-ahead electricity price. Appl. Energy **306**, 118049 (2022)
53. Blohm, I., Antretter, T., Sirén, C., Grichnik, D., Wincent, J.: It's a people's game, isn't it?! A comparison between the investment returns of business angels and machine learning algorithms. Entrep. Theory Pract. **46**(4), 104225872094520 (2022)
54. Gaytan, J.C.T., Ateeq, K., Rafiuddin, A., Alzoubi, H.M., Ghazal, T.M., Ahanger, T.A., Chaudhary, S., Viju, G.K.: AI-Based prediction of capital structure: performance comparison of ANN SVM and LR models. Comput. Intell. Neurosci. **1–13**, 2022 (2022)
55. Karn, A.L., Sachin, V., Sengan, S., Gandhi, I., Ravi, L., Sharma, D.K., Subramaniyaswamy, V.: Designing a deep learning-based financial decision support system for fintech to support corporate customer's credit extension. Malaysian J. Comput. Sci. 116–131 (2022)
56. Kyeong, S., Kim, D., Shin, J.: Can system log data enhance the performance of credit scoring?-Evidence from an Internet Bank in Korea. Sustainability **14**(1), 130 (2021)
57. Mourmouris, J., Poufinas, T.: Multi-criteria decision-making methods applied in health-insurance underwriting. Health Syst. 1–33 (2022)
58. Sharma, K., Dwivedi, Y.K., Metri, B.: Incorporating causality in energy consumption forecasting using deep neural networks. Ann. Oper. Res. (2022)
59. Sujith, A.V.L.N., Qureshi, N.I., Dornadula, V.H.R., Rath, A., Prakash, K.B., Singh, S.K.: A comparative analysis of business machine learning in making effective financial decisions using structural equation model (SEM). J. Food Qual. **1–7**, 2022 (2022)
60. Xiuguo, W., Shengyong, D.: An analysis on financial statement fraud detection for Chinese listed companies using deep learning. IEEE Access **10**, 22516–22532 (2022)
61. Zhang, Z., Xiao, Y., Niu, H.: DEA and machine learning for performance prediction. Mathematics **10**(10), 1776 (2022)
62. Zhu, L., Yan, M., Bai, L.: Prediction of enterprise free cash flow based on a backpropagation neural network model of the improved genetic algorithm. Information **13**(4), 172 (2022)

63. Gravili, G., Hassan, R., Avram, A., Schiavone, F.: Big data and human resource management: paving the way toward sustainability. Eur. J. Innov. Manag. **26**(7), 552–590 (2023)
64. Gu, L.: Optimized backpropagation neural network for risk prediction in corporate financial management. Sci. Rep. **13**(1), 19330 (2023)
65. Gu, R., Zhou, L.: Financial decision management of enterprise cloud accounting based on big data technology. Appl. Math. Nonlinear Sci. **9**(1) (2023)
66. Peng, Y., Li, H.: A rental platform service supply chain network equilibrium model considering digital detection technology investment and big data marketing. Sustainability **15**(13), 9955 (2023)
67. Shen, Q., Zhang, W., Song, M.: RankXGB-Based enterprise credit scoring by electricity consumption in edge computing environment. Comput. Mater. Continua **75**(1), 197–217 (2023)
68. Tong, D.A., Tian, G.: Intelligent financial decision support system based on big data. J. Intell. Syst. **32**(1) (2023)
69. Xu, H., Ge, J., Tong, L.: Application of cloud accounting in enterprise financial forecasting and decision making in the era of big data. Appl. Math. Nonlinear Sci. **8**(1), 3095–3110 (2023)
70. Akour, I., Alzyoud, M., Alquqa, E.K., Tariq, E., Alzboun, N., Sulieman, Alshurideh, M.: Artificial intelligence and financial decisions: Empirical evidence from developing economies. Int. J. Data Netw. Sci. **8**(1), 101–108 (2024)
71. Giannakopoulos, N.T., Terzi, M.C., Sakas, D.P., Kanellos, N., Toudas, K.S., Migkos, S.P.: Agroeconomic indexes and big data: digital marketing analytics implications for enhanced decision making with artificial intelligence-based modeling. Information **15**(2), 67 (2024)
72. Khan, I., Ahmed, W.S., Shad, S., Kumar, C., Usman, M., Jasemi, M., Shafi, K.: Determine intraday trading currency's trend framework evidence from machine learning techniques. Int. J. Bus. Intell. Res. **15**(1), 1–15 (2024)
73. Wenting, L., Mohd, W., Xinlin, J., Na, M., Alam, S.S.: Analyzing the impact on talent acquisition and performance management. J. Organ. End User Comput. **36**(1), 1–30 (2024)
74. Zhang, P.: A study on the management of big data technology in financial decision-making of enterprise cloud accounting. Appl. Math. Nonlinear Sci. **9**(1) (2023)
75. Khan, S., Faisal, S., Thomas, G.: Exploring the nexus of artificial intelligence in talent acquisition: unravelling cost-benefit dynamics, seizing opportunities, and mitigating risks. Probl. Perspect. Manag. **22**(1), 462–476 (2024)
76. Meshref, H.: Predicting loan approval of bank direct marketing data using ensemble machine learning algorithms. Int. J. Circuits Syst. Signal Proc. **14**, 914–922 (2020)
77. Setty, R., Elovici, Y., Schwartz, D.: Cost-sensitive machine learning to support startup investment decisions. Int. J. Intell. Syst. Account. Financ. Manag. **31**(1) (2024)
78. Alshehadeh, A.R., Alia, M.A., Jaradat, Y., Injadat, E., Al-khawaja, H.: Big data analytics techniques and their impacts on reducing information asymmetry: evidence from Jordan. Int. J. Data Netw. Sci. **7**(3), 1259–1266 (2023)
79. Raval, J., Bhattacharya, P., Jadav, N.K., Tanwar, S., Sharma, G., Bokoro, P., Elmorsy, M., Tolba, A., Raboaca, M.S.: RaKShA: a trusted explainable LSTM model to classify fraud patterns on credit card transactions. Mathematics **11**(8), 1901 (2023)

Author Index

H. T. T. Nguyen et al. (eds.), *Data Science in Finance and Accounting*, Studies
in Big Data 181, https://doi.org/10.1007/978-3-032-06179-9

MIX
Papier aus verantwortungsvollen Quellen
Paper from responsible sources
FSC® C105338

If you have any concerns about our products,
you can contact us on
ProductSafety@springernature.com

In case Publisher is established outside the EU,
the EU authorized representative is:
Springer Nature Customer Service Center GmbH
Europaplatz 3, 69115 Heidelberg, Germany

Printed by Libri Plureos GmbH
in Hamburg, Germany